Progress in
PHYSICAL ORGANIC CHEMISTRY

VOLUME 18

Progress in
PHYSICAL ORGANIC CHEMISTRY

VOLUME 18

Editor

ROBERT W. TAFT, *Department of Chemistry
University of California, Irvine, California*

A Wiley-Interscience Publication
John Wiley & Sons, Inc.
New York / Chichester / Brisbane / Toronto / Singapore

An Interscience® Publication
Copyright © 1990 by John Wiley & Sons, Inc.

All rights reserved. Published simultaneously in Canada.

Reproduction or translation of any part of this work
beyond that permitted by Section 107 or 108 of the
1976 United States Copyright Act without the permission
of the copyright owner is unlawful. Requests for
permission or further information should be addressed to
the Permissions Department, John Wiley & Sons, Inc.

Library of Congress Cataloging in Publication Data:
Library of Congress Catalog Card Number:

ISBN 0-471-51706-2

Printed in the United States of America

10 9 8 7 6 5 4 3 2 1

*Dedicated to the memory of
Louis P. Hammett*

Contributors to Volume 18

Marvin Charton
    Department of Chemistry
    Pratt Institute
    Brooklyn, New York

Siegfried Dähne
    Academy of Sciences
    Central Institute of Optics and Spectrometry
    Berlin, German Democratic Republic

Otto Exner
    Institute of Organic Chemistry and Biochemistry
    Czechoslovak Academy of Sciences
    Prague, Czechoslovakia

U. Haldna
    Institute of Chemistry
    Academy of Sciences of the Estonian S.S.R., Akadeemia
    Tee, Tallinn
    U.S.S.R.

Katrin Hoffmann
    Academy of Sciences
    Central Institute of Optics and Spectrometry
    Berlin, German Democratic Republic

Romuald I. Zalewski
    Division of General Chemistry
    Academy of Economics
    Poznan, Poland

Introduction to the Series

Physical organic chemistry is a relatively modern field with deep roots in chemistry. The subject is concerned with investigations of organic chemistry by quantitative and mathematical methods. The wedding of physical and organic chemistry has provided a remarkable source of inspiration for both of these classical areas of chemical endeavor. Further, the potential for new developments resulting from this union appears to be still greater. A closing of ties with all aspects of molecular structure and spectroscopy is clearly anticipated. The field provides the proving ground for the development of basic tools for investigations in the areas of molecular biology and biophysics. The subject has an inherent association with phenomena in the condensed phase and thereby with the theories of this state of matter.

The chief directions of the field are: (a) the effects of structure and environment on reaction rates and equilibria; (b) mechanisms of reactions; and (c) applications of statistical and quantum mechanics to organic compounds and reactions. Taken broadly, of course, much of chemistry lies within these confines. The dominant theme that characterizes this field is the emphasis on interpretation and understanding which permits the effective practice of organic chemistry. The field gains its momentum from the application of basic theories and methods of physical chemistry to the broad areas of knowledge of organic reactions and organic structural theory. The nearly inexhaustible diversity of organic structures permits detailed and systematic investigations which have no peer. The reactions of complex natural products have contributed to the development of theories of physical organic chemistry, and, in turn, these theories have ultimately provided great aid in the elucidation of structures of natural products.

Fundamental advances are offered by the knowledge of energy states and their electronic distributions in organic compounds and the relationship of these to reaction mechanisms. The development, for example, of even an empirical and approximate general scheme for the estimation of activation energies would indeed be most notable.

The complexity of even the simplest organic compounds in terms of physical theory well endows the field of physical organic chemistry with the frustrations of approximations. The quantitative correlations employed in this field vary from purely empirical operational formulations to the approach of applying physical principles to a workable model. The most common procedures have involved the application of approximate theories to approximate models. Critical assessment of the scope and limitations of these approximate applications of theory leads to further development and understanding.

Although he may wish to be a disclaimer, the physical organic chemist attempts to compensate his lack of physical rigor by the vigor of his efforts. There has indeed been recently a great outpouring of work in this field. We believe that a forum for exchange of views and for critical and authoritative reviews of topics is an essential need of this field. It is our hope that the projected periodical series of volumes under this title will help serve this need. The general organization and character of the scholarly presentations of our series will correspond to that of the several prototypes, e.g., *Advances in Enzymology*, *Advances in Chemical Physics*, and *Progress in Inorganic Chemistry*.

We have encouraged the authors to review topics in a style that is not only somewhat more speculative in character but which is also more detailed than presentations normally found in textbooks. Appropriate to this quantitative aspect of organic chemistry, authors have also been encouraged in the citation of numerical data. It is intended that these volumes will find wide use among graduate students as well as practicing organic chemists who are not necessarily expert in the field of these special topics. Aside from these rather obvious considerations, the emphasis in each chapter is the personal ideas of the author. We wish to express our gratitude to the authors for the excellence of their individual presentations.

We greatly welcome comments and suggestions on any aspect of these volumes.

Robert W. Taft

Contents

Free-Energy Relationships in Strongly Conjugated Substituents:
  The Polymethine Approach   1
  By Siegfried Dähne and Katrin Hoffmann

Estimation of the Basicity Constants of Weak Bases by the Target
  Testing Method of Factor Analysis   65
  By Ü. Haldna

Application of Principal Component Analysis in Organic
  Chemistry   77
  By Romuald I. Zalewski

Physicochemical Preconditions of Linear Free-Energy Relationships   129
  By Otto Exner

The Quantitative Description of Amino Acid, Peptide and Protein
  Properties and Bioactivities   163
  By Marvin Charton

Author Index   285

Subject Index   297

Cumulative Index, Volumes 1–18   302

Progress in
PHYSICAL ORGANIC CHEMISTRY

VOLUME 18

Free-Energy Relationships in Strongly Conjugated Substituents: The Polymethine Approach*

BY SIEGFRIED DÄHNE AND KATRIN HOFFMANN
Academy of Sciences of the GDR
Central Institute of Physical Chemistry, Analytical Center
Berlin, German Democratic Republic

CONTENTS

I. Introduction 1
II. Limitations of Linear Free-Energy Relationships in Strongly Conjugated Substituents 3
III. The Ideal Polymethine State 7
   A. Theoretical Foundation of the Ideal Polymethine State 8
   B. Experimental Proof of the Ideal Polymethine State 18
IV. On the Way to a Joint Model between the Ideal Polymethine State and LFERs in Strongly Conjugated Substituents 31
   A. The Quantum-Chemical Method 32
   B. Modeling Intermediate Structures between Polymethines, Polyenes, and Aromatics 37
   C. Limitations of LFER Parameters 46
   D. A Provisional New Parameter in Description of Substituent Effects 55
V. Conclusions 57
   References and Notes 59

I. INTRODUCTION

Since the formulation of Hammett's basic equation (Equation 1) of linear free-energy relationships (LFERs) in the 1930s (1), an inflation of the substituent constants σ is occurring because severe deviations from linearity take place in strongly conjugated substituents:

$$\log \frac{K}{K_0} \sim \log \frac{k}{k_0} \sim P_i = \rho \cdot \sigma \qquad (1)$$

---

*Plenary lecture presented at the 4th European Conference on Correlation Analysis in Organic Chemistry, The Hammett Memorial Symposium, held in Poznán, Poland, July 18–23, 1988.

where K are equilibrium constants of chemical reactions in relation to K_0 of a standard reaction; k, k_0, rate constants; P_i, certain structural parameters of compounds under consideration (electron densities and NMR chemical shifts of certain atoms, bond lengths, bond angles, etc.); and ρ, reaction constants indicating the transmission of substituent effects. For a long period several multiparameter extensions of Hammett's equation (Equation 1) such as Taft's double-substituent parameter (DSP), Equation 2 seemed to be satisfactory (1) as far as

$$P_i = \rho_I \cdot \sigma_I + \rho_R \cdot \sigma_R \tag{2}$$

series of different σ_R constants had been used in case of strong donor (e.g., σ_R^-) and strong acceptor (e.g., $\sigma_R^+, \sigma_R^{C+}$) substituents. However, even this extension proved to be inadequate with increasing accuracy of the available experimental data. Thus, some nonlinear equations were later proposed (2) where the following equation was rendered useful:

$$\sigma_R = \frac{\sigma_R^0}{1 - \varepsilon \sigma_R^0} \tag{3}$$

where ε = electron demand parameter ranging within $-0.6 < \varepsilon < 0.25$. Nevertheless, the situation is still rather unsatisfactory. Thus, for example, Charton (3) as well as Taft and Topsom (4) recently returned to multiparameter treatment using equations such as Equation 4 (4), which contains even four different parameters:

$$-\delta \Delta E^0 = \rho_F \sigma_F + \delta_\chi \sigma_\chi + \rho_\alpha \sigma_\alpha + \rho_R \sigma_R \tag{4}$$

where F = field effect, χ = electronegativity effect, α = polarizability effect, and R = resonance effect. It is worth mentioning that in applications of Equation 4 the main contribution to substituent effects should be the field effect F and the resonance effect R, not the substituent electronegativity and polarizability (4). This conclusion contrasts with the results obtained by the polymethine approach, as will be shown later in this chapter.

The historical development of linear free-energy relationships just now outlined has led to a revival of the basic question for maximum conjugation and maximum delocalization, respectively, in strongly conjugated π-electron systems (5–7). This question had been answered so far solely on the basis of aromaticity. In 1986 Krygowski first proposed (8) that the authors of this paper should try to address the problem of strong conjugation on the basis of the triad theory that had previously been developed in color chemistry. In triad theory it is shown (9) that the totality of conjugated organic compounds can be best described on the assumption that there exists, besides the well-known ideal aromatic state of cyclic molecules and the polyene-like state of chain-shaped molecules, at least one additional, energetically stabilized state with unique features termed "ideal polymethine state," which is realized in chain-shaped molecules, and that, indeed, represents a state maximum conjugation and maximum delocalization, respectively.

Therefore, Section II begins with a brief description of some limitations of correlation analysis in strongly conjugated substituents. Then follows Section III, with discussion of theoretical and experimental proof for justifying the theory of the ideal polymethine state. Finally, in Sections IV and V, initial attempts to link correlation analysis and triad theory will be discussed.

II. LIMITATIONS OF LINEAR FREE-ENERGY RELATIONSHIPS IN STRONGLY CONJUGATED SUBSTITUENTS

In Fig. 1 the most common substituent constants $\bar{\sigma}_R$ are plotted against the σ_R° values. Deviations from the line having a slope of $45°$ indicate the influence of conjugation on σ_R°. It is seen that in the case of strong conjugation exaltations from σ_R° appear that are in some instances much greater than one order of magnitude. Even when using values of ± 1 for the electron demand parameter ε that far exceed the experimentally determined one in the nonlinear Equation 3, many σ_R values are placed beyond the limits given by this equation showing that the problem cannot be solved in this way.

Another phenomenon in correlation analysis is the so-called π-saturation effect when two donor or two acceptor substituents are combined. Figure 2, published by Taft, Topsom, and coworkers (2b) illustrates this effect on the basis of quantum-chemical ab initio calculations on the STO-3G level. The parameter $\sum \Delta q_\pi$ indicates the sum of acceptance (positive values) or release (negative values) of π-electron charges by the substituent from or to the residual molecule. It is shown to be directly proportional to σ_R (12) and thus is assumed to be a measure of the resonance or delocalization effect in molecules.

At the abscissa the π-electron charge at the position of substitution Δq_π^H is given for several molecules. As expected, $\sum \Delta q_\pi$ is the greater the stronger the acceptor strength when acceptor substituents such as —CHO, —NO$_2$, and —CN are combined with donating carbanion residuals (left side of Fig. 2). The same is true when combining donor substituents, such as NH$_2$, —OH, and —F, with accepting carbocation residuals (right side of Fig. 2). However, on combining a donor substituent with a donating residual, there is only a slight or even no release of π-electrons from the donor under consideration to the residual (left side of Fig. 2), which is typical of π-saturation whereas combination of an acceptor substituent with an accepting residual (right side

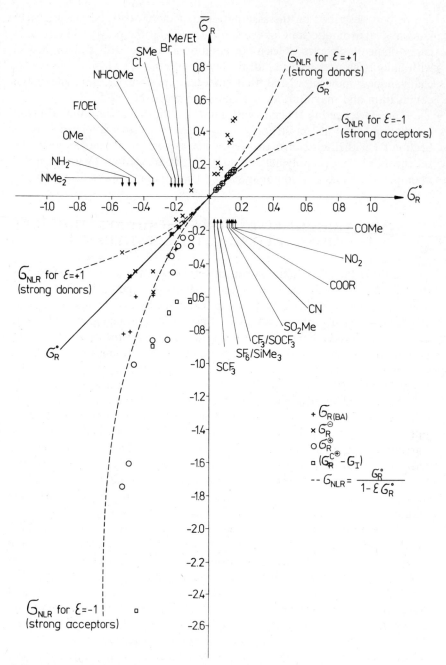

Figure 1 Plot of different substituent constants $\bar{\sigma}_R$ versus σ_R°. The constants σ_R°, $\sigma_{R(BA)}$, σ_R^-, σ_R^+, and σ_I are taken from reference 10a and σ_R^{C+} from reference 11; σ_{NLR} was calculated by Equation 3.

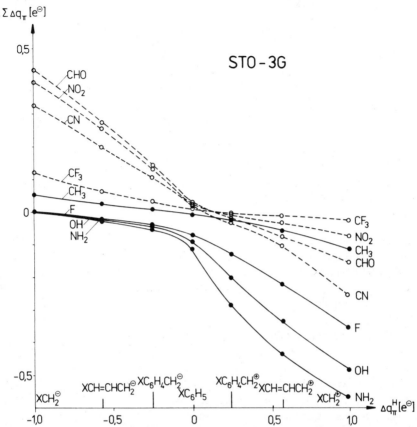

Figure 2 Plot of π-electron charge transfer of a substituent to or from the residual molecule $\Sigma \Delta q_\pi$ versus the π-electron charge density Δq_π^H at the position of substitution calculated by the ab initio STO-3G method with additional geometry optimization of the substituents from Taft et al. (2b).

of Fig. 2) the π-saturation effect is no longer very pronounced. Contrary to the behavior of donor substituents, weak acceptors even may act as donors to strong accepting residuals, as it is indicated at the right side of Fig. 2 by the negative $\Sigma \Delta q_\pi$ values of the acceptor substituents.

From Fig. 2 another conclusion can be drawn concerning the degree of π-electron delocalization. Comparing the residual XCH_2^- with its vinylogue $XCH=CHCH_2^-$ and XCH_2^+ with $XCH=CHCH_2^+$, respectively, which are substituted with one and the same substituent, the $\Sigma \Delta q_\pi$ value of the simple carbanion or carbocation is much higher than that of their vinylogues. Thus, σ_R and hence π-electron delocalization should be greater in the shorter-chain

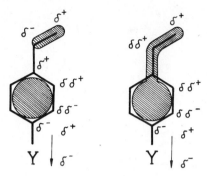

Figure 3 Demonstration of π-polarization induced by an electron-withdrawing substituent Y in substituted benzenes after reference 14.

entities. This effect is even more pronounced if one includes the vinylogous pentadienylium cation having the lowest $\sum \Delta q_\pi$ value (12d). It will be shown in Section III that the result is contrary to experimental experience because there is no doubt that delocalization increases with vinylogous lengthening of polymethine chains.

The strongest curiosity in case of strongly conjugated substituents is the so-called π-polarization effect (13, 14). When π-electron charges in a substituted molecule, as in the *para*-donor–acceptor-substituted benzene in Fig. 3, are drawn into the direction of the Y group by increasing its electronegativity, one should expect that the π-electron system will be polarized as a whole, inducing positive charges in the upper part of the molecule and negative charges in the lower one as shown on the right side of Fig. 3. There is, however, well-defined experimental proof by NMR spectroscopy that the electron density at the C^α atom of the acceptor substituent will be increased when the π-electrons are pulled away and vice versa. To explain this phenomenon, it has been assumed that the π-electron system disintegrates into various bond increments that are separately polarized. Quantum-chemical calculations gave some support of this idea. On positioning a locally separated dipole in the neighborhood of a conjugated π-electron system, the electric field of such a dipole actually produces a certain π-polarization effect in the sense of the formula given at the left side of Fig. 3. The authors themselves admit, however, that this explanation results in an unsatisfactory and artificial impression. If the hypothesis would be correct, the experimental results would lead to the contradictory conclusion that the bond disintegration would be stronger with greater resonance effect, that is, the greater π-electron delocalization within the conjugated molecules.

All these difficulties related to interpreting resonance effects of strongly conjugated substituents provoked Sjoestroem and Wold (15) to question

LFERs as fundamental laws of chemistry and to regard them only as local empirical models of similarity or locally valid linearizations of more complicated relationships. Although this questioning was rejected by Kamlet and Taft in an engaging and vehement manner (16), the crucial problem of the physical origin of LFERs remained unsolved. Only by means of linear solvation energy relationships (LSERs) were Kamlet and Taft able to show that hundreds of physicochemical properties and reactivity parameters of a great variety of solvation effects are based on a few basic physical principles, such as polarity and polarizability properties of the solute and the solvent expressed in terms of refractivity indices and dielectricity constants, as well as specific interactions such as the formation of hydrogen bonds. The same, however, could not be done with LFERs because the physical background of these relationships is well understood only in the case of nonconjugated substituents, preferably on the basis of their field and electronegativity effects, whereas in the case of conjugated substituents, the question of conjugation and π-electron delocalization in respect to field, electronegativity, polarizability, and resonance effects whose parameters are given in Equation 4 is not yet clear. An especially important question to be answered is whether there is a state of maximum π-electron delocalization, because in this case it should be possible to design LFERs a priori as functions between at least two fixed states: that of complete localization and that of complete delocalization of the π-electron system.

III. THE IDEAL POLYMETHINE STATE

In color chemistry a state of maximum delocalization between two localized limiting structures of *para*-donor–acceptor-substituted benzenes was postulated by Ismailski in 1913 (9a,b, 17). He designated this state as a "chromostate," which was further developed by Koenig in his polymethine concept (18) and was included even in later textbooks (e.g., reference 19). In 1966 on the basis of generalization of the polymethine concept (20) we proposed the term "ideal polymethine state" (20a), which is now well accepted in color chemistry (21).

To clarify the relations between polymethines and aromatic compounds, in Fig. 4 we show how the chain-shaped pentamethines (right side) can be converted into *para*-donor–acceptor-substituted benzenes (left side) only by insertion of an ethylene group and vice versa. If we substitute the Y donor of the benzene ring for the hydroxide ion O^-, for instance, we obtain the phenoxide ion, which can be transformed into the anion of the *p*-hydroxybenzaldehyde on substitution the $CH=X$ acceptor group for a aldehyde group $CH=O$. This anion can be more effectively written with a

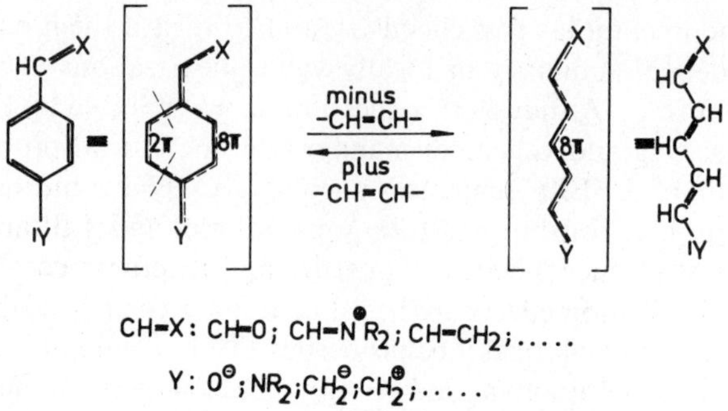

CH=X: CH=O; CH=N⁺R$_2$; CH=CH$_2$; .....

Y: O$^{\ominus}$; NR$_2$; CH$_2^{\ominus}$; CH$_2^{\ominus}$; .....

Figure 4 Commutation between *para*-donor–acceptor-substituted benzenes and pentamethines by removal and insertion of an ethylene fragment. The acceptor substituents of the nonpolar limiting structure are the C=X units and the donors the Y atoms.

fully delocalized π-electron system as indicated in brackets. On cutting off an ethylene fragment, the substituted aromatic molecule is commuted into a pentamethine oxonol and vice versa. On the other hand, the pentamethine oxonol can be formally considered as butadiene substituted in the 1,4-position with a hydroxide anion as donor and a aldehyde group as acceptor. Some more examples of related substituted benzenes and polymethines are grouped in Table 1.

Because of these relationships, substituted aromatic compounds can be theoretically treated not only as substituted aromatic (cyclic) systems but also equally as polymethines branched by ethylene units (22). Whereas the former way of consideration should be preferred in the case of weak conjugation, the latter one is much more convenient in case of strong conjugation, as will be shown later.

The ultraviolet (UV)–visible absorption spectra of *para*-donor–acceptor-substituted benzenes shown in Fig. 5 provide initial arguments that justify this practice. The spectra are continuously shifted to the red region, and their intensity is increased on increasing the donor and the acceptor strength of the substitutents. Finally, on cutting off an ethylene group from the *p*-dimethylaminobenzylidenedimethyliminium cation, yielding the chain-shaped 1,5-bis(dimethylamino)pentamethinium ion, there is, indeed, no essential change of the spectrum, although the pentamethine has two π-electrons less and is no longer "aromatic." In this context it should be mentioned that red-shifting and rise of intensity of absorption bands in the UV–visible region are strong indications of increasing π-electron delocalization as far as the total number of π-electrons remains constant or is even reduced (19).

A. Theoretical Foundation of the Ideal Polymethine State

It is well established from the theory of resonance and mesomerism, respectively, that the state of maximum resonance, that is, of maximum

TABLE 1
Compounds realizable by 1,4 Substitution of Benzene and Butadiene According to Fig. 4

Y \ CH=X	—CH=O	—CH=$\overset{+}{N}R_2$	—CH=CH₂
		para-Substituted Benzenes	
—O⁻	Anion on *p*-hydroxybenzaldehyde	*p*-Dialkylamino-benzylidene-dialkyliminium cation	Anion of *p*-hydroxystyrene
—$\bar{N}R_2$	*p*-Dialkylaminobenzaldehyde		*p*-Dialkylamino styrene
—$\bar{C}H_2^-$			
—CH_2^+			
		Pentamethines (1,4-Substituted Butadienes)	
—O⁻	Pentamethine-oxonol (**3**, n = 2)	Pentamethine merocyanine (**4**, n = 2)	Anion of 1-hydroxyhexatriene
—$\bar{N}R_2$	Pentamethine merocyanine (**4**, n = 2)	Pentamethine streptocyanine (**2**, n = 2)	1-Dialkylamino-hexatriene
—$\bar{C}H_2^-$	Anion of 1-hydroxyhexatriene	1-Dialkylamino hexatriene	Heptatrienylanion
—CH_2^+			Heptatrienylcation

π-electron delocalization, as well as maximum resonance stabilization, is realized in conjugated molecules when two possible limiting structures (**1a** and **1c**) of a conjugated molecule contribute with the same weight to the real structure (**1b**) in the sense of the valence-bond (VB) theory (23). In this case the π-electrons are symmetrically arranged over the conjugated chain.

	1a	1b	1c
Symmetry deviation:	$\Sigma = -1$	$\Sigma = 0$	$\Sigma = +1$
Electronegativity of the substituents:	X < Y	X = Y	X > Y

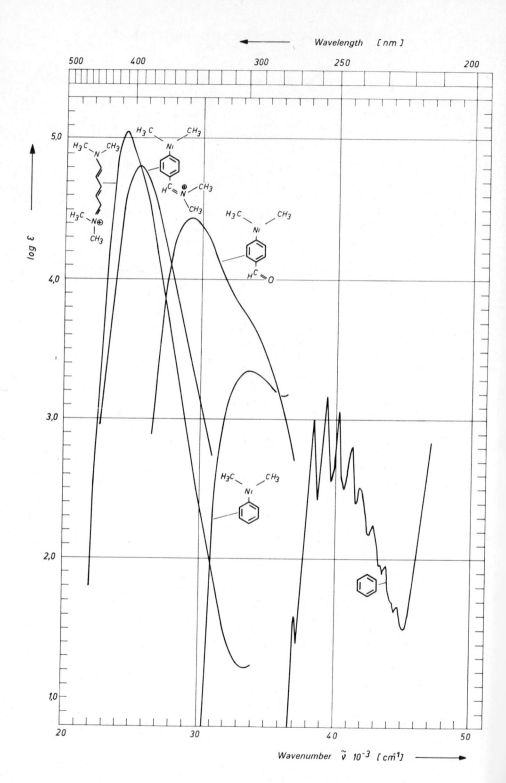

On the other hand, more polyene-like structures (**1a** or **1c**) with a certain localization of the π-electrons at the double bonds are formed if the nonpolar limiting structure (**1a**) or the polar one (**1c**) predominates in the mesomeric equilibrium (**1**). In this case a nonsymmetric distribution of the π-electrons along the molecular chain occurs in respect to the molecule's C_{2v} mirror plane. Since this balance of the π-electrons in donor-acceptor-substituted chromophores is one of the most important measures of its polymethinic character, a new parameter had been introduced (9c, 24), termed *symmetry deviation* $\sum$, which quotes the surplus of π-electron charges at one half of a polymethine chain in respect to its other half. Consequently, $\sum$ equals ± 1 in the limiting structures **1a** and **1c**, respectively, and zero in the symmetric one **1b**.

The symmetry deviation $\sum$ corresponds to the well-known (25) "charge transfer" (CT) in a certain state, but in terms of the numerical values of the transferred charges and in regard to both halves of the molecule under consideration rather than to a substituent. However, in this chapter the expression "charge transfer" in the description of charge shifts within a given state will be avoided because in reality it is not a shift but the normal π-electron distribution of a molecule. We propose to use the term "charge transfer" exclusively for the actual transfer of charges on light excitation, usually designated "charge-transfer transition."

In literature on resonance theory, the nature of the terminal atoms X, Y in conjugated molecules has not been discussed to a greater extend. Usually the hydrocarbon ions (**1**) X, Y = $CH_2^{1/2+}$ or $CH_2^{1/2-}$ are considered as the basic chromophores, which may be perturbed by introducing certain heteroatoms. This is the basis of Dewar's perturbational MO (PMO) theory (25b), which classifies, for instance, the cyanines and oxonols as heteroatom-substituted odd-alternant hydrocarbon ions. Accordingly, the unsaturated hydrocarbon ions were tacitly supposed to possess maximum π-electron delocalization.

In 1966 it turned out, however, by comparing chain-shaped polymethines (**1**) with identical terminal atoms of oxygen, nitrogen, or carbon that chromophores terminated by nitrogen, specifically, the streptocyanines, have unique properties (20) (Fig. 6). For example, using Hückel's MO (HMO) method in quantum-chemical calculations (20a, 26) there is a turning point in the transition energy E_T of the longest wavelength absorption band and a complete π-bond order equalization ($\Delta p = 0$), as in ideal aromatic compounds when the terminal atoms are parametrized adequate to nitrogen ($\alpha_{X,Y} = 1.0\beta$). In contrast to aromatics and polyenes, however, the polymethines exhibit a

Figure 5 Absorption spectra of substituted benzenes and of the 1, 5-bis(dimethylamino)-pentamethinium cation (**2**, $n = 2$; R = Me) in the UV-visible spectral region in ethanolic solution.

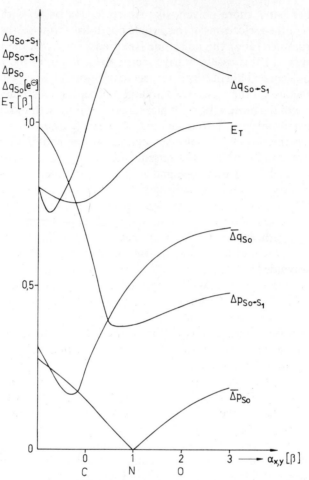

Figure 6 Molecular parameters of the pentamethines (**1**, $n = 2$) calculated by the HMO method in dependence on the Coulomb integral value $\alpha_X = \alpha_Y$ of their terminal atoms (after reference 20a) with $h_{X,Y} = (\alpha_{X,Y} - \alpha_C)/\beta$ and $k_{C=X,C=Y} = \beta_{C=X,C=Y}/\beta_{C=C} = 1.0$; E_T—transition energy; $\Delta p, \Delta q$—alternation of the π-bond orders and π-electron densities, respectively, in the ground state S_0; $\Delta p^{S_0 \rightarrow S_1}, \Delta q^{S_0 \rightarrow S_1}$—changes of the π-bond orders and π-electron densities, respectively on $S_0 \rightarrow S_1$ light excitation. For definition of the parameters, see Section IV.A.

conspicuously strong alternation of their π-electron densities, $\overline{\Delta q}$ in their ground state S_0. This is equal to a strong charge alternation that is reversed in the first excited singlet state S_1. It must be stressed, however, that according to Fig. 6 the maximum of this charge alternation does not coincide with the minimum of the π-bond order alternation $\overline{\Delta p}$ in the nitrogen-terminated polymethines. In the first paper regarding this topic (20a) the delocalization energies E_{deloc} were estimated to be much greater than those of polyenes and to have a maximum in the case of the cyanines. As already mentioned, these findings enabled us to designate this exceptional state "ideal polymethine state." Later it was shown (27) that in ideal polymethines on $S_0 \to S_1$ light excitation a minimum alteration of the π-bond orders $\Delta p^{S_0 \to S_1}$ takes place where contrarily to polyenes no bond contraction but only a slight bond extension occurs. In a like manner, a maximum rearrangement of the π-electron densitites $\Delta q^{S_0 \to S_1}$ can be observed in the ideal polymethines (see Fig. 6).

According to Fabian and Hartmann (28), Dewar's resonance energies RE proved to be very high with ideal streptocyanines in comparison to the constituent polyenic building blocks. However, if one follows the formalism used by the authors and proposed by Hess and Schaad (29), no maximum appears with Dewar's resonance energy at $h_{X,Y} = 1.0$, but the curve is S-shaped, similar to that of the transition energy E_T and of the π-electron density alternation $\overline{\Delta q_{S_0}}$. Because of this rather obscure result it will be necessary to reexamine the problem of resonance in linearly conjugated molecules in the future.

In 1975 Fabian and Hartmann verified the existence of the ideal polymethine state in quantum-chemical terms (30). When the well-known Frost circles of ideal aromatic pericycles consisting of $2N$ atoms and occupied by $2N$ π-electrons are bisected through bonds* into two identical parts that are occupied by $(N + 1)$ and $(N - 1)$ π-electrons, respectively, molecular chains are obtained that correspond to the closed-shell polymethine dyes in odd-numbered fragments and to the open-shell polymethine radicals when N is even-numbered (20c) as illustrated in Fig. 7. Like for the aromatics, the projection of the $2N$ corners perpendicular to their connecting lines of each polygon gives the eigenvalues of the polymethine chains in units of β. Consequently, the orbital energies of polymethine chains exhibit the same pairing properties as do the aromatic and antiaromatic pericycles, with the exception of the unoccupied highest-value MO in case of $(N - 1)$ π-polymethines (left, beside the circles in Fig. 7) and of the occupied lowest-value MO for the $(N + 1)$ π-polymethines (right, beside the circles in Fig. 7). Furthermore, as a result of bisection, the frontier orbitals are no longer degenerate as are those of the aromatic and antiaromatic pericycles.

The bisection procedure can be modeled by factorization of the Hückel determinants of the pericycles. The determinants thus obtained are shown

---

*Bisection procedures by mirror planes crossing the atomic orbitals have been well established in the literature for many years. These operations yield the ideal polyenes, such as ethylene, butadiene, and hexatriene (31).

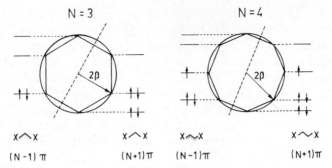

Figure 7 Derivation of the eigenvalues ε_i of $(N \pm 1)$ monomethines and $(N \pm 1)$ dimethine radicals by bisecting Frost's circle of aromatic benzene and antiaromatic cyclooctatetraene after Fabian and Hartmann (30).

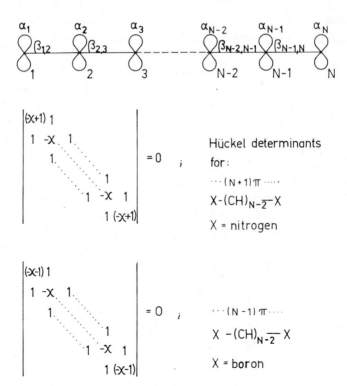

Figure 8 Hückel determinants of $(N + 1)\pi$ and $(N - 1)\pi$ polymethines written in terms of atomic parameters χ after Fabian and Hartmann (30).

in Fig. 8. A chain of $(N-2)$ atoms having identical Coulomb integral values $\alpha = -\chi$, and identical resonance integral values $\beta = 1.0$ necessarily must be terminated by atoms whose Coulomb integral value is increased or decreased by the resonance integral. For carbon atoms within the chain, the terminal atoms consequently must be nitrogen atoms on $(N+1)$ π-electron occupation $\alpha_X = \alpha_C + \beta$ and boron atoms on $(N-1)$ π-electron occupation $\alpha_X = \alpha_C - \beta$. Some analytical expressions of molecular parameters of the polymethine dyes ($N =$ odd) and polymethine radicals ($N =$ even) derived similarly by the factorization procedure are assembled in Table 2. The eigenvalues ε_i really show pairing properties with exception of the lowest-value orbital ($i = 1$). As the resonance integral β is taken constant, the transition energy E_T exclusively depends on the number of π-electrons involved. The reciprocal value of E_T, which is proportional to the absorption wavelength, increases approximately by a constant amount with increasing N. All π-bond orders $p_{r,s}$ of a certain polymethine chain are identical in the ground state S_0 as well as in the first excited singlet state S_1 and are near the value of an ideal aromatic hydrocarbon ($p_{benzene} = 0.6667$). Therefore, their change on $S_0 \to S_1$ light excitation $\Delta_{r,s}^{S_0 \to S_1}$ is small and causes only lowering of the π-bond orders. The π-electron densities q_r in the ground state S_0 alternate along the chain corresponding to position r. On light excitation the π-electron density at each atom is reversed, which means that atoms of low density in the ground state will receive a high π-electron density in the first excited singlet state and vice versa. The π-electron distribution within the frontier MOs based on HMO calculations (26) is reproduced in Fig. 9. Within the inaccuracies of the drawing the same π-electron distribution is obtained with the HMO-β_{SC} calculations mostly used in this chapter.

Later the principal behavior of the geometry and electronic structure of ideal polymethines was confirmed by using more sophisticated quantum-chemical methods such as Pariser–Parr–Pople (PPP) (33a,b), CNDO/2 (33b,c), MINDO/3 (34), and ab initio 6-31G and 4-31G basis sets (35). Moreover, independent arguments for the ideal polymethine state were provided by Dyadyusha et al. (36), who introduced a new parameter ϕ_0 for the electron-donor ability or basicity of substituents, which can be applied to any complicated substituent and expressed by both the Hamiltonian elements of the wavefunctions and the experimentally accessible electrochemical half-wave potentials of the dye molecules. As ϕ_0 is given by trigonometric functions, it is usually measured in degrees, where $\phi_0 < 45°$ means substituents of low electron donor ability and $\phi_0 > 45°$, substituents of high electron ability, whereas the ideal polymethine state is realized when ϕ_0 equals 45°.

Recently Gimarc and others (37) emphasized the finding of strong π-electron density alternation in many rather stable conjugated compounds

TABLE 2

Analytical Solutions for Molecular Parameters of Closed- and Open-Shell Polymethines Consisting of N Atoms and Occupied by $(N+1)$ π-Electrons in One-Electron Approximation Taken from Reference 30b (Orbitals: $i = 1, 2, \ldots N$; Frontier Orbitals: $i = S_0$; $i + 1 = S_1$; Atomic Position: $r = 1, 2, \ldots N$; Neighbored Positions: r, s; Coulomb Integral: α; Resonance Integral: β)

	Closed-Shell Polymethines (N = Odd-Numbered)	Open-Shell Polymethines (N = Even-Numbered)
Eigenvalue	$\varepsilon_i = \alpha + 2\beta \cos\dfrac{\pi(i-1)}{N}$	$\varepsilon_i = \alpha + 2\beta \cos\dfrac{\pi(i-1)}{N}$
Lowest transition energy	$E_T = \Delta\varepsilon^{S_0 \to S_1} = 4\beta \sin\dfrac{\pi}{2N}$	$E_T = \Delta\varepsilon^{S_0 \to S_1} = 2\beta \sin\dfrac{\pi}{N}$ [a]
Reciprocal transition energy	$\dfrac{1}{E_T} = [\Delta\varepsilon^{S_0 \to S_1}]^{-1} \approx \dfrac{N}{2\beta\pi}$	$\dfrac{1}{E_T} = [\Delta\varepsilon^{S_0 \to S_1}]^{-1} \approx \dfrac{N}{2\beta\pi}$ [b]
Coefficients for $i = 1$	$C_{i,r} = \sqrt{\dfrac{1}{N}}$	$C_{i,r} = \sqrt{\dfrac{1}{N}}$
Coefficients for $i = 2, 3, \ldots N$	$C_{i,r} = \sqrt{\dfrac{2}{N}} \cos\dfrac{\pi(2r-1)(i-1)}{2N}$	$C_{i,r} = \sqrt{\dfrac{2}{N}} \cos\dfrac{\pi(2r-1)(i-1)}{2N}$
π-Bond orders in S_0	$p_{r,s} = \dfrac{1}{N} \operatorname{cosec}\dfrac{\pi}{2N}$	$p_{r,s} = \dfrac{1}{N} \cot\dfrac{\pi}{2N}$
Change of π-bond orders during $S_0 \to S_1$ transition	$\Delta p_{r,s}^{S_0 \to S_1} = \dfrac{2}{N} \sin\dfrac{\pi}{2N}$	$\Delta p_{r,s}^{S_0 \to S_1} = \dfrac{1}{N} \sin\dfrac{\pi}{N}$ [c]
π-Electron densities in S_0	$q_r = 1 - (-1)^r \cdot \dfrac{1}{N} \operatorname{cosec}\dfrac{\pi(2r-1)}{2N}$	$q_r = 1 - (-1)^r \cdot \dfrac{1}{N} \cot\dfrac{\pi(2r-1)}{2N}$
Change of π-electron densities during $S_0 \to S_1$ transition	$\Delta q_r^{S_0 \to S_1} = -2(-1)^r \dfrac{1}{N} \sin\dfrac{\pi(2r-1)}{2N}$	$\Delta q_r^{S_0 \to S_1} = +(-1)^r \cdot \dfrac{1}{N} \sin\dfrac{\pi(2r-1)}{2N}$ [c]

[a] The semioccupied orbital is nonbonding in nature. Because of the pairing properties of the remaining orbitals, the transition energies from and to the semioccupied orbital are identical.
[b] For a more detailed treatment of the behavior of the light absorption of polymethine radicals, including LHP–PPP calculations, see reference 32.
[c] Mean values of the Δq_r and $\Delta p_{r,s}$ changes for the two possible lowest-energy transitions that are identical. (see Footnote *a*, above.)

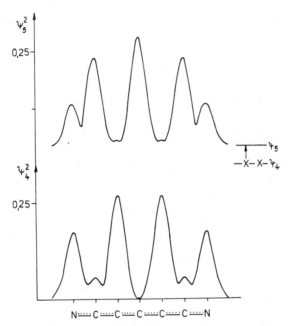

Figure 9 π-Electron density distribution of an ideal pentamethine cyanine (**1**, $n = 2$) in its HOMO (ψ_4) and LUMO (ψ_5) calculated by the HMO formalism. The π-electron densities are nearly identical with that calculated by the HMO-β_{SC} method.

and presumed that charge alternation should cause a certain stabilization effect ["charge stabilization rule" (37a)]. Although the molecules considered by these authors are by definition polymethine-like compounds, the typical polymethinic π-electron density alternation alone cannot be an essential measure of maximum delocalization, as it had been shown for some time (20a) (see Fig. 6) and later emphasized (38) that there are examples where charge alternation may increase when π-electrons are again partly localized at certain bonds.

Another unique property of ideal polymethines is their high mean π-electron polarizability $\bar{\alpha}_\pi$ in the ground state, which greatly exceeds that of polyenes and aromatic compounds of comparable size (39). However, whereas $\bar{\alpha}_\pi$ of the latter is strongly enhanced on $S_0 \to S_1$ light excitation, it remains nearly constant with ideal polymethines (40).

The basic theoretical problem of resonance stabilization and π-electron delocalization in ideal polymethines has been already discussed for many years. As mentioned earlier, the delocalization energy (20a) and Dewar's resonance energy (28) are relatively high in the case of ideal polymethine cyanines. Nevertheless, Shaik et al. (6, 41) doubted that the driving force of π-electron delocalization in both cyclic and chain-shaped molecules is maximum resonance stabilization. On the basis of ab initio investigations [at levels of STO-3G, 6-31G, and 6-311G with extensive configuration interaction (CI) calculations] they concluded that π-electron delocalization is only a by-product of the geometric constraint imposed by suitable symmetry conditions of the rigid σ-electron frames of the molecules. To distinguish between the topologically caused resonance stabilization of aromatic compounds and the resonance stabilization of polymethine-like chains, it proved useful (35) to consider the different contributions of the kinetic energy and the potential energy to molecular stabilization. Using ab initio split-valence basis sets 6-31G and 4-31G, respectively, Ichikawa et al. showed that ideal polymethines are preferentially stabilized by lowering the potential energy of their π-electron system (35), whereas ideal aromatic compounds possess π-electrons of strikingly low kinetic energy (42). Experimental evidence of the features of substances having an approximately ideal polymethine structure will be presented in the following section.

B. Experimental Proof of the Ideal Polymethine State

Since the formulation of the ideal polymethine state in 1966, much work has been done to prove experimentally the predictions of the unique properties of ideal polymethines. Although much material has been provided regarding polymethines with more complicated structures the following survey will focus mainly on the most simple ones, the so-called streptopolymethines or $1,\omega$-substituted polymethinium ions, where terminal substitutions of **1b**, n by amino groups ($X = Y = -NR_2^{1/2+}$) yields the polymethinestreptocyanines (**2**, n) by hydroxide anions ($X = Y = -O^{1/2-}$), the polymethine oxonols (**3**, n), and by both one amino group and one hydroxide anion the polymethine merocyanines (**4**, n). The relationship between these simple polymethines and

the $1,\omega$-substituted polyenes was explained in Fig. 4 and Table 1. Other terminologies often used are vinylogous formamidinium ions with the cyanines **2**, n, anions of the ω-hydroxypolyenals (e.g., 5-hydroxyglutaconal-dehyde sodium, 7-hydroxysorbinaldehyde sodium) with the oxonols (**3**, n) and vinylogous formamides or ω-aminopolyenals with the merocyanines (**4**, n).

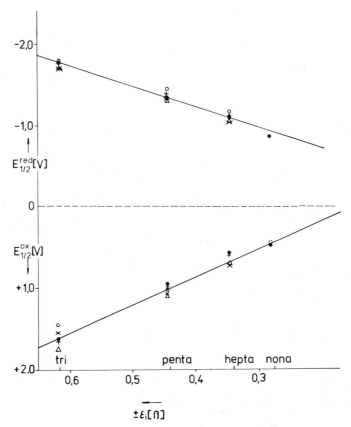

Figure 10 Experimentally determined oxidation and reduction half-wave potentials $E^{ox}_{1/2}$, $E^{red}_{1/2}$ (solvent acetonitrile vs. saturated calomel electrode) of polymethine streptocyanines (**2**, $n = 2$) versus frontier orbital energies ε_i calculated by the HMO-β_{SC} method. Substituents NR$_2$ (43): ● dimethylamino; △ monomethylamino; ○ pyrrolidyl; + piperidyl; × morpholyl. HMO parameters: $h_N = 1.0$; $k_{C=X} = k_{C-X} = 1.0$.

The eigenvalues ε_i (HMO-β_{SC} values in β units) of the frontier MOs linearly correlate with the respective anodic and cathodic half-wave potentials of simple polymethine cyanines as shown in Fig. 10(43). The correlation functions read

$$E^{ox}_{1/2}[V] = 3.478\varepsilon_{HOMO} - 0.537 \qquad \begin{array}{l} n = 17 \\ SD = 0.437 \\ r = 0.9853 \end{array} \qquad (5)$$

$$E^{\text{red}}_{1/2}[V] = 2.720\varepsilon_{\text{LUMO}} + 0.144 \qquad \begin{array}{l} n = 17 \\ SD = 0.348 \\ r = 0.9711 \end{array} \qquad (6)$$

where the following statistical parameters are used throughout the text (10, 14): number of measuring points n, standard deviation SD and correlation coefficient r.

There is also a linear correlation between the lowest transition energy E_T of polymethinecyanine dyes (which corresponds to the difference of their frontier orbital energies ε_i) and the difference between their oxidation and reduction half-wave potentials $\Delta E_{1/2}$, whereby all functions have a common origin (43a).

Because molecules will undergo intramolecular redox reactions if their oxidation half-wave potential is more negative than their reduction potential, it was derived from the correlation function that cyanine dyes having a chain length greater than 15 methine atoms are no longer stable. The same should be true with the vinylogous pyrylium cyanines whose half-wave potentials were published by Tolmachev et al. (44) giving the following correlation:

$$\Delta E_{1/2}[V] = E^{\text{ox}}_{1/2} - E^{\text{red}}_{1/2} = 1.289\, E_T - 0.999 \qquad \begin{array}{l} n = 13 \\ SD = 0.488 \\ r = 0.9940 \end{array} \qquad (7)$$

where the transition energy E_T of the longest wavelength absorption band is given in electron volts. In this case $\Delta E_{1/2}$ equals zero when the light absorption amounts to about 1600 nm, corresponding to, for example, a heptadecamethine thiapyrylium dye.

The problem of resonance stabilization of polymethines is of special interest. From the experimentally accessible enthalpies of protonation reactions of polymethines, in comparison to their polyenic or aromatic derivatives, resonance energies of about 4 kcal mole^{-1} of triphenylmethane dyes (45) and about 25 kcal mole^{-1} of simple trimethine cyanine entities (46) were estimated. From the increase in the experimentally determined heats of combustion of homologous series, in comparison to the first entity of the series under consideration [e.g., formamide in the simple merocyanines (**4**, n) and benzene in the polyacenes], the resonance energy of vinylene groups and their π-electrons, respectively, were determined (47). In this way the polyacenes yield a resonance energy per π-electron of 2, 5 kcal mole^{-1}, whereas, as expected, the polyenes give no additional resonance stabilization on vinylogous lengthening. For the ideal streptocyanines (**2**, n), the resonance energy per π-electron could be theoretically estimated to be about 2.0 kcal mole^{-1}.

As the molecular structure of merocyanines (**4**, *n*) is intermediate between that of ideal polymethines and of ideal polyenes (48), the experimentally obtained resonance energy per π-electron of about 0.6 kcal mole^{-1} is quite resonable.

The theoretically expected constant vinylene shift of polymethines is in agreement with the well-known fact that the maximum of their longest wavelength absorption band λ_{max} is red-shifted by about 100 nm on vinylogous lengthening. An inspection of the available data resulted in the revised Lewis–Calvin rule according to the following general equation (49)

$$\lambda_{max}[\text{nm}] = k(n+q)^a \qquad (8)$$

where the exponent a describes the nature of the basic molecular structure. It is 1.0 in case of ideal polymethines and approximates ≈ 0.37 as the structure becomes more polyene-like. The increment q describes the extraconjugation of additional substituents and of branching effects of the proper molecular chain in units of π-electron pairs and can be compared with the "effective length" of substituents, which were theoretically derived by Dyadyusha et al. (36). The constant k is given by the hypothetical absorption wavelength of one pair of π-electrons. It amounts to about 100 nm with ideal polymethines.

In Fig. 11 the longest wavelength absorption band of the bis(dimethylamino)polymethinium cations **2**, *n*(R = Me) is shown. Simultaneous with the vinylene shift of 104.5 nm, the transition probability, that is, the absorption intensity, is increased and the half-width of the absorption band is reduced. According to the Franck–Condon principle, the latter is a reliable indication of diminishing vibrational coupling between the π-electron system and the residual molecular frame. This implies, however, that the π-electron delocalization is the stronger the longer the polymethine chain is, in contrast to the prediction of the $\sum \Delta q_\pi$ parameter (see Fig. 2), which is reduced on vinylogous lengthening of molecular chains (2b). According to this point of view, the $\sum \Delta q_\pi$ parameter cannot be used generally as a measure of π-electron delocalization and resonance effects in LFERs.

Experimental proof of π-bond order equalization in ideal polymethines is provided by X-ray structural analyses of several polymethine streptocyanines (**2**, *n*) with trimethine chains (48, 50a), pentamethine chains (48, 50b), and heptamethine chains (50c), as well as the acetylacetonates having trimethine oxonol fragments (**3**, $n = 1$) (48) and the trimethine oxonol itself (51), which all possess an all-*trans* structure. Nearly unperturbed trimethine fragments in all-*trans* configuration are also realized in 2,5-donor-substituted *p*-benzoquinones, such as the 2,5-diamino-benzoquinone-1, 4-diiminium dication (52), which has two cyanine units (**2**, $n = 1$), and the dianion of the 2,5-dihydroxy benzoquinone-(1,4), which has two oxonol units (48). Heptamethine fragments in all-*cis* configuration contain the 7-hydroxytropolonate anion

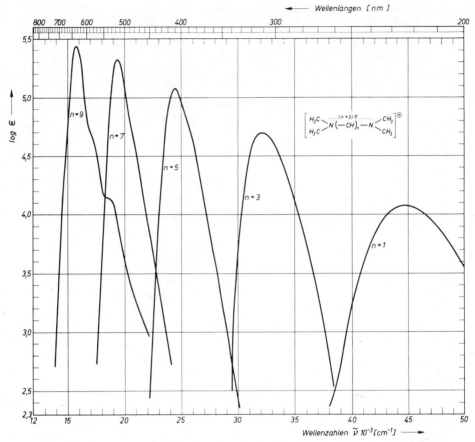

Figure 11 Longest wavelength absorption band of 1,ω-bis(dimethylamino)polymethinium cations (**2**, n; R = Me) in the UV–visible region in ethanolic solution.

and its amino- and thia-derivatives (48). Rather realistic bond lengths of the ideal polymethine state can now be estimated from the numerous experimental results. The mean value of 48 independently determined C⋯C bond lengths amounts to $(1{,}383 \pm 0{,}024)$ Å and that of 28 C⋯N bonds, to $(1{,}314 \pm 0{,}013)$ Å. Using the empirical functions (Equations 9 and 10) between bond lengths and HMO π-bond orders determined by Häfelinger (53), the π-bond orders of all C⋯C and C⋯N bonds indeed are close to the value given by the analytical function presented in Table 2.

$$r_{r,s}^{\text{CC,HMO}-\beta_{\text{SC}}}[\text{Å}] = 1.548 - 0.233\, p_{r,s} \qquad \begin{array}{l} n = 136 \\ SD = 0.016\,\text{Å} \\ r = 0.9080 \end{array} \qquad (9)$$

$$r_{r,s}^{CN,HMO}[\text{Å}] = 1.460 - 0.209\, p_{r,s} \quad \begin{array}{l} n = 308 \\ SD = 0.021\,\text{Å} \\ r = 0.7900 \end{array} \tag{10}$$

Further confirmation of the π-bond order equalization in the polymethine streptocyanines has been recently provided by the interpretation of their infrared (IR) and Raman spectra in connection with force-field calculations (54). Also, the NMR spectroscopically determined barrier of hindered internal rotation around the $C \cdots NR_2$ group (55) in terms of activation free energy $\Delta G^{\#}$ in the order of 17 kcal mole^{-1} (55c) agrees well with the expected π-bond order equalization in vinylogous streptocyanines. Nearly the same value of the rotational barrier around polymethinic $C \cdots C$ bonds was estimated (55c) from the kinetics of the photochemical induced *cis–trans* isomerization process (56).

The mean value of the $C \cdots C$ bond lengths is remarkably smaller than the ideal aromatic bond length of 1,395 Å. It seems reasonable to assume that this bond length contraction is brought about by additional Coulomb attraction between the neighbored positively and negatively charged carbon atoms, as a consequence of the π-electron density alternation in polymethines, indicating that the polymethinic bonds are partly ionic in nature. Kulpe earlier assumed for the same reason (57a) that the anomalous short $C=O$ bond length in polymethines is caused by Coulomb attraction forces between oppositely charged atoms. The adequate Coulomb repulsion of equally charged, neighbored atoms had been unambiguously proved in coupled polymethines such as the quinone dyestuffs observed by the same author (20b, 57b).

More sophisticated quantum-chemical methods, such as ab initio calculations with valence split basis sets 6-31G and 4-31G (35), PPP calculations (33a), and even the HMO-β_{SC} calculations (9c), give some indication that there should be slight deviation in full π-bond order equalization with symmetric polymethine cyanines. Within the limits of the standard deviation the X-ray results (48) as well as the H—H coupling constants of the ^{1}H NMR spectra (58a) seem to confirm the theoretically predicted behavior.

Quite another reason for slight deviation from π-bond order equalization is found in the crystalline state of symmetric cyanines when the anions of the dyes are nonsymmetrically positioned in respect to the mirror plane of the cationic dye molecules. In this case the π-electron system is polarized in direction to one possible polyenic limiting structure (see structure **1a**), causing small π-bond order alternation (48, 50b). Because of the exceptionally high polarizability of ideal polymethines, this is quite understandable.

The theoretically anticipated charge alternation in polymethine-like molecules was first proved by means of ^{1}H NMR spectroscopy by Scheibe et al. (58a) and Daehne et al. (58b) and later confirmed by means of ^{13}C NMR measurements (59). According to Equation 11, the measured ^{1}H and ^{13}C NMR chemical shifts δ are well correlated with each other (59b).

$$\delta(^{13}\text{C})[\text{ppm}] = 23.13\,\delta(^1\text{H}) - 15.27 \qquad \begin{array}{l} n = 33 \\ SD = 32.6 \\ r = 0.9670 \end{array} \qquad (11)$$

On the other hand, the chemical shifts satisfactorily correlate with the quantum-chemical accessible π-electron densities q_r (33a, 59b, 60). The correlation function (Equation 12) reads for the results of PPP calculations as an example (59b):

$$\delta(^{13}\text{C})[\text{ppm}] = 318 - 184\,q_r^{\text{PPP}} \qquad \begin{array}{l} n = 33 \\ SD = 32.6\,\text{ppm} \\ r = 0.9609 \end{array} \qquad (12)$$

Nearly the same results are obtained using CNDO/2 (33c) or ab initio (35) calculations.

As to LFERs, charged substituents have usually been omitted because of strong deviations from linearity (60, 61). It should be emphasized, however, that there are no significant deviations from the correlation function given above in respect to the cationic cyanines (**2**, n), the anionic oxonol dyes (**3**, n), and the neutral merocyanines (**4**, n). It must be concluded, therefore, that the reason for the deviation in LFERs is not the ionic charge of special substituents, but their strong conjugation effect, which is more pronounced in the ions only because of their more symmetric π-electron structure.

As will be shown later, thorough investigations of the electronic structure of the neutral merocyanines (**4**, n) revealed that they possess a more or less polymethinic structure in dependence on the polarity of the solvent used. This structure involves a rather high dipole moment, where the charges at the positive end and at the negative end, respectively, are of the same order of magnitude as the terminal charges of the cyanines or the oxonols (33c, 62) causing, for instance, that the protonation of merocyanine-like molecules occurs at the oxygen atom and not at the terminal amino group (63). This may explain why the methine chemical shifts of charged and neutral polymethinic entities behave in nearly the same manner.

The strong π-electron alternation in polymethine-like molecules explains why they are subject to electrophilic and nucleophilic substitution reactions much more readily than are substituted aromatic compounds. Electrophilic nitration, bromination, diazotation, Vilsmeyer formylation, and related phenomena at atomic centers of high π-electron density are common reactions in color chemistry (64). Electrophilic deuterium exchange reactions also easily take place (65). For instance, methylene blue is instantaneously deuterated at the 2,7 positions, even in neutral solution at room temperature (66). With the polymethine streptocyanines (2, n) and merocyanines (4, n), the logarithm of the rate constant k_D of the H/D exchange reaction at pH 5.0 is linearly related to the quantum-chemical localization energy LE of the atoms under consideration (in β-units on HMO basis) (67) according to

$$\log k_D = 18.77 LE + 16.65 \qquad \begin{array}{l} n = 7 \\ SD = 1.35 \\ r = 0.8883 \end{array} \qquad (13)$$

The longer the polymethine chain, the higher the H/D exchange rate. Expectedly, the neutral merocyanines are more easily deuterated than the positively charged cyanines.

Nucleophilic reactions at polymethine atoms of low π-electron density are well known with the addition of hydroxyl ions at the *meso* position of di- and triphenylmethane dyes to form the colorless carbinol bases. This reaction, however, has much more general meaning in that all Lewis bases are more or less easily added to the nucleophilic centers of polymethines (68).

In the case of heptamethine cyanines and longer vinylogues, the nucleophilic attack of primary or secondary amines even results in a C⋯C splitting of the chain, forming enamine and polymethines of shorter chain length (69), which supports the conclusion made from the bond length contraction that polymethinic bonds are partly ionic in nature. Also, intramolecular cycloaddition reactions between electrophilic and nucleophilic atomic centers appear in streptopolymethines forming after additional elimination of amine or water pyridine derivatives in the case of pentamethines (70), benzaldehyde and cinnamic aldehyde, respectively, in the case of hepta- and nonamethines (71). Obviously such electrocyclic reactions are controlled by the orbital symmetry of the atoms involved according to the Woodward–Hoffmann rule.

The reversal of the π-electron densities in the first excited singlet state after light excitation has been detected also by the expected interchange of the electrophilic and nucleophilic reaction centers. For instance, the photochemically induced electrophilic deuteration of *para*-donor–acceptor-substituted benzenes takes place at atoms that have low π-electron density in the ground state (72). The unique polymethinic π-electron density alternation in molecules also produces second-order effects that are related to hybridization changes of the methine atoms according to their different π-electron densities. The significant shortening of the C⋯C bonds in comparison to aromatic ones was already mentioned. The initial proof of hybridization changes caused by the π-electron alternation was provided by X-ray structural analyses of streptopolymethines in the crystalline state (48, 50). Kulpe et al. (73) showed clearly that there is a significant alternation of the C—C—C bond angles within polymethine chains between about 117° and 128° that resembles the π-electron density alternation. Obviously, deviation of the π-electron density from unity causes rehybridization effects in the σ-electron frame. As the experimentally determined bond angles θ_{ij} are related to the hybridization parameter λ according to Equation 14, it was observed that the square of the hybridization parameter λ^2 linearly depends on the total electron density Q_r at atoms r calculated, for example, by the CNDO/2 method (Equation 15):

$$\lambda^2 = -\frac{1}{\cos}\theta_{ij} \qquad (14)$$

$$\lambda^2 = -0.964 Q_r + 2.001 \cong 2 - Q_r \qquad (15)$$

These equations show that atoms with high π-electron density have higher p character of the sp^2 hybridized state, so that the bond angles θ_{ij} become smaller, whereas at atoms with low π-electron density, the s character of the atomic orbital is increased and hence the θ_{ij} values increase. Thus, it follows that the overall shape of cyanine and merocyanine chains, which is expected to be banana-like because of the one-sided positioned *cis*-configurated R substituents at the terminal nitrogen atoms (structure **2**, *n*), is given a more elongated, planar form in its ground state (50c, 74).

The results are in agreement with the principle of the maximum overlap of atomic orbitals (75) and have been confirmed in the liquid state by Radeglia (59b), who showed that the C—H coupling constants of the ^{13}C NMR spectra of simple cyanines, oxonols, and merocyanines alternate in the same manner as do the bond angles. However, their absolute values additionally depend on the charge of the molecules investigated.

The alternating hybridization changes of the carbon atoms in ideal polymethine chains expectedly influence the IR and Raman spectra in similar manner. Especially the C—H bending frequency depends strongly on the π-electron density and varies between 760 and 820 cm^{-1} in monomethines and trimethines, assignments of which were first given on the basis of force-field calculations by Mitzinger et al. (54a, b) and were recently confirmed with the 1,5-bis(dimethylamino)pentamethinium perchlorate by Tanaka et al. (54c).

As ring strain and other steric effects as well as electronegativity effects of the substituents are absent within the central part of polymethine chains, the relationships derived are much easier to interpret than are the bond length and bond angle dependencies of substituted aromatic compounds in LFERs published by Domenicano, Krygowski, and others (76). Also in the interpretation of the electronic structure of donor–acceptor-substituted ethylenes (77) the π-electron-induced changes of hybridization must be taken into account.

As with the reversal of the electrophilic and nucleophilic reaction centers on $S_0 \to S_1$ light excitation, one should expect a reversal of the bond angles, too, which should cause a geometric bending of the polymethine chains in their first excited singlet state (78). This prediction was confirmed recently by Tanaka et al. (54c) with preresonance Raman spectroscopical measurements. The totally symmetrical bending modes of the central C—C—C fragment of the pentamethine streptocyanine (**2**, $n = 2$, R = Me) give the most intense Raman bands at 248 and 1131 cm^{-1}, which means that the displacement of the potential minimum of the normal coordinate upon light-excitation into the first excited singlet state is exceptionally large at the central C—C—C angle.

As polymethines have longer wavelength absorption than do aromatic and polyenic compounds having the same number of π-electrons (9), they should, expectedly also have the largest π-electron polarizability. This was confirmed by measuring the molar refractivity of vinylogous streptopolymethines (79). The increase of the mean π-electron polarizability $\bar{\alpha}_\pi$ of homologous polyacene, polyene, and polymethine series in dependence on the number of π-electrons can be seen in Fig. 12. The contribution of the σ-electron frame of the molecules to the measured total polarizability was eliminated by incremental values. In agreement with the theoretical results presented in Section III. A, the polymethine cyanines (**2**, *n*) are, indeed, much more easily polarizable than are polyenic and aromatic compounds. Again the merocyanines (**4**, *n*) exhibit a behavior that is intermediate in terms of dependence on the polarity of the solvent used.

The exceptionally high π-electronic polarizability of polymethines results in some interesting consequences. It explains, for example, the previously mentioned induction of a nonsymmetric polyene-like structure in crystals of symmetrically structured polymethines only by nonsymmetric positioning of the anions in respect to the dye cations. Obviously, the high polarizability also explains why polymethine dyes have an uncommonly strong tendency to form dimeric and polymeric molecular aggregates already in the ground state (80), which is assumed to be caused by strong dispersion forces. As aromatic and polyenic compounds have the same or even a larger π-electronic polarizability in their first excited singlet state as do the polymethines in their ground state (see Section III.A), it is understandable that only after light excitation do those molecules produce molecular aggregates, which are well described in literature as excimers and exciplexes (81).

The high π-electronic polarizability of polymethines also causes a strong dependence of their π-electron structure on the polarity of the solvent used, even when the dye molecules are completely symmetric (82). Thus the solvent-induced changes of the chemical shifts and of the vicinal H—H

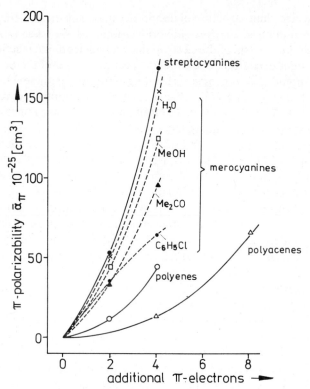

Figure 12 Increase of the mean π-electron polarizability $\bar{\alpha}_\pi$ of polyacenes, polyenes, merocyanines (**4**, n; R = Me), and streptocyanines (**2**, n; R = Me) on homologous enlargement. The reference substances are benzene, ethylene, dimethylformamide, and tetramethylformamidinium ion, respectively. The solvent used with the merocyanines is indicated on the curves.

coupling constants in the ^{1}H and ^{13}C NMR spectra of the streptocyanines (**2**, n) assume values of the same order of magnitude as that usually found with strongly solvatochromic molecules having a high dipole moment such as the merocyanines (**4**, n) and donor–acceptor-substituted aromatic compounds (see Table 1).

Strong evidence of the unique properties of ideal polymethines comes from the solvent-dependent experimental behavior of the positively solvatochromic merocyanines (**4**, n). The vinylene shift of their longest wavelength absorption band on vinylogous lenthening is much higher in polar solvents than in nonpolar ones (83). According to the revised Lewis–Calvin rule (Equation 8) (49), the exponent a amounts to 0.72 in water and 0.47 in n-hexane, indicating a typical polyene-like behavior in nonpolar solvents and a polymethine-like one in polar solvents. The exponent a is well correlated with the solvent polarity parameter E_T after Reichardt (21c) according to Equation 16 (49), which proves that it is the overall solvent polarity that changes the structure of the merocyanine between the ideal polymethine state ($a \approx 1.0$) and the ideal polyene state ($a \approx 0.37$). The solvent-dependent correlation functions calculated according to Equation 8 show that the increment q of additionally conjugating π-electron pairs is zero in polar solvents but -1 in nonpolar ones, which means that in the latter case the conjugation of the polymethine chain is clearly restricted. The same result is obtained with the solvent dependence of the oscillator strength f of the longest wavelength absorption band of the pentamethine merocyanine (**4**, $n = 2$), which amounts to $f = 0.97$ in water, close to the value of the corresponding streptocyanine (**2**, $n = 2$) ($f = 1.04$ in ethanol), whereas it has only a polyene-like value of $f = 0.64$ in n-hexane (84).

$$E_T[\text{kcal mole}^{-1}] = 129.7a - 29.2 \qquad \begin{array}{l} n = 15 \\ SD = 10.7\,\text{kcal mole}^{-1} \\ r = 0.9896 \end{array} \qquad (16)$$

Well-defined proof of solvent-dependent geometric changes in merocyanines (**4**, n) between the polyene-like structure (**1a**, n) and the more polymethine-like structure (**1b**, n) is given by NMR and IR spectroscopy. The bond orders estimated by the H—H coupling constants of the ^{1}H NMR spectrum alternate much more in nonpolar solvents than in polar ones (85):

$$\Delta J\,[\text{Hz}] = 6.63 - 0.0575 E_T \qquad \begin{array}{l} n = 10 \\ SD = 0.62\,\text{Hz} \\ r = 0.9391 \end{array} \qquad (17)$$

This equation holds for trimethine merocyanine (**4**, $n = 1$; $R = \text{Me}$), for instance (83), where ΔJ is the difference of the coupling constants between the C_1–C_2 and C_2–C_3 bond and E_T is again Reichardt's polarity parameter in kcal mole^{-1}. The adequate increase of the bond C$\cdots$N order with increasing polarity of the solvent is indicated by the growing activation free energy of the hindered internal rotation around the C$\cdots$N bond (86), whereas the analogous decrease of the C$\cdots$O bond order can be seen in the IR spectrum with the frequency of the C=O stretching vibration (87). Somewhat more complicated are the ^{15}N chemical shifts of the NMR spectra, which depend on both the C$\cdots$N bond order and the π-electron density at the nitrogen atom. Nevertheless, these measurements are also in good agreement with the expected solvent-dependent structural changes (88).

Parallel to the alterations of the bond lengths with varying solvent polarity in the merocyanines, strong changes in the π-electron densities occur,

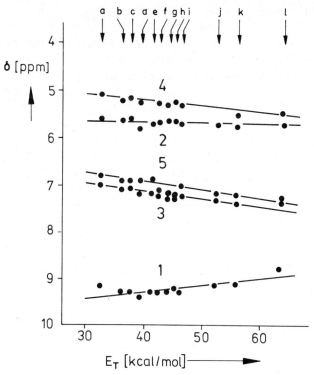

Figure 13 The ^{1}H NMR chemical shifts δ of the methine protons of the pentamethine merocyanine (**4**, $n = 2$; R = Me) after Radeglia et al. (87) versus the solvent polarity parameter E_T after Reichardt (21c). The numbering starts at the carbonyl methine atom. Solvents used: (*a*) *n*-hexane; (*b*) dioxane; (*c*) tetrahydrofurane; (*d*) chloroform; (*e*) methylene chloride; (*f*) acetone; (*g*) pyridine; (*h*) dimethylsulfoxide; (*i*) acetonitrile; (*j*) ethanol; (*k*) methanol; (*l*) water.

as indicated by the ^{1}H and ^{13}C NMR spectra (59, 83). The ^{1}H NMR chemical shifts δ of the pentamethine merocyanine (**4**, $n = 2$) in dependence on Reichardt's E_T scale are shown in Fig. 13 as an example.

As with the symmetric polymethines in the merocyanines a strong alternation of the bond angles have been observed (48, 89), proving its polymethine-like character. On the other hand, according to the X-ray structural results (48), their bond lengths alternate much more than is to be expected for ideal polymethines. All these results do not fit simple resonance considerations because, as in structures **1**, *n*, only a linear polarization of the π-electron system accompanied by an increase of the molecule's dipole moment is to be expected when the solvent polarity is increased.

For this reason, and on the basis of CNDO/2 and VESCF calculations, a microstructural model of solvatochromism was developed (62, 90) that gives a consistent explanation of the experimental results found. From that model it follows that the dipole moment of strongly conjugated donor–acceptor-substituted polyenes such as the merocyanines (**4**, *n*) must be strongly influenced by the solvent polarity already in the ground state. Such changes have been hitherto neglected in other models of solvatochromism (21c, 91) as well as in LFERs, although the behavior of bond length, bond angle, and electron density parameters gave some evidence of the special features of molecules substituted with strongly conjugated substituents (77, 92). It will be shown in Section IV that the structural and solvent-induced changes in such substances have a strong impact on LFERs.

The microstructural model of solvatochromism includes not only the case of positively solvatochromic dyes where the molecular structure is varying between its nonpolar from **1a** and its symmetric one **1b** but also the case of negatively solvatochromic dyes, such as **5** and **6**, which are assumed to posses,

in nonpolar solvents a symmetric structure **1b** or even a more polar structure in the direction of **1c**. As this structure is further stabilized with increasing solvent polarity, the molecules are, indeed, further transformed into the polyenic structure **1c** accompanied with a blue-shift of the light absorption (93). Also in this case ^{1}H and ^{13}C NMR spectroscopical measurements gave evidence that the model well describes the experimental facts of solvent-dependent changes of the electronic structure (94).

IV. ON THE WAY TO A JOINT MODEL BETWEEN THE IDEAL POLYMETHINE STATE AND LFERs IN STRONGLY CONJUGATED SUBSTITUENTS

In LFERs aromatic and polyenic groups and their substituents are considered to have primarily localized π-electron structures that become increasingly delocalized. Thus the general concept of LFERs is related to O. N. Witt's classical model of color and constitution, which considers molecules to consist of several building blocks such as the aromatic and polyenic chromophores and the auxochromic (i.e., donating) and antiauxochromic (i.e., accepting) substituents (17). It has been already mentioned, that the stronger the delocalization of the π-electron system, the greater the deviations from linearity in free-energy relationships.

On the other hand, the theory of the ideal polymethine state starts with fully delocalized π-electronic structures represented by symmetric limiting structures, which become more localized with stronger perturbations of the symmetry of the π-electron distribution along a molecular chain. As illustrated in Fig. 4 and Table 1, polymethines can also be formally considered to consist of aromatic and polyenic units that are substituted with donor and acceptor substituents whose individuality, however, is completely lost.

To combine both theories, it will be thus necessary to investigate the molecular alterations when fully delocalized, ideal polymethine structures are progressively transformed into localized polyenic and/or aromatic ones.

A. The Quantum-Chemical Method

Fundamental relationships between polymethinic, polyenic, and aromatic structures should be best recognized on the basis of quantum-chemical calculations. In present-day quantum chemistry, usually ab initio methods are used in describing LFERs (2, 12, 13, 95), which doubtlessly are most appropriate for precisely calculating the molecular structure of any concrete compound. However, if only the tendency of certain parameters on structural changes should be estimated, such methods—as well as semiempirical all-valence electron methods and even methods based on π-electron level, which take into account π-electron repulsion effects such as the PPP method—are too sophisticated.

The ab initio methods allow the treatment of only actually existing atoms and substituents and, as to semiempirical methods, require the arbitrary variation of more than one parameter for modeling molecular changes, which is rather difficult to be seen through. For this reason and as mentioned in a preceding study (9c), the one-electron Hückel MO (HMO) treatment on π-electron level has been used in connection with the β_{SC} procedure proposed by Golebiewski et al. (96) in order to cover the influence of polyenic bond orders in a better way. Test calculations showed that the same basic relationships can be achieved by including the ω technique to take into account the strong π-electron density changes, or even more accurately by PPP calculations. The advantages of the HMO-β_{SC} procedure are:

1. The restriction on the π-electron level excludes a priori the influence of the σ-electrons on molecular parameters and hence allows the study of pure resonance effects.

2. With the exception of some vectorial quantities such as transition moment and molecular polarizability, the essential molecular parameters such as π-bond orders and π-electron densities are exclusively dependent on the topology, but not on the geometry of the molecules. In this way "through-bond" effects along "molecular lines of force" (97) can be investigated without any interference with "through-space" or field effects.

3. The more or less arbitrary variation of the resonance integral β of bonds with participating heteroatoms according to Equation 18 can be eluded because this is covered by the β_{SC} formalism:

$$\beta_{C=X} = k_{C=X} \cdot \beta_{C=C} \qquad (18)$$

Thus the Coulomb integral value of the heteroatoms $\alpha_{X,Y}$ is the only parameter to be varied in the calculations according to

$$\alpha_{X,Y} = \alpha_C + h_{X,Y} \cdot \beta_{C=C} \qquad (19)$$

where $h_{X,Y}$ represents the π-orbital electronegativity of the terminal heteroatoms X and Y. It was shown that the $h_{X,Y}$ values can be directly derived from the valence state electronegativities introduced by Hinze and Jaffé (98) by setting $\alpha_{X,Y} = -\chi_{X,Y}$ with due allowance for the number of the $2p$ electrons supplied by the atom under consideration (99). Thus the values $h_{X,Y}$ are a crude measure of the atom's overall electronegativity. Roughly speaking, $h_X = 0$ corresponds to carbon atoms, $h_X = 1.0$ to nitrogen atoms, $h_X = 2.0$ to oxygen atoms, $h_X = 3.0$ to fluorine atoms, and $h_X = -1.0$ to boron atoms, confirming the primarily used values mentioned in Section III.A. As it is usual, α_C is set to be zero and hence h_X is equal to the Coulomb integral value of the heteroatom α_X. The latter is often taken within the text instead of h_X. Besides the well-known quantities used in quantum chemistry (25, 31), some new parameters describing important molecular features will be used in addition. These parameters were defined in previous studies also (9c, 27).

π-Bond order alternation:

$$\overline{\Delta p} = \frac{1}{N-1} \sum_{1}^{N-1} |\Delta p|$$

where N is the number of the atomic orbitals and Δp the difference between two neighbored π-bond orders $p_{r,s}$ in a polymethine chain.

π-Bond order equalization:

$$\frac{1}{\overline{\Delta p}}$$

π-Electron density alternation:

$$\overline{\Delta q} = \frac{1}{N} \sum_{r=1}^{N} |\Delta q|$$

where Δq is the difference between the π-electron density q_r of two neighbored atoms.

Sum of π-bond order changes on $S_0 \to S_1$ light excitation:

$$\Delta p^{S_0 \to S_1} = \sum_{1}^{N-1} |p_{r,s}^{S_0} - p_{r,s}^{S_1}|$$

where it is sometimes appropriate to additionally distinguish between bond length extension and bond length contraction by summing up separately the positive and negative differences (90).

Sum of π-electron density changes on $S_0 \to S_1$ light excitation:

$$\Delta q^{S_0 \to S_1} = \sum_{1}^{N} |q_r^{S_0} - q_r^{S_1}|$$

This parameter is identical to the total charge alteration on light excitation abbreviated, as CA in reference 100, where CA is the sum of the light-induced charge transfer CT plus the light-induced charge resonance CR within molecules (100b).

Symmetry deviation:

$$\Sigma = \sum_{r=1}^{1/2(N-1)} q_r - \sum_{r=1/2(N+3)}^{N} q_r$$

The parameter was explained in connection with structure **1** in Section III.A

Electronegativity difference of the terminal atoms:

$$\Delta \alpha_X = \alpha_X^{\Sigma=0} - \alpha_X$$

where $\alpha_X^{\Sigma=0}$ is that Coulomb integral value of the terminal atom X, which is necessary to symmetrize a molecule having a definite reaction site Y. Negative values mean that the electronegativity of the atom X is smaller than the electronegativity of the reaction site Y and vice versa. The quantity $\Delta\alpha_X$ can be taken as another measure of deviations from the symmetrical π-electron distribution of ideal polymethine chains. Therefore, it is more or less proportional to the symmetry deviation in the ground state Σ_{S_0}.

To ensure that the relationships derived from HMO-β_{SC} calculations are compatible with ab initio results, the outcomes of both methods are compared in Figs. 14 and 15. Figure 14 corresponds to Fig. 2 published by Topsom, Taft, et al. (2b), showing the relationship between the π-electron charge transfer $\Sigma\Delta q_\pi$ from a substituent to the residual molecule in dependence on the π-electron charge at the substituted atom of the residue. The curves in both figures are, indeed, very similar in appearance. As typical of the HMO method,

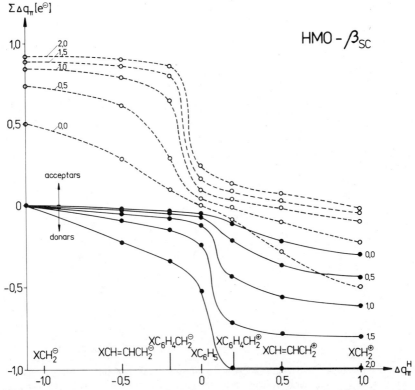

Figure 14 Plot of the π-electron charge transfer $\Sigma\Delta q_\pi$ of a substituent to and from the residual molecule versus the π-electron charge density, q_π^H, at the position of substitution in analogy to Fig. 2 but calculated by the HMO-β_{SC} method.

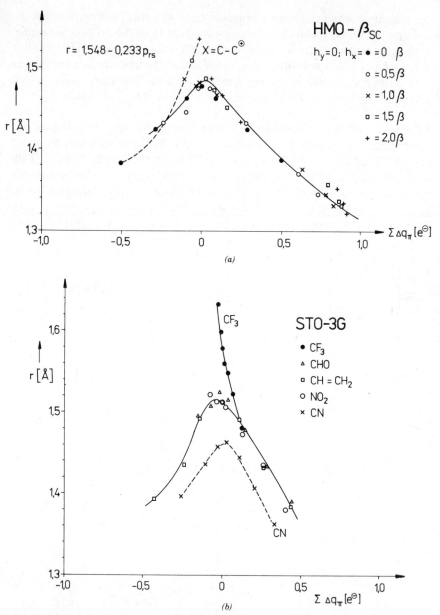

Figure 15 Plot of the bond length r between the substituent and the residual molecule versus the substituent's π-electron charge transfer parameter $\sum \Delta q_\pi$ for acceptor substituents that can act as donors on combination with very strong acceptors. STO-3G calculations from reference 2b; HMO-β_{SC} results from this work. The given bond lengths in the HMO-β_{SC} presentation were calculated from the π-bond orders $p_{r,s}$ by the empirical formula in Equation 9.

the calculated π-electron densities and hence also the charge transfer $\sum \Delta q_\pi$ are too large in comparison to experimental values. By using a scaling factor of about 0.5 in Fig. 14 and by comparing the results with Fig. 2, the Coulomb integral values of the heteroatoms used in the HMO-β_{SC} calculations can be directly assigned to the substituents used in the ab initio calculations. In this way the values indicated before ($h_C = 0.0$; $h_N \approx 1.0$; $h_O \approx 2.0$) are, indeed, realistic.

Because only π-electron charges are compared in the examples shown in Figs. 2 and 14, a comparison of π-bond orders and bond lengths, respectively, in respect to dependence on the π-electron charge-transfer parameter $\sum \Delta q_\pi$ is made in Fig. 15, with the acceptor substituents being able to donate or to accept π-electrons as shown before. In case of the HMO-β_{SC} calculations the π-bond orders $p_{r,s}$ were transformed to bond lengths r according to Häfelinger's empirical formula (Equation 9). Apart from the exceptional behavior of the trifluoromethyl-substituted compounds in the STO-3G results and of the acceptor-substituted carbocations in the HMO-β_{SC} results, both diagrams are very similar and justify the application of HMO-β_{SC} calculations for evaluation of principal relationships between LFERs of strongly conjugated compounds and ideal polymethine features.

B. Modeling Intermediate Structures between Polymethines, Polyenes, and Aromatics

The electronic structure of the ideal pentamethine streptocyanines (**1b**, $n = 2$; $X = Y = NR_2^{1/2+}$) having maximum π-electron delocalization is gradually transformed into substances with more or less localized polyenic structures (**1a, 1c**) through substitution of only one terminal nitrogen atom, $X = N$, or one $C=N$ group for substituents with varied electronegativity. This is modeled in Fig. 16. It is seen that all molecular parameters feature extremes in the symmetric state as previously outlined in reference 9c. For instance, transition energy E_T and π-bond order alternation $\overline{\Delta p}$ have a minimum, the transition moment μ_T and the π-electron density alternation $\overline{\Delta q}$ yield a maximum, the symmetry deviation $\sum_{S_0}$ equals zero, and the frontier orbital energies ε_{HOMO} and ε_{LUMO} exhibit at least a bending point. For comparison with LFERs, the $\sum \Delta q_\pi$ parameters of the donor substituent Y and of the acceptor group $C=X$ of the 1,4-substituted butadiene fragment calculated as in reference 2b are additionally included in Fig. 16. The disadvantage of those parameters is evident because with exception of their S shape, the $\sum \Delta q_\pi$ curves continuously increase and decrease, respectively, with increasing electronegativity of the terminal X atom without having a zero point in the symmetric, ideal polymethine structure like the symmetry deviation $\sum_{S_0}$. Another shortcoming was mentioned in Section II demonstrating that $\sum \Delta q_\pi$ is diminished on vinylogous lengthening of the polymethine chain (structures **1**), although there

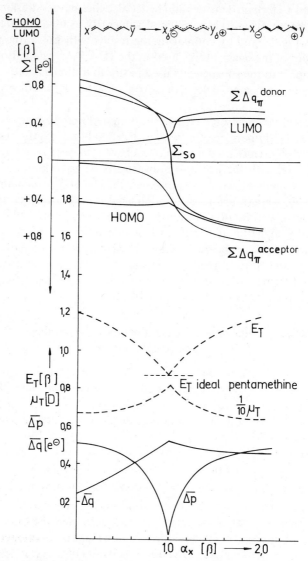

Figure 16 Symmetrization of the 1,4-donor–acceptor-substituted butadiene models **1a** and **1c**, $n = 2$, after Daehne and Moldenhauer (9c) by variation of the Coulomb integral value α_x, yielding the ideal pentamethine **1b**. See text for explanation of the parameters.

is strong experimental evidence (Section III.B) that resonance and delocalization within polymethines is higher the longer the polymethine chain. Therefore, the most appropriate parameters in the description of the electronic structure of chain-shaped molecules obviously are π-bond order alternation $\overline{\Delta p}$ and symmetry deviation $\sum$. Both parameters are close to zero or even zero in the ideal polymethine state. As for the polyene–polymethine intermediate structures shown in Fig. 16, maximum π-electron density alternation also seems to be well suited to represent ideal polymethinic behavior. It has been already demonstrated, however, that this is no longer true when symmetric polymethines having terminal atoms of varying electronegativity are compared with each other (see Fig. 6). Also, comparison of vinylogous polymethine chains reveals that charge alternation decreases with increasing chain lengths whereas π-electron delocalization increases in the same order. For that reason Gimarc's charge stabilization rule (37) mentioned in Section III. A cannot be the intrinsic property of maximum resonance and maximum π-electron delocalization, respectively.

It should be mentioned in this connection that the results of Fig. 16 also reflect the solvatochromic behavior of polymethine dyes. If a dye molecule in the gaseous state or in nonpolar solvents possesses a more or less nonpolar electronic structure in the ground state tending toward the limiting structure **1a** at the left side of Fig. 16 its electronic structure will be changed toward the direction of the symmetric structure **1b** with rising solvent polarity because the polar solvent molecules will stabilize the more polar electronic structure. This means that the π-bond order equalization, π-electron density alternation, and dipole moment will increase and the light absorption will be red-shifted, exhibiting positive solvatochromism in agreement with Fig. 16.

The same stabilization effect occurs when an isolated dye molecule having already in its ground state a polar structure in the direction of **1c** is given in solvents of rising polarity. The dipole moment will increase further. But according to Fig. 16, a blue-shift of the light absorption connected with rising π-bond order alternation must occur, yielding negative solvatochromism. The basic relationships between the structural parameters and solvent-dependent light absorption has been modeled in detail in the microstructural model of solvatochromism on the basis of PPP and CNDO/2 calculations (62, 90), which excellently confirm the conclusions suggested now by the HMO-β_{SC} picture.

To give a more convincing impression of the importance of π-electron localization at certain bonds, which is easily perceptible at their π-bond orders, the $p_{r,s}$ values of the pentamethine chromophor (**1**, $n = 2$) are plotted in Fig. 17 against the electronegativity difference of the terminal atoms $\Delta\alpha_X$. The symmetric structure is realized when $\Delta\alpha_X$ equals zero. In agreement with the

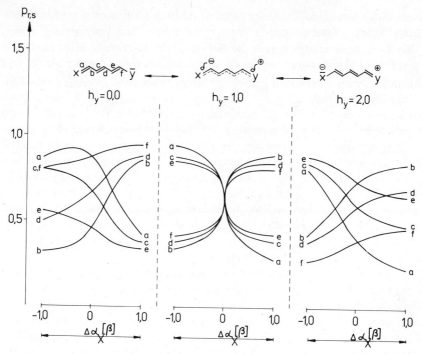

Figure 17 HMO-β_{SC} π-bond orders of pentamethine chains (**1**, $n = 2$) versus the electronegativity difference of the terminal atoms $\Delta\alpha_X$.

results of simple HMO calculations shown in Fig. 6, complete π-bond order equalization, that is, full π-electron delocalization, is realized only with the ideal pentamethine streptocyanine ($h_Y = 1.0$; $\Delta\alpha_X = 0$). Neither in the heptatrienyl anion ($h_Y = 0$; $\Delta\alpha_X = 0$) nor in the pentamethine oxonol ($h_Y = 2.0$; $\Delta\alpha_X = 0$) the π-bond orders can be fully equalized.

The complete connection between π-bond order equalization $1/\overline{\Delta p}$ in the ground state and the electronegativity of the substituents is presented in Fig. 18. The three-dimensional plot represents $\log(1/\overline{\Delta p})$ in dependence on both the electronegativity difference of the terminal atoms $\Delta\alpha_X$ and the electronegativity of the reaction site h_Y. The unique property of ideal polymethines having maximum delocalization is clearly seen with the maximum of $1/\overline{\Delta p}$ at $\Delta\alpha_X = 0$ and $h_Y = 1.0$. It is seen likewise that the strength of delocalization depends not only on the symmetry of the π-electron distribution along a polymethine chain but also on the Coulomb integral value of the reaction site h_Y, which represents the electron demand of the

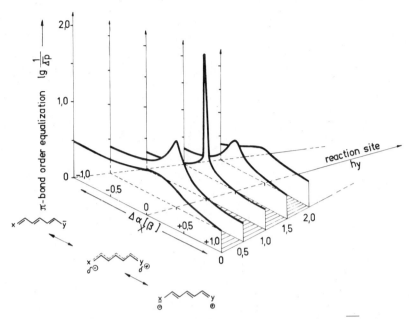

Figure 18 Three-dimensional plot of logarithm of π-bond order equalization $1/\overline{\Delta p}$ in the ground state S_0 of pentamethine chains (**1**, $n = 2$) versus the electronegativity difference of the terminal atoms $\Delta\alpha_X$ and versus the Coulomb integral value h_Y of the reaction site Y.

reaction site. Therefore, Fig. 18 reveals the physical background of the findings in LFERs (2) that the σ_R constants are a function of the electron demand of the reaction site. If one should try to linearize the functional relationships from a principal point of view this will be possible only between the limiting structures **1a** and **1b** or **1b** and **1c**, or between the reaction sites $h_Y = 0$ and $h_Y = 1.0$, or $h_Y = 1.0$ and $h_Y = 2.0$ or in any other desired direction starting with the maximum π-bond order equalization in the ideal polymethine state at $\Delta\alpha_X = 0$ and $h_Y = 1.0$. But linearization can never be realized over the full range of the possible structures shown. The analogous relationship concerning the symmetry deviation $\sum_{S_0}$ of pentamethines (**1**, $n = 2$) is given in Fig. 19. By definition, $\sum$ equals zero if the structures are symmetric. In this case the ideal polymethinic behavior is disclosed only by the strongest changes of $\sum$ on variations of $\Delta\alpha_X$ near to values of $\Delta\alpha_X = 0$ and $h_Y = 1.0$. Obviously, this is due to the exceptional high ground-state π-electronic polarizability of molecular structures in the ideal polymethine state.

In order to model intermediate structures between polymethinic and aromatic compounds, the same considerations have been applied with the *para*-donor–acceptor substituted benzenes **7** (see Fig. 4). Figure 20 shows the

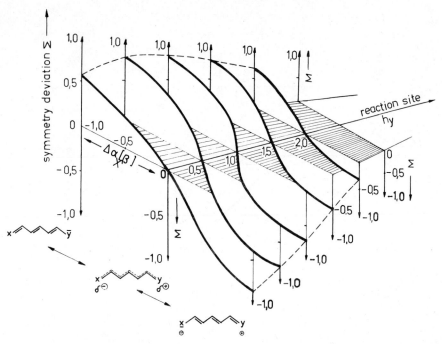

Figure 19 Three-dimensional plot of symmetry deviation Σ in the ground state S_0 of pentamethine chains (**1**, $n = 2$) versus the electronegativity difference of the terminal atoms $\Delta\alpha_X$ and versus the Coulomb integral value h_Y of the reaction site Y.

outcomes obtained in analogy to the 1,4-donor–acceptor-substituted butadienes (**1**, $n = 2$) presented in Fig. 16.

Remarkably, an aromatic structure with a nonsymmetric distribution of the π-electron system along the pentamethine chain ($\Sigma_{S_0} \approx 0.6$) is realized when the terminal heteroatoms X and Y are identical ($\alpha_X = \alpha_Y = 1.0\beta$). Following a proposal made by Dyadyusha et al. (101), this structurally

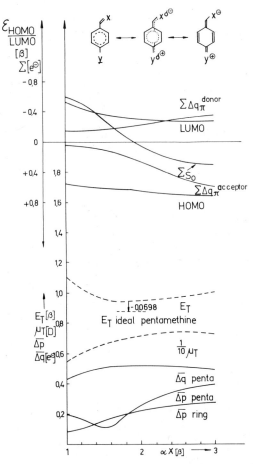

Figure 20 Symmetrization of *para*-donor–acceptor-substituted benzenes (**7**) by variation of the Coulomb integral value α_X, after Daehne and Moldenhauer (9c). See text for explanation of the parameters.

induced asymmetry of the π-electrons should be designated "topological deviation" or "topological asymmetry." To symmetrize the model it is necessary to enhance the acceptor strength of the C=X substituent, that is, to increase the Coulomb integral value of the X atom. According to Fig. 20, symmetrization with $\sum_{S_0} = 0$ is achieved at $\alpha_X = 1.85\beta$. At this point all the extremes observed in the nonbranched polymethine chains **1** (see Fig. 16), such as minimum transition energy E_T, and maximum transition moment μ_T, are present but no longer so distinctly marked. Only the minimum of π-bond order alternation $\overline{\Delta p}$ is slightly shifted to the left side of Fig. 20 because the aromatic benzene ring realized under this condition possesses minimum π-bond order alternation $\overline{\Delta p_{\text{ring}}}$ likewise. Similar to the polymethines **1** shown in Fig. 16, the $\sum \Delta q_\pi$ parameters of LFERs do not indicate the state of maximum delocalization at $\sum_{S_0} = 0$.

The π-bond order equalization $1/\Delta p$ and the symmetry deviation $\sum$ of the *para*-donor–acceptor-substituted benzenes in the ground state, in dependence on both the electronegativity difference of the terminal atoms $\Delta \alpha_X$ and the reaction site h_Y, are depicted in Figs. 21 and 22. In Fig. 21 the π-bond order equalization is by far no longer as pronounced as in the nonbranched polymethine chains shown in Fig. 18. However, the behavior is basically the same: maximum π-bond order equalization with ideal cyanines ($h_Y = 1.0$) near the symmetric state with $\Delta \alpha_X = 0$. (The shift of the maximum in direction to the nonpolar aromatic limiting structure **7a** is due to the π-bond order equalization within the benzene ring as explained immediately above.) Worth mentioning is the fact that substituted benzenes (**7**) whose donor electronegativity is much greater than nitrogen ($h_Y \geq 1.5$ in Fig. 21) are symmetrized only

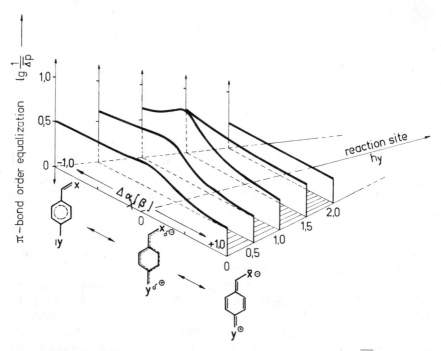

Figure 21 Three-dimensional plot of logarithm of π-bond order equalization $1/\overline{\Delta p}$ in the ground state S_0 of *para*-donor–acceptor-substituted benzenes (**7**) versus the electronegativity difference of the terminal atoms $\Delta \alpha_X$ and versus the Coulomb integral value h_Y of the reaction site Y.

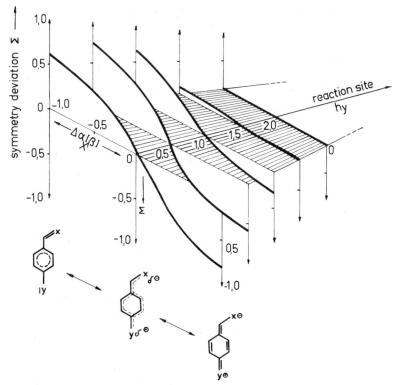

Figure 22 Three-dimensional plot of symmetry deviation Σ in the ground state S_0 of para-donor–acceptor-substituted benzenes (**7**) versus the electronegativity difference of the terminal atoms $\Delta\alpha_X$ and versus the Coulomb integral value h_Y of the reaction site Y.

when the C=X acceptor strength by far exceeds the experimentally realizable values (see Fig. 29, later), thus resulting in nearly constant $\overline{\Delta p}$ values with $h_Y = 2.0$. This independence of the electronic structure from variation of the C=X acceptor strength in molecules with donors of high electronegativity is even better illustrated with the symmetry deviation Σ_{S_0} at $h_Y = 2.0$ shown in Fig. 22. Here the strongest polymethine-like changes of the π-electron distribution occur when the reaction site has an electronegativity corresponding $h_Y = 0.5$.

It is beyond the scope of this contribution to consider excited-state properties of the investigated model compounds. It should be mentioned, however, that the unique features of molecules in the ideal polymethine state are also reflected in characteristic properties of their first excited singlet state (100). The model considerations prove, for instance, that ideal polymethine chains exhibit maximum π-electron reversal, that is, maximum charge resonance CR, and minimum π-bond order alterations on $S_0 \to S_1$ light excitation as well (100b). In addition in nonsymmetric, polymethine-like chains, there occurs on light excitation a charge transfer CT from both the nonpolar, polyenic structure **1a** and from the highly polar structure **1c** in the direction of the symmetric, ideal polymethine structure **1b**, meaning that the dipole moment will increase if one starts on the nonpolar side between structures **1a** and **1b**, but will decrease on starting with molecules having a more polar structure between **1b** and **1c**, in agreement with the behavior of positively and negatively solvatochromic dyes. In other words, nearly ideal polymethine models have the tendency to symmetrize their π-electron system in their first excited singlet state.

On the other hand, for the *para*-donor–acceptor-substituted benzenes (**7**), the light-induced charge transfer will be directed only in the forward direction; that is, it will increase the dipole moment in any case regardless of whether light excitation starts with molecules having a more nonpolar structure such as **7a** or a more polar one such as **7b** or **7c** (100b). Furthermore, there are substituted aromatic molecules, such as the *para*, *para'*-donor–acceptor-substituted stilbenes and azobenzenes, which cannot be symmetrized in the ground state even on changing the electronegativity of the substituents (100a). On the basis of these results, a new systematization and classification of unsaturated organic compounds has been proposed (100).

C. Limitations of LFER Parameters

The physically founded limitations of the charge-transfer parameter $\sum \Delta q_\pi$ and the σ_R constants in the description of free-energy relationships require a reexamination of the utility of other parameters used in LFERs.

As to correlation analysis, many publications are concentrated on ^{13}C NMR chemical shifts of the *para*-positioned carbon atom of substituted benzenes. The behavior of the π-electron density q_π at this position in dependence on the electronegativity difference of the terminal atoms $\Delta\alpha_X$ is presented by means of the HMO-β_{SC} calculations in Fig. 23 for both the pentamethines (**1**, $n = 2$) and the *para*-donor–acceptor-substituted benzenes (**7**). The q_π values decrease when the π-electrons are shifted from the donor Y to

Figure 23 Calculated π-electron density at the C^5 position of pentamethine chains (**1**, $n = 2$) and at the corresponding C^4_{para} position of the *para*-donor–acceptor-substituted benzenes (**7**) in dependence on the electronegativity difference of the terminal atoms $\Delta\alpha_X$. The h_Y values used for the reaction site Y are indicated on the curves. The values at the ordinate correspond to the ideal pentamethine fragment with $h_Y = 1.0$. For reasons of clarity, the upper two curves have been vertically shifted by $+0.3$ and $+0.6$ units and the two lower curves by -0.3 and -0.6 units, respectively.

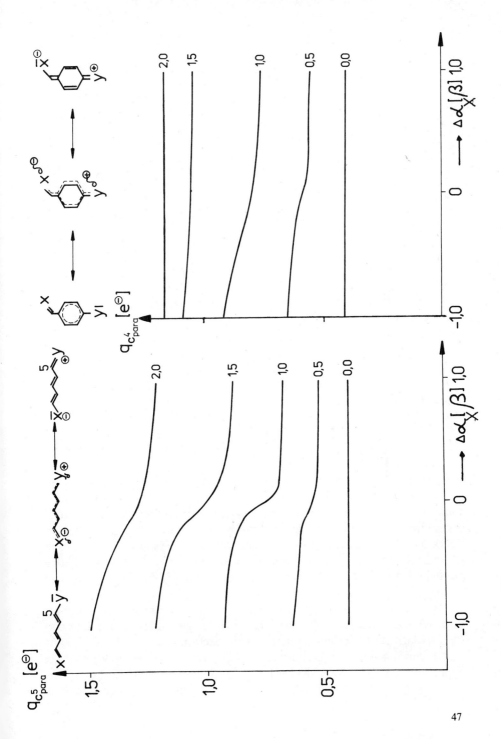

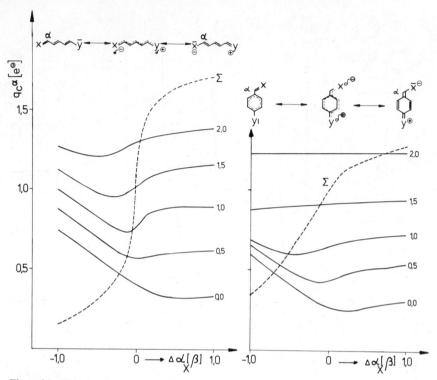

Figure 24 Calculated π-electron density at the α-position of the C=X acceptor group in pentamethines (**1**, $n = 2$) and in *para*-donor–acceptor-substituted benzenes (**7**) plotted in the same manner as explained in the legend of Fig. 23. The dotted curve Σ represents the symmetry deviation in the ground state S_0 in case of ideal pentamethine fragments with $h_Y = 1.0$.

the acceptor C=X, that is, with increasing $\Delta\alpha_X$ values. Apart from the weak turning point of the functions with $h_Y = 0.5 \cdots 1.5$ near the ideal polymethine state with $\Delta\alpha_X = 0$ there is no extreme value that may characterize this ideal state. Furthermore, no dependence of the π-electron density on $\Delta\alpha_X$ occurs in the case of the carbanion-substituted derivatives ($h_Y = 0$) or with benzenes in the case of reaction sites of high electronegativity ($h_Y = 2.0$). Thus, this parameter does not satisfactorily describe the unique features of the ideal polymethine state.

Much more interesting is the behavior of the π-electron density at the α position of the acceptor substituent because this gives rise to formulation of the π-polarization mechanism described in Section II.

Figure 24 exactly represents what has been experimentally observed. When the π-electrons are drawn from the donor reaction site Y into the direction of the acceptor group C=X by increasing the Coulomb integral value of the X-atom, the π-electron density at the α position of the pentamethine (left side of Fig. 23) is decreased at first. This abnormal behavior occurs with each reaction site having Coulomb integral values between $h_Y = 0$ and $h_Y = 2.0$. The overcompensation of the π-electron shift toward a "normal" behavior, that is, the expected increase in the π-electron density at C^α with increasing dipole moment, begins the earlier the more electronegative the reaction site is. Especially with the carbanions ($h_Y = 0$) there must be a relatively high predominance of the C=X group electronegativity to produce a reversal of the π-electron shift very similar to the behavior of the substituted benzenes on the right side of Fig. 24. Only reaction sites having an electronegativity equivalent to $h_Y = 1.5$ and higher produce from the very beginning a small increase in the π-electron density at C^α in accordance with the increase of $\sum$ also shown in Fig. 24.

The influence of $\Delta\alpha_X$ variations on the π-electron density at each position is investigated in detail in Fig. 25 for the pentamethine cyanine (**1**, $n = 2$) and the *para*-donor–acceptor-substituted benzene (**7**) with reaction sites $h_Y = 1.0$. Fitting the π-electron shift from the donor substituent Y to the acceptor group C=X, the π-electron density at the terminal atoms X and Y is increased and decreased, respectively. The curves at α position correspond to that shown in Fig. 24 with $h_Y = 1.0$, which were explained above. However, the π-electron density functions at any other position within the molecule's carbon chain do not behave as was expected in the case of a linear polarization of the π-electron system according to structures **1a↔1c** and **7a↔7c**. The reason is quite obvious and can be seen in Fig. 25 by the function $\overline{\Delta q}$ of the π-electron density alternation along the chain; there is a maximum π-electron density alternation near $\Delta\alpha_X = 0$, that is, near the ideal polymethine state. Therefore, the seemingly anomalous behavior of π-electron densities in the π-polarization effect is the consequence of the formation tendency of the ideal polymethine state, and quite normally it is the stronger with greater π-electron delocalization.

It was mentioned several times within the text that contrary to the unique features in the ideal polymethine state with its symmetric π-electron structure (**1b**), the behavior of the dipole moment should follow the linear polarization scheme illustrated by structures **1a↔1c** and **7a↔7c**. Figure 26 demonstrates the expected behavior. The dipole moments in the ground state of **1** and of **7** in dependence on the electronegativity difference of the terminal atoms, $\Delta\alpha_X$, are continuously ascending with increasing values and exhibit no extreme in the symmetric case at $\Delta\alpha_X = 0$.

The first authors who used bond length alternation in LFERs to describe the quinoid character of substituted benzenes were Taft, Topsom et al. (102) as well as Krygowski et al. (76b). The latter defined the parameter $\sum_\Delta = 2b - a - c$ for the lengths a, b, c of three neighbored bonds within six-

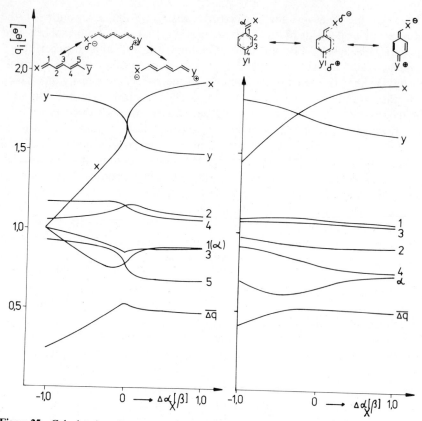

Figure 25 Calculated π-electron densities of pentamethines (**1**, $n=2$) and of *para*-donor–acceptor-substituted benzenes (**7**) at the positions indicated at the formulas in dependence on the electronegativity difference of the terminal atoms $\Delta\alpha_X$.

membered aromatic cycles. The application of this parameter to intermediate structures between polyenes (**1a, 1c**) and polymethines (**1b**) seems to be very useful, as can be seen in Fig. 27. With ideal polymethines (reaction site $h_Y = 1.0$) the reversal of the sign of Σ_Δ from positive to negative values at $\Delta\alpha_X = 0$ exactly coincides with the minimum π-bond order alternation $\overline{\Delta p}$. In the case of polymethines terminated with atoms of electronegativity lower than that of nitrogen ($h_Y \to 0$), the sign reversal of Σ_Δ is shifted to somewhat negative $\Delta\alpha_X$ values and the opposite occurs in the case of polymethines with terminal atoms of higher electronegativity ($h_Y \to 2.0$). Nevertheless, the appearance of Σ_Δ and $\overline{\Delta p}$ is in principle the same. This is no longer true, however, with the *para*-donor–acceptor substituted benzenes **7** shown in Fig. 28. Here, the Σ_Δ

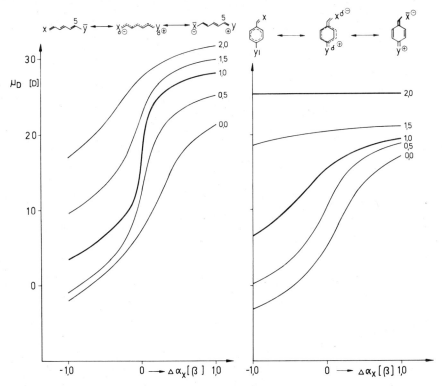

Figure 26 Calculated dipole moment in the ground state $\mu_D(S_0)$ (in Debeye units) of the pentamethines ($1, n = 2$) and the *para*-donor–acceptor-substituted benzenes (7) in dependence on the electronegativity difference of the terminal atoms $\Delta\alpha_X$. The h_Y values used for the reaction site Y are indicated on the curves. The values on the ordinte correspond to the ideal pentamethine fragment with $h_Y = 1.0$. For reasons of clarity, the upper two curves have been vertically shifted by $+ 5.0$ and $+ 10.0$ units and the two lower curves by $- 5.0$ and $- 10.0$ units, respectively.

parameter remains positive over the whole $\Delta\alpha_X$ range. The minimum π-bond order alternation $\overline{\Delta p}$ within the nitrogen terminated pentamethine chain near the ideal polymethine state ($h_Y = 1.0; \Delta\alpha_X = 0$) is indicated by only a very weak turning point of the Σ_Δ function, which even disappears at reaction sites of higher electronegativity ($h_Y \rightarrow 2.0$). Therefore, the Σ_Δ parameter cannot be used as universal parameter in the description of LFERs.

Another shortcoming of LFER parameters is the transmittivity of molecular substituent effects expressed in terms of the reaction constant ρ in regard to a reference system ρ_0. The transmittivity is believed to be a constant of each structural unit such as *p*-phenylene, butadiene, and various substituted naphthalenes (103). However, if we consider conjugation in a

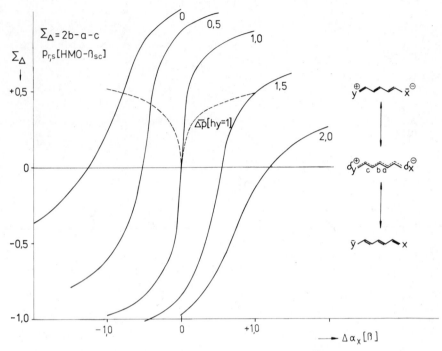

Figure 27 Bond length alternation Σ_Δ in the pentamethine chain $(\mathbf{1}, n = 2)$ defined by Krygowski et al. (76b) in terms of π-bond orders $p_{r,s}$ in dependence on the electronegativity difference of the terminal atoms $\Delta\alpha_X$. The Coulomb integral value h_Y of the reaction site Y is indicated on the curves. The values on the abscissa correspond to the ideal pentamethine fragment with $h_Y = 1.0$. For sake of clarity, the curves with $h_Y = 0.5$, $h_Y = 0$, $h_Y = 1.5$, and $h_Y = 2.0$ have been horizontally shifted by -0.5, -1.0, $+0.5$, and $+1.0$ units, respectively. The dotted line represents π-bond order alternation $\overline{\Delta p}$ in the case of the ideal pentamethine with $h_Y = 1.0$.

molecule as a whole in the sense of the polymethine approach, the problem of transmittivity of certain structural fragments becomes invalid. Only the strengths of conjugation or the scale of π-electron delocalization can be dealt with, which depends on the respective substituents. To demonstrate this, in Figs. 16–19 we have illustrated the π-bond order alternation $\overline{\Delta p}$ in connection with symmetry deviation Σ, which shows that maximum delocalization and thus maximum transmittivity is realized only in the ideal polymethine chains, that is, in the nitrogen-terminated streptocyanines $\mathbf{1}$, $n(h_X = h_Y = 1.0)$ having symmetric π-electron distribution. Localization increases and hence transmittivity decreases if the π-electron structure is no longer symmetric $(\alpha_X \neq \alpha_Y)$ and also if the electronegativity of the terminal substituents in the symmetric case $(\alpha_X = \alpha_Y)$ deviates from $h_X = h_Y = 1.0$ toward substituents of

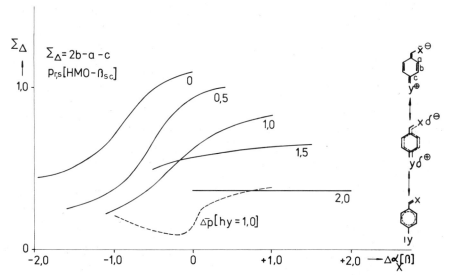

Figure 28 Bond length alternation Σ_Δ in the *para*-donor–acceptor-substituted benzene (7) defined by Krygowski et al. (76b) in terms of π-bond orders $p_{r,s}$ in dependence on the electronegativity difference of the terminal atoms $\Delta\alpha_X$. The Coulomb integral value h_Y of the reaction site Y are indicated on the curves. The values on the abscissa correspond to the ideal pentamethine fragment with $h_Y = 1.0$. For sake of clarity, the curves with $h_Y = 0.5$; $h_Y = 0$; $h_Y = 1.5$; $h_Y = 2.0$ have been horizontally shifted by -0.5, -1.0, $+0.5$, and $+1.0$ units, respectively. The dotted line represents π-bond order alternation Δp in the case of the polymethine-like chromophor with $h_Y = 1.0$.

higher electronegativity such as oxygen ($h_Y \to 2.0$) or toward those of lower electronegativity such as carbon ($h_Y \to 0$). In other words, transmittivity is not a constant of a certain structural unit but a function of the electronegativity of its terminal substituents.

The problem is even more complicated when a branched polymethine chain in the substituted structural units is realized, as with the substituted benzenes, naphthalenes, and other homocyclic and heterocyclic compounds. For example, in Fig. 29 the symmetry deviation of the *para*-donor–acceptor-substituted benzenes (7) with different donor atoms Y is shown in dependence on the electronegativity of the acceptor group C=X. With identical electronegativity of terminal atoms X and Y, all structures exhibit topologically induced asymmetry in favor of π-electron localization at the donor atom. The molecules are symmetrized on successive increase in the electronegativity of the acceptor group C=X. However, symmetrization becomes more difficult with greater electronegativity of the donor atom. Specifically, in the case of benzylanion (with $h_Y = 0$), a surplus of only 0.40β units at the acceptor side is

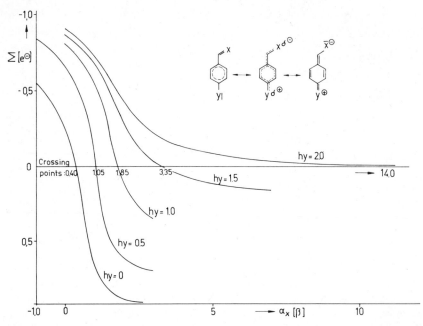

Figure 29 Symmetry deviation in the ground state Σ of *para*-donor–acceptor-substituted benzenes (7) with different reaction sites Y versus the Coulomb integral value of the acceptor atom α_X.

sufficient for symmetrization, whereas a surplus of 1.85β units is needed to symmetrize *para*-acceptor-substituted anilines (with $h_Y = 1.0$). With reaction sites of $h_Y = 1.5$ and $h_Y = 2.0$, the electronegativity of the acceptor atom X must be raised to 3.35β and 14.0β, respectively, to achieve $\Sigma = 0$. As such high values cannot be verified experimentally, it is obvious that the π-electrons in a phenoxide anion ($h_Y \approx 2.0$) cannot be symmetrized any longer irrespective of the degree of the acceptor's electronegativity. The result shows again that the transmittivity through *p*-phenylene is a strong function of the electronegativity of the reaction site.

In summary, it must be said, however, that the limitations of the LFER parameters outlined here are not so important as it might appear at first sight because nearly every substituent parameter considered previously covers only the range between the nonpolar limiting structures *a* and the symmetric polymethine-like limiting structures *b* of structures **1** and **7**, as examples.

Within this range of validity the relationships between molecular parameters and substituent effects can be approximated by more or less well fitted linear functions. An inspection of Figs. 16–27 obviously confirms this possibility. Therefore, hitherto there was no need to look over the borderline of

LFERs. However, if one should try to understand their physical background, to explain the exaltations and deviations of the σ_R constants, and to model free-energy relationships of highly polar substances, such as the negatively solvatochromic dyes **5** and **6**, it would crucial to combine correlation analysis with the theory of the ideal polymethine state.

D. A Provisional New Parameter in Description of Substituent Effects

One of the most important parameters in the characterization of the ideal polymethine state is the symmetry deviation $\sum$, which approaches zero in this state. To realize such a symmetric π-electron distribution along donor–acceptor-terminated molecular chains it is necessary to vary the donor strength or the acceptor strength, that is, the electronegativity, of the terminal substituents until $\sum = 0$ is attained. If the electronegativity of one terminal substituent would be fixed, the electronegativity—or in terms of the HMO-β_{SC} method, the h_Y value—of the opposite terminal substituent that brings $\sum$ to zero would then be a new substituent parameter.

The initial results of this method are compiled with structurally modified model compounds in Table 3. Considering at first the horizontal lines, it turns out that symmetrization becomes more difficult when the acceptor strength of the substituent increases. Consequently, the sequence of the provisional parameter corresponds to the σ_R substituent parameter scale. Particularly the sudden increase in the values from the carbonyl group to the nitroso group in the first three rows of Table 3 is in full agreement with experimental experience (104). Also, the fluoromethyl group exhibits significantly lower acceptor strength than does the nitroso group, although both substituents have the same sum of the Coulomb integral values of the constituent atoms, namely, 3.0β.

This result is remarkable in so far as it clearly demonstrates that the π-electron distribution is governed not only by the electronegativity of the atoms involved but also by their position within the molecules, proving that the topology of the heteroatom substitution also plays an important part in the realization of the electronic structure of conjugated molecules.

In the vertical columns of Table 3 the models are grouped into two sections. Both sections show that branching of the polymethine resonance, as a consequence of the insertion of aromatic ring systems, reduces the transmittivity of conjugation between the terminal substituents because the donor strength of the Y group increases (i.e., its h_Y value becomes more negative) when more aromatic cycles are included within the polymethine chain.

The influence of branching polymethine chains by monomethine units can be studied by comparing both sections in Table 3. Taking the first column,

TABLE 3
Provisional Parameter Describing the Donor Strength of a Substituent Y with Respect to a Fixed Acceptor Group C=X Connected to the Donor in Terms of the Donor's Electronegativity Value h_Y[a]

Substituent / Model (–C=X)	–CH=CH₂	–CH=ṄR₂	–CH=O	–CH₂F	–N=O
IY~~~X	0	+1.0	+2.0	+3.0	∞
IY–⟨⟩–X	−0.25	+0.4	+1.0	+1.6	∞
IY–⟨⟩–⟨⟩–X	−0.95	−0.15	+0.4	+0.6	+50.0

Substituent / Model (–C(4π)(X)(X))	–C(=CH₂)(CH₂⁻)	–C(=ṄR₂)(NR₂)	–C(=O)(O⁻)		–N(=O)(O⁻)
IY~~~(X)(4π)(X)	+0.3	+0.7	+1.35	—	∞
IY–⟨⟩–(X)(4π)(X)	+0.2	+0.4	+0.8	—	+3.1
IY–⟨⟩–⟨⟩–(X)(4π)(X)	−0.25	−0.2	+0.15	—	+1.55

[a] The numbers indicate the value by which is necessary in order to symmetrize the π-electron system. Infinity sign means that symmetrization cannot be realized.

which represents the hydrocarbon ions ($\alpha_X = 0$), branching obviously facilitates conjugation because in the second section containing the branched systems the donor strength of the Y substituent is lower (i.e., its h_Y value is more positive) than in the first section. However, in the more polymethine-like models of the second and third columns, the opposite effect is observed inasmuch as the donor strength of the Y substituent is higher (i.e., its h_Y value is more negative) in the branched model compounds. This is in full agreement with the experimental observation that conjugation through *para*-substituted acetophenones and benzophenones is reduced when the carbonyl group is replaced by carboxylic acid or carbamide groups or related substituents (13a, 105).

To give some practical recommendations from Table 3, benzaldehyde (second line, third column), for instance, should be substituted with a *para*-amino group ($h_Y = 1.0$) to realize a nearly ideal polymethine system with symmetric π-electron distribution. In order to perform the same symmetric π-electron distribution within the anion of the benzoic acid (fifth line, third column), however, one must slightly increase the donor strength of the *para*-amino group (perhaps by using the dimethylamino group). A much higher donor strength of the *para* substituent will be necessary if one should try to realize a polymethine-like π-electron structure in the stilbene aldehyde (third line, third column) or in the anion of the stilbene carboxylic acid (sixth line, third column). In the styrene analogue of stilbene (third line, first column) it must be even a boron atom ($h_Y = -0.95$) to symmetrize the π-electron distribution within the polymethine chain. On the other hand, the nitro-substituted benzene and stilbene (fifth and sixth lines, fifth column) should be combined with a donor group Y of rather high electronegativity such as the phenoxide anion, for example, to obtain a nearly polymethine-like π-electron structure. Because of the infinite h_Y values obtained for the nitroso and nitro group, respectively, in lines 1–4 of column 5, one must conclude that it will be impossible to verify nearly ideal polymethine structures in these models.

V. CONCLUSIONS

There is enough experimental proof and theoretical confirmation to justify a revision of LFERs for strongly conjugated substituents. The guideline for clarifying the complex relationships is the triad theory. This theory depicts the interrelationships between three ideal states—the aromatic, polymethinic, and polyenic states—which make up the immense variety of existing molecules. Thus, triad theory provides the physicochemical basis for an understanding of LFERs.

The present work yields the following results:

1. By limiting quantum-chemical calculations to the π-electron level, without considering electron repulsion effects, only the overall electronegativity of substituents is taken into account. The results show that the resonance and polarizability behavior of conjugated molecules is governed mainly by both the electronegativity of the substituents and the topology of their atoms, proving that the influences of field and through-space effects and

any other variables related to the σ-electron framework as well as to the molecule's geometry have been overestimated in the past.

2. The LFER models are suitable for describing the influences of substituents only with respect to changes between the nonpolar and the symmetric limiting structure of a molecule on one side, or that between the symmetric limiting structure and the highly polar one on the other side (see structures **1** and **7**). In the sense of the classical resonance theory, the symmetric structures **1b** and **7b** are resonance hybrids, whose canonical structures **1a** and **1c** contribute with equal weight to mesomerism. Difficulties arise, however, with nonsymmetric models, such as **7**, where the weight of structures **7a** and **7c** for realizing the symmetric one, **7b**, depends on the topological deviation (101) of the molecule. Within these limits molecular properties can, indeed, be approximated by more or less accurately linearized functions. However, LFERs necessarily will be reversed when the substituent-induced structural changes pass over the symmetric resonance structures, **1b** or **7b**, because these structures are energetically stabilized and possess maximum π-bond order equalization. For that reason the measures of resonance hitherto used in LFERs such as the charge shift from the substituent to the residual molecule $\Sigma \Delta q_\pi$ does not fully reflect reality.

3. Apart from the truism of resonance theory explained in paragraph 2 (above), one must consider that the state of utmost resonance and π-electron delocalization in linearly conjugated molecules is realized in neither hydrocarbon ions consisting exclusively of carbon atoms nor carbon chains terminated by oxygen or other atoms of rather high electronegativity, but only in the cyanines and related compounds that are terminated by nitrogen atoms or other substituents of comparable electronegativity. This state of maximum π-electron delocalization is designated the "ideal polymethine state."

4. As a consequence of the formation tendency of the ideal polymethine state, there is a strong tendency of π-electron density alternation for strongly conjugated substituents, explaining in a simple way the π-polarization effect.

5. One precondition of polymethine formation is a conjugated chain of N atoms that is occupied by a surplus or a deficit of about one π-electron, that is, by $(N \pm 1)$ π-electrons. Therefore, combinations of two donor substituents through an aromatic or polyenic fragment does not fulfil this condition as two additional π-electrons are provided for conjugation. In this case the well-described behavior, the so-called π-saturation effect, results. On combination of two strong acceptor substituents, however, the π-electron occupation tends toward the deficit situation of less than one π-electron per N chain atoms, and hence acceptors may act as donors on combination with extremely strong acceptor substituents forming a polymethine-like structure.

6. Contrary to the assumption underlying LFER models, the transmit-

tivity of any conjugated system is not a constant of the system but depends strongly on the nature of the reaction site in proportion to the degree of the ideal polymethine state that is realized.

Our present knowledge of LFERs in connection with the polymethine approach suggests the need for many future tasks. The most important one will be to characterize the ideal polymethine state, namely, the state of maximum π-electron delocalization, from a modern point of view in terms of experiments and quantum theory. It is true that some basic relationships between three ideal states have been now formulated. Much research is still needed, however, until it will be possible to describe these relationships quantitatively by means of appropriate functions as has been done, for instance, with regard to linear solvation energy relationships. The tools necessary for solving this task are available; for example, ^{13}C, ^{15}N, and ^{17}O NMR spectroscopic measurements provide information about the precise π-electron distribution in a molecule. Infrared and Raman spectroscopy, NMR coupling constants, and X-ray structure analyses provide detailed information about the bond lengths and bond angles. Ionization potentials of substituents can be determined by means of ESCA (electron spectroscopy for chemical analysis) measurements. The relative position of molecular frontier orbitals is available by the electrochemical half-wave potentials. Ab initio calculations provide molecular data with almost as much precision as one requires. Very much has been done in this context, but it is the field of polymethinic compounds with maximum π-electron delocalization that has been hitherto neglected to a large extent.

Finally, the question raised by Sjoestroem and Wold (15): "Are LFERs only local empirical rules of limited validity or fundamental laws in chemistry?" should be answered. By modifying the quotation made by Kamlet and Taft (16a), it can be said "that the controversial and spirited dialogue on this subject has now generated sufficient heat that some light is shed on the origin of LFERs." LFERs do, indeed, reflect fundamental laws in chemistry, but they have well-defined limitations in validity, which are provided by the polymethine approach within the bounds of triad theory.

REFERENCES AND NOTES

1. J. Shorter, Chem. unserer Zeit., 19, 197 (1985); J. Shorter, Progr. Phys. Org. Chem., 17
2. (a) J. Bromilow, R. T. C. Brownlee, D. J. Craik, M. Sadek, and R. W. Taft, J. Org. Chem., 45, 2429 (1980); (b) W. Reynolds, P. Dais, D. W. MacIntyre, R. D. Topsom, S. Marriott, E. von Nagy-Felsobuki, and R. W. Taft, J. Am. Chem. Soc., 105, 378 (1983).
3. M. Charton, Progr. Phys. Org. Chem., 16, 287 (1987).

4. R. W. Taft, R. D. Topsom, *Progr. Phys. Org. Chem.*, *16*, 1 (1987).
5. I. Agranat and A. Skancke, *J. Am. Chem. Soc.*, *107*, 867 (1985).
6. S. S. Shaik, P. C. Hiberty, J. M. Lefour, and G. Ohanessian, *J. Am. Chem. Soc.*, *109*, 363 (1987).
7. J. Aihara, *Bull. Chem. Soc. Jpn.*, *60*, 2268 (1987).
8. M. Krygowski, private communication in letters from December 30, 1986 and February 23, 1987.
9. (a) S. Daehne, *Wiss. Z. T. U. Dresden*, *20*, 671 (1971); *Chem. Abstr.*, *77*, 138811u (1972); (b) S. Daehne, *Science*, *199*, 1163 (1978); (c) S. Daehne and F. Moldenhauer, *Progr. Phys. Org. Chem.*, *15*, 1 (1985).
10. (a) S. J. Ehrenson, R. T. C. Brownlee, and R. W. Taft, *Progr. Phys. Org. Chem.*, *10*, 1 (1973); (b) S. J. Ehrenson, *J. Org. Chem.*, *44*, 1793 (1979).
11. H. C. Brown, M. Periasamy, and K. T. Liu, *J. Org. Chem.*, *46*, 1646 (1981).
12. (a) R. T. C. Brownlee and R. W. Taft, *J. Am. Chem. Soc.*, *92*, 7007 (1970); (b) S. Marriott, and R. D. Topsom, *J. Chem. Soc. Perkin II*, 1045 (1985); (c) R. D. Topsom, *Progr. Phys. Org. Chem.*, *16*, 85 (1987); (d) S. Marriott, A. Silvestro, and R. D. Topsom, *J. Chem. Soc. Perkin II*, 457 (1988); *J. Mol. Struct. (Theochem)*, *184*, 23 (1989).
13. (a) J. Bromilow, R. T. C. Brownlee, D. J. Craik, P. R. Fiske, J. E. Rowe, and M. Sadek, *J. Chem. Soc. Perkin II*, 753 (1981); (b) R. T. C. Brownlee and D. J. Craik, *J. Chem. Soc. Perkin II*, 760; (1981); (c) W. Reynolds, A. Gomes, A. Maron, D. W. MacIntyre, R. G. Maunder, A. Tanin, H. E. Wong, G. K. Hamer, and I. R. Peat, *Can. J. Chem.*, *61*, 2367, 2376 (1983); (d) S. Marriott, and R. D. Topsom, *J. Chem. Soc. Perkin II*, 113 (1984).
14. D. J. Craik and R. T. C. Brownlee, *Progr. Phys. Org. Chem.*, *14*, 1 (1983).
15. M. Sjoestroem and S. Wold, *Acta Chem. Scand.*, *B35*, 537 (1981).
16. (a) M. J. Kamlet and R. W. Taft, *Acta Chem. Scand.*, *B39*, 611 (1985); (b) M. J. Kamlet, R. M. Doherty, G. R. Famini, and R. W. Taft, *Acta Chem. Scand.*, *B41*, 589 (1987).
17. S. Daehne, *Z. Chem.*, *10*, 133, 168 (1970).
18. W. Koenig, *J. prakt. Chem.*, /2/, *112* (2), 1 (1926).
19. H. A. Staab, *Einführung in die theoretische organische Chemie*, 3rd ed., Verlag Chemie, Weinheim, 1962.
20. (a) S. Daehne, D. Leupold, *Ber. Bunsenges. Phys. Chem.*, *70*, 618 (1966); (b) S. Daehne and D. Leupold, *Angew. Chem.*, *78* 1029 (1966); *Angew. Chem. Int. Edn. Engl.*, *5*, 984 (1966); (c) S. Daehne, *Z. Chem.*, *5*, 441 (1965).
21. (a) J. Griffiths, *Rev. Progr. Color. Relat. Top.*, *11*, 37 (1981); *14*, 21 (1984); (b) S. Tokita, *Senryo to Yakuhin*, *27*, 164 (1982); *Chem. Abstr.*, *98*, 88399u (1983); (c) C. Reichardt, *Solvents and Solvent Effects in Organic Chemistry*, 2nd ed., Verlag Chemie, Weinheim, 1988.
22. (a) H. Hartmann, *J. Signalaufzeichnungsmater.*, *7*, 101, 181 (1979); *Chem. Abstr.*, *92*, 57970v (1980); *91*, 212555p (1979); (b) S. Daehne, H. Hartmann, *Zhur. nauchn. prikl. Fotogr. Kinematogr.*, *31*, 312 (1986).
23. T. Foerster, *Z. Elektrochem.*, *45*, 548 (1939).
24. S. Daehne, *Z. Chem.*, *21*, 58 (1981).
25. (a) H. Suzuki, *Electronic Absorption Spectra and Geometry of Organic Molecules*, Academic Press, New York, 1967; (b) R. C. Dougherty and M. J. S. Dewar, *PMO Theory of Organic Chemistry*, Plenum Press, New York, 1975.
26. In reference 20a a completed HMO method taking into account non-neighbored resonance integrals was used, which is now no longer practiced. Therefore, in Figs. 6 and 9 the functions shown were recalculated by the original HMO method.

27. S. Daehne, D. Leupold, K. D. Nolte, in *Dye Sensitization: Symposium*, Bressanone 1967, W. F. Berg, U. Mazzucato, M. Meier, and G. Semerano, Eds., Focal Press, London, 1970, p. 136; *Chem. Abstr.*, *78*, 64409h (1973).
28. J. Fabian and H. Hartmann, *Z. Chem.*, *13*, 263 (1973).
29. B. A. Hess and L. J. Schaad, *J. Am. Chem. Soc.*, *93*, 305 (1971).
30. (a) J. Fabian and H. Hartmann, *J. Mol. Struct.*, *27*, 67 (1975); (b) J. Fabian and H. Hartmann, *Theor. Chim. Acta*, *36*, 351 (1975).
31. B. Heilbronner and H. Bock, *Das HMO Modell und seine Anwendung*, Verlag Chemie, Weinheim, 1970.
32. J. Fabian and S. Daehne, *J. Mol. Struct. (Theochem)*, *92*, 217 (1983).
33. (a) R. Radeglia, E. Gey, K. D. Nolte, and S. Daehne, *J. prakt. Chem.*, *312*, 877 (1970); M. Klessinger, *Theor. Chim. Acta*, *5*, 251 (1966); (b) G. G. Dyadyusha, I. V. Repyakh, and A. D. Kachkovskii, *Teor. Eksp. Khim.*, *20*, 398 (1984); *Chem. Abstr.*, *101*, 132445u (1984); *21*, 138 (1985); *Chem. Abstr. 103*, 38686m (1985); *22*, 347 (1986); *Chem. Abstr.*, *105*, 178632g (1986); G. G. Dyadyusha and I. V. Repyakh, *Teor. Eksp. Khim.*, *24*, 129 (1988); (c) R. Radeglia, E. Gey, T. Steiger, S. Kulpe, R. Lueck, M. Ruthenberg, M. Stierl, and S. Daehne, *J. prakt. Chem.*, *316*, 766 (1974); K. D. Nolte and S. Daehne, *J. prakt. Chem.*, *318*, 993 (1976).
34. J. Fabian and A. Mehlhorn, *J. Mol. Struct. (Theochem)*, *109*, 17 (1984).
35. H. Ichikawa, J. Aihara, and S. Daehne, *Bull. Chem. Soc. Jpn.*, *62*, 2798 (1989).
36. G. G. Dyadyusha and A. D. Kachkovskii, *J. Inf. Rec. Mater.*, *13*, 95 (1985); *Chem. Abstr.*, *104*, 7201r (1986); *Teor. Eksp. Khim.*, *17*, 393 (1981); *Chem. Abstr.*, *95*, 152134w (1981); *Chem. Abstr.*, *15*, 152 (1979); *Chem. Abstr.*, *91*, 6392e (1979); *Ukr. Khim. Zh.*, *44*, 948 (1978); *Chem. Abstr.*, *89*, 199073s (1978); *Chem. Abstr.*, *41*, 1176 (1975); *Chem. Abstr.*, *84*, 46033a (1976).
37. (a) B. M. Gimarc, *J. Am. Chem. Soc.*, *105*, 1979 (1983); (b) B. M. Gimarc and J. J. Ott, *J. Am. Chem. Soc.*, *108*, 4298 (1986); J. J. Ott and B. M. Gimarc, *J. Am. Chem. Soc.*, *108*, 4303 (1986); (c) J. Klein, *Tetrahedron*, *39*, 2733 (1983); *44*, 503 (1988); *J. Am. Chem. Soc.*, *110*, 4634 (1988).
38. S. Daehne, *Wiss. Z. Tech. Univ. Dresden*, *29*, 101 (1980).
39. K. D. Nolte, thesis, Academy of Sciences of the GDR (German Democratic Republic), Berlin, 1972.
40. I. Kanev and G. Neykov, *J. Chim. Phys. Phys.-Chim. Biol.*, *79*, 115 (1982); S. Daehne, unpublished results.
41. S. S. Shaik and R. Bar, *Nouv. J. Chim.*, *8*, 411 (1984).
42. H. Ichikawa and Y. Ebisawa, *J. Am. Chem. Soc.*, *107*, 1161 (1985).
43. (a) S. Daehne and O. Guertler, *J. prakt. Chem.*, *315*, 786 (1973); (b) O. Guertler and S. Daehne, *Z. phys. Chem. (Leipzig)*, *255*, 501 (1974).
44. A. D. Kachkovskii, M. A. Kudinova, B. I. Shapiro, N. A. Derevjanko, L. G. Kurkina, and A. I. Tolmachev, *Dyes and Pigments*, *5*, 295 (1984).
45. G. Schwarzenbach, *Z. Elektrochem.*, *47*, 40 (1941).
46. D. Lloyd and D. R. Marshall, *Chem. Ind.*, *1972*, 335; *Angew. Chem.*, *84*, 447 (1972); *Angew. Chem. Int. Edn. Engl.*, *11*, 404 (1972).
47. S. Daehne, H. J. Rauh, R. Schnabel, and G. Geiseler, *Z. Chem.*, *13*, 70 (1973); H. J. Rauh, R. Schnabel, G. Geiseler, and S. Daehne, *Z. Phys. Chem. (Leipzig)*, *255*, 651 (1974).
48. S. Daehne and S. Kulpe, *Structural Principles on Unsaturated Organic Compounds, with Special Reference vo X-Ray Structure Analyses of Coloured Substances*, Abh. Akad. Wiss. DDR, Abt. Math. Naturwiss., Tech., Akademieverlag, Berlin, 1977, Vol. 8N, pp. 1–128; *Chem. Abstr.*, *89*, 162951a (1978).

49. S. Daehne and R. Radeglia, *Tetrahedron*, 27, 3673 (1971).
50. (a) B. B. Johnson, P. Gramaccioni, and W. T. Simpson, *Mol. Cryst. Liq. Cryst.*, 28, 99 (1973); J. Dale, O. I. Eriksen, and P. Groth, *Acta Chem. Scand.*, B41, 653 (1987); (b) J. O. Selzer and B. W. Matthews, *J. Phys. Chem.*, 80, 631 (1976); F. Chentli-Benchikha, J. P. Declercq, G. Germain, and M. Van Meerssche, *Cryst. Struct. Commun.*, 6, 421 (1977); J. Dale, S. Krueger, and C. Roemming, *Acta Chem. Scand.*, B38, 117 (1984); M. Honda, C. Katayama, and J. Tanaka, *Acta Cryst.*, B42, 90 (1986); (c) S. Kulpe and B. Schulz, *Z. Chem.*, 18, 146 (1978); S. Kulpe, R. J. Kuban, B. Schulz, and S. Daehne, *Cryst. Res. Technol.*, 22, 375 (1987); P. Groth, *Acta Chem. Scand.*, B41, 547 (1987).
51. P. Groth, *Acta Chem. Scand.*, A41, 117 (1987).
52. K. Elbl, C. Krieger, and H. A. Staab, *Angew. Chem.*, 98, 1024 (1986); *Angew. Chem. Int. Edn. Engl.*, 25, 1023 (1986).
53. G. Haefelinger, *Tetrahedron*, 26, 2469 (1970); *Chem. Ber.*, 103, 2941 (1970).
54. (a) L. Mitzinger, thesis, Academy of Sciences of the GDR, Berlin, 1973; (b) L. Mitzinger and S. Daehne, *J. prakt. Chem.*, 324, 458 (1982); (c) N. Sano, M. Shimizu, J. T anaka, and M. Tasumi, *Bull. Chem. Soc. Jpn.*, in press.
55. (a) J. Ranft and S. Daehne, *Helv. Chim. Acta*, 47, 1160 (1964); (b) G. Scheibe, C. Jutz, W. Seiffert, and D. Grosse, *Angew. Chem.*, 76, 270 (1964); *Angew. Chem. Int. Edn., Engl.*, 3, 374 (1964); (c) J. Dale, R. G. Lichtenthaler, and G. Teien, *Acta Chem. Scand.*, B33, 141 (1979).
56. F. Doerr, J. Kotschy, and H. Kausen, *Ber. Bunsenges. Phys. Chem.*, 69, 11 (1965).
57. (a) S. Kulpe, *Z. Chem.*, 20, 377 (1980); *Angew. Chem.*, 93, 283 (1981); *Angew. Chem. Int. Edn. Engl.*, 20, 271 (1981); (b) S. Kulpe and S. Daehne, *Acta Cryst.*, B34, 3616 (1978).
58. (a) G. Scheibe, W. Seiffert, H. Wengenmayr, and C. Jutz, *Ber. Bunsenges. Phys. Chem.*, 67, 560 (1963); (b) S. Daehne and J. Ranft, *Z. Phys. Chem. (Leipzig)*, 224, 65 (1963).
59. (a) R. Radeglia, G. Engelhardt, E. Lippmaa, T. Pehk, K. D. Nolte, and S. Daehne, *Org. Magn. Res.*, 4, 571 (1972); (b) R. Radeglia, *J. prakt. Chem.*, 315, 1121 (1973).
60. A. Pross and L. Radom, *Progr. Phys. Org. Chem.*, 13, 1 (1981).
61. S. Marriott, W. F. Reynolds, and R. D. Topsom, *J. Org. Chem.*, 50, 741 (1985).
62. K. D. Nolte and S. Daehne, *Adv. Mol. Relax. Interact. Processes*, 10, 299 (1977).
63. A. R. Katritzky and R. A. Y. Jones, *Chem. Ind.*, 722 (1961).
64. J. Kučera and Z. Arnold, *Collect. Czech. Chem. Commun.*, 32, 1704 (1967); R. Brehme and H. E. Nikolajewski, *Tetrahedron*, 25, 1159 (1969); 32, 731 (1976); D. Lloyd and H. McNab, *Angew. Chem.*, 88, 496 (1976); *Angew. Chem. Int. Edn. Engl.*, 15, 459 (1976).
65. G. Scheibe, C. Jutz, W. Seiffert, and D. Grosse, *Angew. Chem.*, 76, 270 (1964); *Angew. Chem. Int. Edn. Engl.*, 3, 306 (1964); A. R. Butler, D. Lloyd, and D. R. Marshall, *J. Chem. Soc. (B)*, 795 (1971); N. Bacon, W. O. George, and B. H. Springer, *Chem. Ind. (London)* (II), 1377 (1965); H. E. A. Kramer, *Liebigs Ann. Chem.*, 696, 15, 28 (1966).
66. S. Daehne, unpublished results.
67. R. Radeglia, M. Waehnert, S. Daehne, and H. Boegel, *J. prakt. Chem.*, 320, 539 (1978).
68. G. Scheibe, G. Buttgereit, and E. Daltrozzo, *Angew. Chem.*, 75, 1023 (1963); *Angew. Chem. Int. Edn. Engl.*, 2, 1019 (1963); G. Buttgereit and G. Scheibe, *Ber. Bunsenges. Phys. Chem.*, 69, 301 (1965).
69. H. E. Nikolajewski, S. Daehne, B. Hirsch, and E. A. Jauer, *Angew. Chem.*, 78, 1063 (1966); *Angew. Chem. Int. Edn. Engl.*, 5, 1044 (1966).
70. T. Zincke, *Liebigs Ann. Chem.*, 330, 361 (1904); W. Koenig, *J. prakt. Chem.*, 70 (2), 19 (1904); J. Joussot-Dubien and J. Houdard, *Tetrahedron Lett.*, 1967, 4389.

71. H. Althoff, B. Bornowski, and S. Daehne, *J. prakt. Chem.*, *319*, 890 (1977).
72. E. Havinga, R. O. De Jongh, and M. E. Kronenberg, *Helv. Chim. Acta*, *50*, 2550 (1967); E. Havinga and M. E. Kronenberg, *Pure Appl. Chem.*, *16*, 137 (1968); S. Daehne, *Wiss. Z. Tech. Hochsch. "Carl Schorlemmer", Leuna-Merseburg*, *24*, 455 (1982); *Chem. Abstr.*, *98*, 106448r (1983).
73. S. Kulpe, A. Zedler, S. Daehne, and K. D. Nolte, *J. prakt. Chem.*, *315*, 865 (1973),
74. S. Kulpe, and I. Seidel, *Z. Chem.*, *20*, 300 (1980); *Cryst. Res. Techn.*, *19*, 947 (1984).
75. R. Radeglia, *J. Mol. Struct.*, *17*, 47 (1973).
76. (a) A. Domenicano, A. Vaciago, and C. A. Coulson, *Acta Cryst.*, *B31*, 221, 1630 (1975); A. Domenicano, P. Mazzeo, and A. Vaciago, *Tetrahedron Lett.*, 1029 (1976); (b) M. Krygowski, *J. Chem. Res.*, 238 (1984), 120 (1987); J. Maurin, and T. M. Krygowski, *J. Mol. Struct.*, *158*, 359 (1987); T. M. Krygowski and G. Haefelinger, *J. Chem. Res.*, 348 (1986); T. M. Krygowski, G. Haefelinger, and J. Schuele, *Z. Naturforsch.*, *41B*, 895 (1986); R. Anulewicz, G. Haefelinger, T. M. Krygowski, C. Regelmann, and G. Ritter, *Z. Naturforsch.*, *42B*, 917 (1987).
77. S. Marriott and R. D. Topsom, *J. Mol. Struct. (Theochem)*, *109*, 305 (1984); *139*, 101 (1986).
78. S. Daehne, S. Kulpe, K. D. Nolte, and R. Radeglia, *Photogr. Sci. Eng.*, *18*, 410 (1974); S. Daehne, S. Kulpe, and K. D. Nolte, *Z. Chem.*, *13*; 422 (1973).
79. S. Daehne and K. D. Nolte, *J. Chem. Soc., Chem. Comm.*, 1056 (1972); K. D. Nolte and S. Daehne, *J. prakt. Chem.*, *318*, 643 (1976).
80. H. Kobischke and S. Daehne, *Photogr. Sci. Eng.*, *16*, 173 (1972); B. Grimm and S. Daehne, *J. Signalaufzeichnungsmater.*, *1*, 339 (1973); *Chem. Abstr.*, *81*, 162058z (1974); S. Kulpe, S. Daehne, B. Ziemer, and B. Schulz, *Photogr. Sci. Eng.*, *20*, 205 (1976); S. Daehne, B. Bornowski, B. Grimm, S. Kulpe, D. Leupold, and M. Naether, *J. Signalaufzeichnungsmater.*, *5*, 277 (1977); *Chem. Abstr.*, *87*, 203055r (1977); N. Sano and J. Tanaka, *Bull. Chem. Soc. Jpn.*, *59*, 843 (1986); N. Sano, M. Shimizu and J. Tanaka, *J. Raman Spectr.*, *20*, 399 (1989).
81. M. Gordon and W. R. Ware, Eds., *The Exciplex*, Academic Press, New York, 1975.
82. M. Waehnert, S. Daehne, R. Radeglia, A. M. Alperovich, and I. I. Levkojev, *Adv. Mol. Relax. Interact. Processes*, *11*, 263 (1977); M. Waehnert and S. Daehne, *J. Signalaufzeichnungsmater.*, *4*, 403 (1976).
83. R. Radeglia anf S. Daehne, *J. Mol. Struct.*, *5*, 399 (1970).
84. S. Daehne, F. Schob, and K. D. Nolte, *Z. Chem.*, *13*, 471 (1973).
85. (a) H. E. A. Kramer and R. Gompper, *Z. Phys. Chem. NF* *43*, 292 (1964); (b) G. Scheibe, W. Seiffert, G. Hohlneicher, C. Jutz, and H. J. Springer, *Tetrahedron Lett.*, *1966*, 5053.
86. R. Radeglia, *Z. Phys. Chem. (Leipzig)*, *235*, 335 (1967).
87. S. Daehne, D. Leupold, H. E. Nikolajewski, and R. Radeglia, *Z. Naturforsch.*, *20b*, 1006 (1965).
88. R. Radeglia, R. Wolff, B. Bornowski, and S. Daehne, *Z. Phys. Chem. (Leipzig)*, *261*, 502 (1980); *J. prakt. Chem.*, *323*, 125 (1981).
89. S. Kulpe and B. Schulz, *Z. Chem.*, *17*, 377 (1977); *Krist. Techn.*, *11*, 707 (1976); *14*, 159 (1979); S. Kulpe, *Z. Chem.*, *16*, 56 (1976).
90. S. Daehne and K. D. Nolte, *Acta Chim. Acad. Sci. Hung.*, *97*, 147 (1978).
91. N. S. Bayliss and E. G. McRae, *J. Phys. Chem.*, *58*, 1002, 1006 (1954); W. Liptay, in *Excited States*, Vol. 1, E. C. Linn, Ed., Academic Press, New York, 1974, pp. 129ff.
92. G. P. Ford, A. R. Katritzky, and R. D. Topsom, in *Correlation Analysis in Chemistry*, N. B. Chapman, N. Bellamy, and J. Shorter, Eds., Plenum Press, New York, 1978.
93. P. Jacques, *J. Phys. Chem.*, *90*, 5535 (1986).

94. (a) H. G. Benson and J. N. Murrell, *J. Chem. Soc., Faraday Trans. II*, 68, 137 (1972); (b) M. Waehnert and S. Daehne, *J. prakt. Chem.*, 318, 321 (1976); (c) A. Botrel, A. Le Beuze, P. Jacques, and H. Strub, *J. Chem. Soc., Faraday Trans. II*, 80, 1235 (1984).
95. W. J. Hehre, R. W. Taft, and R. D. Topsom, *Progr. Phys. Org. Chem.*, 12, 159 (1976).
96. A. Golebiewski and J. Nowakowski, *Acta Phys. Polon.*, 25, 647 (1964).
97. D. J. Craik, R. T. C. Brownlee, and M. Sadek, *J. Org. Chem.*, 47, 657 (1982).
98. J. Hinze and H. H. Jaffé, *J. Am. Chem. Soc.*, 84, 540 (1962); *J. Phys. Chem.*, 67, 1501 (1963); J. Hinze, *Fortschr. Chem. Forsch.*, 9, 448 (1967/1968).
99. N. C. Baird, M. A. Whitehead, *Can. J. Chem.*, 44, 1933 (1966); P. R. Wells, *Progr. Phys. Org. Chem.*, 6, 11 (1968); D. Bergmann and J. Hinze, *Struct. Bonding (Berlin)*, 66, 145 (1987).
100. (a) S. Daehne, in "Progress and Trends in Applied Optical Spectroscopy," *Teubner Texte zur Physik*, Vol. 13, D. Fassler, K. H. Feller, and B. Wilhelmi, Eds., Teubner, Leipzig, 1987, p. 178; (b) S. Daehne and I. Ritschl, in "Recent Developments in Molecular Spectroscopy; *Proceedings Xth National Conference on Molecular Spectroscopy*, Blagoevgrad, Bulgaria, B. Jordanov, N. Kirov and P. Simova, Eds., World Scientific, Singapore, New Jersey, London, Hong Kong, 1989, p. 549.
101. G. G. Dyadyusha, M. N. Ushomirskii, M. S. Lynbich, E. B. Lifshits, And R. A. Gerstein, *Teor. Eksp. Khim.*, 21, 641 (1985).
102. E. von Nagy-Felsobuki, R. D. Topsom, S. Pollack, and R. W. Taft, *J. Mol. Struct. (Theochem)*, 88, 255 (1982).
103. M. Charton, *Progr. Phys. Org. Chem.*, 10, 81 (1973); M. Charton, in *Correlation Analyses in Chemistry*, N. B. Chapman, N. Bellamy, and J. Shorter, Eds., Plenum Press, New York, 1978, pp. 176ff.
104. G. Butt, R. D. Topsom, B. G. Gowenlock, and J. A. Hunter, *J. Mol. Struct. (Theochem)*, 164, 399 (1988).
105. J. F. Gal, S. Geribaldi, G. Pfister-Guillouzo, and D. G. Morris, *J. Chem. Soc. Perkin II*, 103 (1985).

Estimation of the Basicity Constants of Weak Bases by the Target Testing Method of Factor Analysis

BY Ü. HALDNA

Institute of Chemistry
Estonian Academy of Sciences
Tallinn, Estonia, U.S.S.R

The basicity of weak organic bases in aqueous solutions is quantitatively estimated by the respective basicity constant

$$K_{BH^+} = \frac{a_{H^+} \cdot a_B}{a_{BH^+}} = \frac{c_{H^+} \cdot c_B}{c_{BH^+}} \cdot \frac{f_{H^+} \cdot f_B}{f_{BH^+}} \qquad (1)$$

The basicity constant K_{BH^+} is referred to a standard state of infinite dilution in water, where the activity coefficients approach unity. As a rule, extrapolation from the strong acid solutions, in which the base studied is remarkably protonated to the standard state, is carried out by the excess acidity method (1, 2):

$$\log \frac{c_{BH^+}}{c_B} - \log c_{H^+} = m^* \cdot X + pK_{BH^+} \qquad (2)$$

where c_{H^+} is the molar concentration of the solvated protons, X is the excess acidity, m^* is the solvation parameter, and $pK_{BH^+} = -\log K_{BH^+}$. The indicator ratio of base B, $I = c_{BH^+}/c_B$, is calculated from the experimental data. If the classical method, the absorption spectroscopy in the ultraviolet (UV)-visible region is used, the indicator ratio is given by the equation

$$\frac{c_{BH^+}}{c_B} = \frac{\varepsilon_B - \varepsilon}{\varepsilon - \varepsilon_{BH^+}} \qquad (3)$$

---

*A plenary lecture given at the IVth European Conference on the Correlation Analysis in Organic Chemistry, Poznán, Poland, July 18–23, 1988.

where ε_B and ε_{BH^+} are the molar extinction coefficient values at the wavelength λ for the unprotonated (B) and protonated (BH$^+$) forms, respectively, and ε is the molar extinction coefficient for a mixture of B and BH$^+$ forms at the same wavelength. A well-behaved base yields values of I that do not depend on the wavelength (λ) chosen. In this case there may be isobestic points in the measured sets of spectra but no spectral shifts in the positions of absorption bands are observed. Consequently, in such a simple case, there is no need for a sophisticated treatment of the measured set of spectra. For a number of weak bases, however, shifts in the absorption bands with increasing strong acid concentration are observed (3, 4). This is a typical situation when the basicity of carbonyl compounds in aqueous sulfuric acid solutions is studied. In such a case the indicator ratio I values depend strongly on the wavelength chosen and as a result, the pK_{BH^+} value cannot be obtained unambiguously.

In order to overcome the difficulties caused by these "medium effects," a number of empirical correction methods have been suggested. However, none of them has yielded a solution to the problem (3, 4). The situation described is illustrated in Fig. 1, which represents a digital simulation model generated by a computer using the prefixed pK_{BH^+} and m^* values in Equation 2 (pK_{BH^+} = -2.40, $m^* = 0.90$). To obtain the set of spectra presented in Fig. 1, we took two

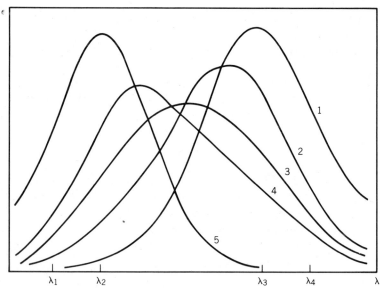

Figure 1 Simulated set of spectra for applying the empirical correction methods (5, 6). The parameters used were pK_{BH^+} = -2.40 and $m^* = 0.90$. Sulfuric acid % (w/w): 1–5%, 2–33%, 3–40%, 4–45%, 5–86%.

Gaussian profiles, one for the B form and the other for the BH^+ form, and made them shift along the wavelength axis toward shorter wavelengths with increasing sulfuric acid concentration. For both forms, shifts were assumed to be linear, with the H_2SO_4 concentration given as w/w% of acid. The molar extinction coefficient values for the mixtures of B and BH^+ forms were calculated by the equation

$$\varepsilon_\lambda = \alpha \cdot \varepsilon_{\lambda, BH^+} + (1 - \alpha) \cdot \varepsilon_{\lambda, B} \tag{4}$$

where $\varepsilon_{\lambda, B}$ and $\varepsilon_{\lambda, BH^+}$ are the ordinate values at the wavelength λ for the shifting profiles of B and BH^+ forms, respectively. The protonation degree

$$\alpha = \frac{c_{BH^+}}{c_B + c_{BH^+}} = \frac{I}{1 + I} \tag{5}$$

was obtained by the use of the indicator ratio ($I = c_{BH^+}/c_B$) values for the acid solution considered (Equation 2). The model set of spectra was treated by the empirical methods of Stewart and Granger (5) and Davis and Geissman (6) to recover the pK_{BH^+} and m^* values used in generating these spectra. Applying the Stewart–Granger method to four arbitrarily chosen wavelengths (see Fig. 1), the following pK_{BH^+} and m^* values were obtained:

$\lambda_1: -2.31 \pm 0.22$, 1.12 ± 0.10; $\lambda_2: -2.21 \pm 0.12$, 1.24 ± 0.06;
$\lambda_3: -2.09 \pm 0.18$, 1.30 ± 0.10; $\lambda_4: -1.85 \pm 0.26$, 1.28 ± 0.16.

(The confidence limits are presented at a risk level of 0.05.) All these values are rather different from those used in generating the digital model set of spectra, specifically, -2.40 and 0.90. The Davis–Geissman method was applied to the pair of wavelengths (λ_2, λ_3), and as a result, we obtained $pK_{BH^+} = -2.24 \pm 0.05$, $m^* = 0.82 \pm 0.03$.

Summarizing the results obtained by the model set of spectra, we can say that there is no confidence in calculating the pK_{BH^+} and m^* values from the shifting spectra using well-known empirical methods. Probably a similar conclusion was made by Edward and Wong in 1977, who suggested the use of factor analysis in the basicity studies (7). The method suggested was a version of the principal component analysis (PCA) described in detail by Simonds (8). In references 7 and 9–12 the results obtained by this method were found to be satisfactory. But the application of Simonds's version of PCA to amide protonation was unsuccessful: the coefficient of the first vector did not reach a stable final value in sulfuric acid solutions where the base is practically wholly in the BH^+ form; this coefficient continues to vary slowly with increasing acid concentration (13). This failure of Simonds's method is not surprising because it yields an abstract factor analysis solution to the problem only. The abstract vectors and their coefficients are useful for a short-circuit reproduction of the initial data matrix but may have no real physical or chemical meaning (14). Applying the digital simulation models, the ability of the Simonds method to recover the shapes of vectors and the coefficients used to generate data matrixes have been estimated (15). It was found that the results obtained by the Simonds method should be treated with caution because in a number of cases an agreement between the initial and recovered shapes of vectors and their coefficients was rather poor (15). This could be anticipated as the Simonds method (8) is, in fact, PCA; in other words, first the maximum possible variation is assigned to the first eigenvector, then the maximum amount of the remainder is assigned to the second, and so on. Thus, the Simonds method shows a tendency to overestimate the first vector on account of all others, as well as the the second on account of the remainders, and so forth.

The failure of the Simonds method (8) does not mean that factor analysis should be disqualified as a tool in protonation studies. For further use of factor analysis in protonation studies, the abstract factor analysis solution must be transformed into a more meaningful one. The mathematical bases for such transformation are summarized (14) by the sequence of equations

$$[D] = [R]_{PCA} \cdot [C]_{PCA} = [R]_{PCA} \cdot [T] \cdot [T]^{-1} \cdot [C]_{PCA} = [R]_{TFA} \cdot [C]_{TFA} \quad (6)$$

where the spectral data matrix is $[D]$; $[R]$ and $[C]$ are the row and column matrixes, respectively. The subscript PCA indicates that the matrix was obtained by PCA. The transformed matrixes have the subscript TFA. The $[T]$ is the transformation matrix and $[T]^{-1}$, its inverse.

Two distinctly different approaches—abstract rotation and target testing (TT)—may be used to transform the PCA solution. The first of them, abstract rotation, transforms abstract PCA matrixes into other abstract matrixes and is therefore not useful for our purposes. The second approach, target testing, is a unique method for testing potential factors one at a time. Target testing enables us to evaluate ideas concerning the nature of factors and thereby develop physically significant models for the data considered (14). For this reason we decided to apply the target testing approach to the RCA solution in order to estimate the meaningful values of pK_{BH^+} and m^* for weak bases whose UV–visible spectra show a complicated pattern of behavior in strongly acidic solutions.

The program, written in FORTRAN IV, is based on the mathematical procedures given by Malinowski (14). In the first, the PCA stage, the covariance about the origin of data is used. The iterations for decomposition of the covariance matrix are repeated until no element of the eigenvector considered changed by more $1 \times 10^{-4}\%$ in two subsequent steps. (The FORTRAN IV double-precision mode was applied.)

The number of abstract factors NF required in principal component analysis was chosen using the root-mean-square value for the data matrix $[D]$:

$$\mathrm{rms}(D) = \left[\frac{1}{NR \cdot NC} \sum_{i=1}^{i=NR} \sum_{k=1}^{k=NC} d_{i,k}^2 \right]^{0.50} \tag{7}$$

and for the residual matrix

$$[E] = [D] - [R]_{\mathrm{PCA}} \cdot [C]_{\mathrm{PCA}} \tag{8}$$

the root-mean-square element is

$$\mathrm{rms}(E) = \left[\frac{1}{NR \cdot NC} \cdot \sum_{i=1}^{i=NR} \sum_{k=1}^{k=NC} e_{i,k}^2 \right]^{0.50} \tag{9}$$

where NR is the number of rows and NC is the number of columns. The NF was increased until $\mathrm{rms}(E) \cdot [\mathrm{rms}(D)]^{-1} < 0.02$ was reached. The value 0.02 was chosen as the higher limit for the relative rms experimental error. This procedure yielded $NF \leqslant 5$.

For the target testing we must have a test vector quantitatively representing the model under test. The effect of protonation on the absorption spectra is given by Equation 4. Keeping Equation 4 in mind, we decided to study three cases using different test vectors $\bar{\mathbf{R}}_1$ and data matrixes $[D]$.

In case 1 we used

$$\bar{\mathbf{R}}_1 = \alpha \tag{10}$$

where α is the protonation degree (see Equation 5). In case 1 the data matrix $[D]$ consisted of the measured spectra; that is, each element of $[D]$ was an $\varepsilon_{\%\mathrm{acid}, \lambda}$ value. We are aware of the fact that this choice is arbitrary because in Equation 4 only one term is proportional to α. Nevertheless, we were interested in checking the TT method of factor analysis to deal with such a complicated case.

In case 2 we used the same $\bar{\mathbf{R}}_1 = \alpha$ (Equation 10), but the data matrix $[D]$ consisted of differences

$$d_{i,j} = \varepsilon_{\%\mathrm{acid}, \lambda} - \varepsilon_{\mathrm{B}, \%\mathrm{acid}, \lambda} \tag{11}$$

because Equation 4 may easily be transformed into

$$\varepsilon - \varepsilon_B = \alpha \cdot (\varepsilon_{BH^+} - \varepsilon_B) \tag{12}$$

As ε_B we applied the ε values for the solution with the lowest strong acid concentration considered; thus, the spectrum of the base in pure water may be subtracted from all other spectra.

In case 3 we made use of

$$\bar{\bar{R}}_1 = 1 - \alpha \tag{13}$$

and the data matrix $[D]$ consisted of differences

$$d_{i,j} = \varepsilon_{\%\text{acid},\lambda} - \varepsilon_{BH^+,\%\text{acid},\lambda} \tag{14}$$

because from Equation 4 it is easy to obtain

$$\varepsilon - \varepsilon_{BH^+} = (1 - \alpha)(\varepsilon_B - \varepsilon_{BH^+}) \tag{15}$$

For ε_{BH^+} we used the ε values measured in the most concentrated strong acid solution considered; thus, the spectrum of the base in the concentrated strong acid solution is to be subtracted from all other spectra.

The test vector $\bar{\bar{R}}_1$ is in each case constructed by using the protonation degree of the base studied (Equation 5); thus, the components of the test vector $\bar{r}_{1,1}, \bar{r}_{2,1}, \bar{r}_{3,1}, \ldots, \bar{r}_{NR,1}$ are the α (or $1 - \alpha$ in case 3) values for the acid solutions applied in recording the spectra. In terms of the TT method of factor analysis our goal is to find such a dependence, $\alpha = \varphi$ (% strong acid), which most suits to the measured spectra. It is evident from Equations 2 and 5 that $\alpha = \varphi$ (% strong acid) is determined by two variables: pK_{BH^+} and m^*. Thus we decided to generate a large number of the values of pK_{BH^+}, m^* pairs for constructing the respective number of dependences $\alpha = \varphi$ (% strong acid), which, in turn, will be used in Equation 10 and/or 13. Our program uses all combinations of pK_{BH^+} and m^* values given by the researcher. The number of the included pK_{BH^+} values (NPK) and m^* values (NM) was NPK ≤ 25 and NM ≤ 25, respectively. Thus the total number of $\bar{\bar{R}}_1$ vectors tested (NT) was NT ≤ 625. Each $\bar{\bar{R}}_1$ was subjected to the target testing (14):

$$\bar{R}_1 = [R]_{PCA} \cdot T_1 \tag{16}$$

where the transformation vector

$$T_1 = [\lambda]_{PCA}^{-1} \cdot [R]_{PCA}^T \cdot \bar{\bar{R}}_1 \tag{17}$$

The $[\lambda]_{PCA}$ is the diagonal matrix with eigenvalues $\lambda_1, \lambda_2, \ldots$ obtained in principal component analysis on the main diagonal.

If the suspected test vector $\bar{\mathbf{R}}_1$ is a real one, that is, fits the measured set of UV–visible spectra, regeneration according to Equation 16 will be successful: each element of the predicted vector $\bar{R}_1$ will be equal to the corresponding element of $\bar{\mathbf{R}}_1$ within experimental error. The program calculates and stores for each $\bar{\mathbf{R}}_1$ tested the value of

$$z = \sum_{i=1}^{i=NR} (\bar{r}_{i,1} - \bar{r}_{i,1})^2 \qquad (18)$$

where $\bar{r}_{i,1}$ and $\bar{r}_{i,1}$ are the elements of the predicted vector $\bar{\mathbf{R}}_1$ and the test vector $\bar{\mathbf{R}}_1$, respectively.

When all the tests have been performed, the smallest z value is chosen. The pK_{BH^+} and m^* values used to obtain the minimal z value are considered to be the best of those tested for describing the protonation equilibrium studied. Of course, if in the run the pK_{BH^+} and m^* were varied in large steps, a new run with smaller pK_{BH^+} and m^* steps (in a closer range) would be necessary in order to obtain the pK_{BH^+} and m^* values within the commonly used precision of ± 0.01 in the corresponding units. The parameters for the new run are determined by the researcher on the basis of the four best combinations of the pK_{BH^+} and m^* values printed out by the program.

When the most suitable combination of the pK_{BH^+} and m^* values has been found, the program estimates how well this combination allows reproduction of the measured spectral data matrix. For this purpose a new row matrix $[R]_{TFA}$ is constructed by replacing in the $[R]_{PCA}$ the first column by $\bar{\mathbf{R}}_1$ corresponding to the minimal z value (Equation 18). This substitution is based on the assumption that the first, largest factor, can be associated with the spectral change accompanying protonation. Next the general transformation matrix is calculated by

$$[T] = [\lambda]_{PCA}^{-1} \cdot [R]_{PCA}^T \cdot [R]_{TFA} \qquad (19)$$

The column matrix in the new coordinate system is obtained by

$$[C]_{TFA} = [T]^{-1} \cdot [C]_{PCA} \qquad (20)$$

The program calculates the reconstructed data matrix

$$[D]_{TFA} = [R]_{TFA} \cdot [C]_{TFA} \qquad (21)$$

and compares it to the initial data matrix $[D]$. Of course, the content of $[D]$

depends on the case selected: in case 1 the matrix $[D]$ consists of the measured ε, in case 2 $[D]$ consists of $\varepsilon - \varepsilon_B$, and in case 3 $[D]$ consists of $\varepsilon - \varepsilon_{BH^+}$. The rest,

$$[ERR] = [D]_{TFA} - [D] \qquad (22)$$

shows how $[D]_{TFA}$ and $[D]$ coincide. As a rule, the root-mean-square element of $[ERR]$ was found to be in the range of 2–4% from the same element of $[D]$.

It should be emphasized that the test performed by Equations 21 and 22 is not a unitary transformation and does not force the model constructed, using the pK_{BH^+} and m^* values, to fit the data. Consequently, this is a unique and severe test of the model (14).

When the program described was completed, we first applied it to a number of digital simulation models in order to check whether the program enables us to recover the pK_{BH^+} and m^* values used to generate the data matrixes. The digital models used consisted of two, three, or four Gaussian profiles. In run 1 we calculated the elements of data matrix by the equation

$$d_{i,j} = \alpha_i(Y1_{i,j} + Y2_{i,j}) + (1 - \alpha_i)(Y3_{i,j} + Y4_{i,j}) \qquad (23)$$

where $d_{i,j}$ is an element of spectral data matrix $[D]$ for the sulfuric acid concentration i and wavelength j; α_i is the degree of protonation (Equation 5); and $Y1_{i,j}$, $Y2_{i,j}$, $Y3_{i,j}$, $Y4_{i,j}$ are the ordinates of shifting Gaussian profiles. (In some models the Gaussian half-widths were also linear with acid concentration.) All these Gaussian profiles were shifted independently from each other along the abscissa axes to longer or shorter wavelengths but linearly with the sulfuric acid concentration in w/w% − s. In run 2 we calculated

$$d_{i,j} = \alpha_i \cdot Y1_{i,j} + (1 - \alpha_i)Y3_{i,j} + Y2_{i,j} + Y4_{i,j} \qquad (24)$$

Thus, the two shifting Gaussian profiles ($Y2_{i,j}$ and $Y4_{i,j}$) represent the "medium effects," which do not depend on protonation but vary linearly with acid concentration. Using different pairs of pK_{BH^+} and m^* values in the ranges −1.58 to −3.30 and 0.45–0.90, respectively, altogether 22 digital simulation models were generated. Two typical examples of the models tested are presented in Fig. 2. All the digital models generated were subjected to the treatment by the program in order to recover the pK_{BH^+} and m^* values used. The respective results are summarized in terms of

$$\Delta pK = pK_{BH^+}(\text{found}) - pK_{BH^+}(\text{used in model}) \qquad (25)$$

and

$$\Delta m = m^*(\text{found}) - m^*(\text{used in model}) \qquad (26)$$

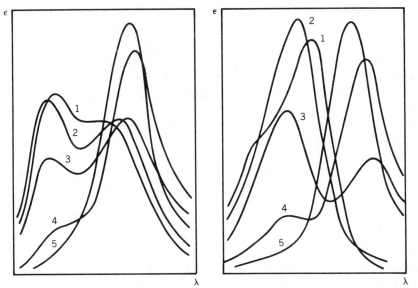

Figure 2 Samples of computer-generated digital models used for recovering pK_{BH^+} and m^* values by the method suggested ($pK_{BH^+} = -2.44$, $m^* = 0.67$). Sulfuric acid % (w/w): 1–5%, 2–25%, 3–48%, 4–60%, 5–95%. On left—model constructed using Equation (24). On right—model constructed using Equation (23).

The ΔpK and Δm values obtained are listed in Table 1, which shows that the method suggested allows us to recover the pK_{BH^+} and m^* values with reasonably small errors. The best results are obtained in case 3, where in the data matrix $[D]$ we have the differences $\varepsilon - \varepsilon_{BH^+}$; thus, the spectrum in the most concentrated acid solution is subtracted from each spectrum at a lower acid concentration. In this case we can expect that the ΔpK value is mostly in the range of -0.05 to $+0.15$ and Δm in the range of -0.05 to $+0.01$, respectively.

In experimentation with digital models we learned that when applying the suggested method at least two limitations exist. First, if the "medium effects" are very large compared to the protonation effects, the absolute values

TABLE 1
Mean Values of ΔpK (25) and Δm (26)

Mean Value[a]	Case 1 ε in $[D]$	Case 2 $\varepsilon - \varepsilon_B$ in $[D]$	Case 3 $\varepsilon - \varepsilon_{BH^+}$ in $[D]$
ΔpK	-0.02 ± 0.12	0.08 ± 0.27	0.05 ± 0.09
Δm	0.01 ± 0.08	0.03 ± 0.21	-0.02 ± 0.03

[a]With standard error for $n = 22$.

of the errors ΔpK and Δm both increase remarkably. For instance, this is the case when the profiles $Y2$ and $Y4$ in Equation 24 make up more than a third of the profiles $Y1$ and $Y3$, which reflects the changes caused by protonation.

The second point is that the absolute values of the errors ΔpK and Δm will increase if the "medium effect" depends on the acid concentration in a way similar to that in the protonation process. In this case the orthogonality between the two effects considered is poor, leading to some "overpumping" of the "medium effect" into the protonation process: the method is not able clearly to distinguish between the "medium effect" and changes in the spectra reflecting the protonation of the base any more.

Keeping this in mind, we decided to use the suggested method for estimation of the pK_{BH^+} and m^* values of some substituted benzamides. The respective spectra in aqueous sulfuric acid solutions have been recorded by Cox and Yates (13) and were available as supplementary material to their paper. The pK_{BH^+} and m^* values of the weak bases considered have also been estimated by Cox and Yates. They developed and used rather a sophisticated method for treating the shifting spectra based on the excess acidity (X function) (13). Thus we can compare the pK_{BH^+} and m^* values calculated by the suggested target testing method of factor analysis (16) and the corresponding values obtained by the X-function method. The results are presented in Table 2.

The Table 2 shows that the corresponding pK_{BH^+} and m^* values are rather close to each other. This fact may be taken as a mutual validation of the

TABLE 2
Comparison of Results for Amides Obtained by Excess Acidity Method (13) and TFA Method (Case 1) (16)

Base	Obtained by TFA			
	pK_{BH^+}	m^*	ΔpK^a	Δm^b
Benzamide	−1.38	0.55	−0.10	0.03
N-Ethylbenzamide	−1.50	0.51	−0.07	0.01
N-Isopropylbenzamide	−1.56	0.58	+0.01	0.04
N-Isobutylbenzamide	−1.66	0.40	+0.10	−0.01
N-sec-Butylbenzamide	−1.45	0.52	−0.08	0.13
N-t-Butylbenzamide	−1.34	0.65	0.02	0.01
N-Benzylbenzamide	−1.61	0.45	−0.15	0.21
α-Methyl-N-Benzylbenzamide	−1.44	0.30	−0.08	0.15
N,N-Diethylbenzamide	−1.08	0.54	−0.03	0.05
N,N-Diisopropylbenzamide	−0.74	0.23	+0.12	0.48
Benzoylpiperidine	−0.94	0.32	0.00	0.21

$^a \Delta pK = pK_{BH^+}$ (excess acidity) $- pK_{BH^+}$ (TFA, case 1).
$^b \Delta m = m^*$ (excess acidity) $- m^*$ (TFA, case 1).

two methods considered. It should be noted that the method used by us and that of Cox and Yates are completely different, being based on different approaches to the problem.

Summarizing the work done so far, it should be emphasized that the suggested target testing method of factor analysis is by no means a panacea. Its usefulness would be proved by further study and applications.

ACKNOWLEDGMENT

I would like to express my gratitude to Mr. A. Murshak for the technical assistance in using the EC-1052 computer installed in the Institute of Cybernetics of the Academy of Sciences of the Estonian S.S.R.

REFERENCES

1. R. A. Cox and K. Yates, *J. Am. Chem. Soc.*, *100*, 3861 (1978).
2. R. A. Cox and K. Yates, *Can. J. Chem.*, *61*, 2225 (1983).
3. L. P. Hammett, *Physical Organic Chemistry*, McGraw-Hill, New York, 1970, p. 373.
4. J. I. Berstein and J. L. Kaminski, *Spectrophotometric Analysic in Organic Chemistry* (in Russian), Khimia, Leningrad, 1975, p. 147.
5. R. Stewart and M. R. Granger, *Can. J. Chem.*, *39*, 2508 (1961).
6. C. T. Davis and T. A. Geissman, *J. Am. Chem. Soc.*, *76*, 3507 (1954).
7. J. T. Edward and S. C. Wong, *J. Am. Chem. Soc.*, *99*, 4229 (1977).
8. J. T. Simonds, *J. Opt. Soc. Am.*, *53*, 968 (1963).
9. R. A. Cox, C. R. Smith, and K. Yates, *Can. J. Chem.*, *57*, 2952 (1979).
10. Ü. Haldna, A. Murshak, and H. Kuura, *Org. Reactiv. (Tartu)*, *17*, 4 (64), 384 (1980).
11. Ü. Haldna, A. Murshak, and H. Kuus, *Org. Reactiv. (Tartu)*, *17*, 3 (63), 313 (1980).
12. R. I. Zalewski, *J. Chem. Soc. Perkin II*, 1637 (1979).
13. R. A. Cox and K. Yates, *Can. J. Chem.*, *59*, 1560 (1981).
14. E. R. Malinowski and D. G. Howery, *Factor Analysis in Chemistry*, Wiley, New York, 1980.
15. Ü. Haldna and A. Murshak, *Comput. Chem.*, *8*, 39 (1984).
16. A. Ebber, Ü. Haldna, and A. Murshak, *Org. Reactiv. Tartu*, *23*, 1 (81), 40 (1986).

Application of Principal Component Analysis in Organic Chemistry

BY ROMUALD I. ZALEWSKI
Division of General Chemistry
Academy of Economics
Poznań, Poland

CONTENTS

I. Introduction 77
II. Principal Component Analysis 78
    A. Principal Component Model 78
    B. Dimensionality of Data 80
    C. Data Pretreatment 81
    D. Goodness-of-fit Residuals 82
    E. "Abstract" Solution 83
    F. "Real" Solution 84
    G. Mathematical and Geometric Foundation of PCA 87
III. Application of PCA in Organic Chemistry 89
    A. Substituent and Conformation Effects on ^{13}C NMR Spectra of Aromatic Compounds, Cyclohexanes, and Related Compounds 89
        1. Unsaturated and Aromatic Systems 90
        2. Alicyclic Systems 92
    B. UV Spectra in Basicity Studies 101
    C. Selected Spectroscopic and Chromatographic Problems 107
        1. IR and Raman Spectroscopy 107
        2. Gas and Liquid Chromatography 108
        3. Mass Spectrometry 109
    D. Classification of Solvents 110
    E. Separation of Inductive and Resonance Substituent Effects and Related Phenomena 114
    F. Acidity of Nonideal Solvents 117
    G. Reactivity of Organic Molecules 120
    References 123

I. INTRODUCTION

Principal component analysis (PCA) was first established by Pearson (1) for application in psychology and developed to its present stage by Hotelling

(2). Simultaneously Thurstone began to formulate factor analysis (FA) (3). FA is closely related, but not identical, to PCA. The two methods are often confused and the names are incorrectly used.

The power of PCA has been rediscovered in various scientific fields besides psychology. As a result, redundant terminology of PCA may be found in the literature: FA, singular-value decomposition in numerical analysis (4) and Karhunen–Loeve expansion (5) in electrotechnics. In physical sciences the terms "eigenvector analysis" and "characteristic vector analysis" (6) are in frequent use. A variant of PCA in France is called "correspondence analysis."

Many good textbooks including discussion of PCA have been published (7–10). PCA was introduced to chemistry in the 1960s by Malinowski, and a large number of applications have been described (11, 12).

II. PRINCIPAL COMPONENT ANALYSIS

A. Principal Component Model

At the end of experimental work on samples in chemistry the final table of data emerges. A large table of data is very difficult for the human mind to comprehend. A routine is to compare between two rows or two columns simultaneously, using linear regression or a plot. The graphical examination is useless beyond three dimensions. Lines, planes, or surfaces from one side and equations describing them are frequently used by the chemist in the form of LFFR (13), structure–activity relationships, (14), and other branches of regression analysis and similarity models (15). The limits of regression analysis are overcome by PCA.

The starting point of PCA is a table of data, a data matrix $\mathbf{D}$, consisting of r rows and k columns. The rows and columns are called *row* (or *column*) *vectors*. In rows we can place data characteristic for "objects" corresponding to chemical samples. For each object we collect numbers (data) in columns representing variables. Figure 1 gives a graphical representation of a data matrix of format $k \times r$. The main results of PCA of a data matrix are:

1. Examination of the relationship between column or/and row vectors.
2. Finding new orthogonal vectors.
3. Finding the rank of the data matrix.
4. Reduction of dimensionality in the data matrix.
5. Calculation of principal components.
6. Calculation of loadings for principal components. These results may be used by researchers for simplification of the data matrix from redundant information, modeling, detection of outliers, selecting of better sets of

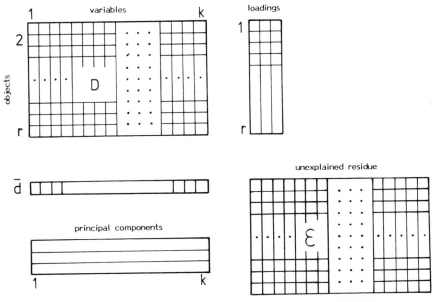

Figure 1 Block diagram of the principal component analysis (PCA).

variables or objects, classification or discrimination, and prediction. Depending on the problem at hand, some or all of these applications should be considered at the same time.

In general, almost any data matrix can be processed by PCA without any starting conditions, leading to principal components, their loadings, the matrix of unexplained residuals, and other features, part of which are shown in Fig. 1.

Each data point is then represented by

$$d_{r,k} = \overline{d_k} + p_{i,k}l_{r,1} + p_{2,k}l_{r,2} + \cdots + \varepsilon_{r,k} \tag{1}$$

or in matrix notation

$$\mathbf{D} = \mathbf{P}\cdot\mathbf{L} + \varepsilon \tag{2}$$

in which p signifies the principal component, l signifies loading, and ε is unexplained residue.

Such results are sufficient to construct a PC model of the data matrix, represented by Equation 1 or Fig. 2.

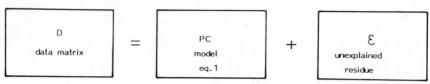

Figure 2 The data matrix is a combination of the PCA model and unexplained residue (noise). The number of principal components used must be sufficient to reduce the noise below certain limits.

The PC model thus extracted may be used for prediction of new data for the same system.

B. Dimensionality of Data

The main task in constructing the PC model is to decide how many components to use in it. The total number of components is k or r (whichever is less), but in further analysis the first two or three are examined. In many applications this serves well to clean the data of typing errors, sampling errors, or similar. In more serious work the correct number of components **A** is essential and some precise criterion is needed to estimate it.

Often the number of components extracted is such that the variance of residuals unexplained by the model ($\varepsilon_{r,k}$) is equal to or less than the error of measurement in **D**. This assumption may be misleading in the case when all systematic chemical and physical variations of data are not explained by the PC model.

Malinowski developed a procedure on the basis of the theory of error (11,16) and using only the primary set of eigenvectors. The criterion is based on the rate of decrease of the remaining residual sum of squares but is not clear theoretically.

A similar criterion was that of dividing each eigenvàlue by the succeeding one and finding the value of **A** where this ratio reaches a maximum (17).

Cartwright (18) used submatrix analysis to determine dimensionality of spectroscopic data and to select a subset of spectra with lowest dimensionality. The method is simple only when three components are present and the number of objects is not very large. With r spectra, eigenvalues of a full matrix and $(r - 1)$ submatrices are calculated and the set of spectra is selected with the lowest number of significant eigenvalues. The spectra of higher dimensionality are thus selected.

A very effective method to find dimensionality of a data matrix is cross-validation developed by Wold (19). The idea of this procedure is to keep part of the data out of the PCA calculation and than predict the data through the developed model. The quality of prediction is calculated as squared differences between predicted and experimental data. The procedure is repeated until each data element is removed only once. Cross-validation (CV) starts with $A = 0$ and stops with A_n if A_{n+1} gives a worse result. The other version of CV was developed using the singular-value decomposition technique (20).

Both versions of CV lead to too few components rather than too many. This is an advantage; data are not overestimated and false conclusions or models are not created.

C. Data Pretreatment

The principal component values and magnitudes depend on the data values, magnitudes, distribution, and structure. Data pretreatment may than play an important role and must be explicitly described.

Data pretreatment is not necessary when the data table consists of homogenous data reflecting similar properties [e.g., a set of ultraviolet (UV)–visible or infrared (IR) absorbances] at fixed frequencie for mixtures of different concentrations or recorded at various temperatures, acidity, and so on. Most frequently a data matrix consists of nonhomogenous data (e.g., UV–visible, IR absorbances, mass spectra abundances, physical data of various origins). The numerical values of variables could be very small or very large and differ by order of 10^n. Also, variability of magnitude for objects may vary from small (e.g., density, refraction index, $E_{T(30)}$) to very large (e.g., reaction rate, molecular absorption coefficient, solubility). Mixing such types of data in one matrix will lead to errenous results and the derived PC model may be false.

The other source of false results is the presence of outliers that severely influence the model. It is strongly recomended to find and correct or eliminate outliers before creating the final PC model. There are several possible indices showing outliers. Outliers having critical influence will appear on plots between principal components or loadings and will cause large unexplained residuals in columns or rows. Outliers could be found among samples or variables.

The other cases are irrelevant variables or samples visible after isolation of outliers. Such data are of no or little effect on the PC model and could be removed, especially if they are increasing the cost of computation or experiment. It is, however, better to leave them in the body of data.

It is practical to create a training data set of objects beside a test set. After developing the PC model on the training set, the test may be applied to check the validity of the model.

The format of the data matrix is not restricted. It is, however, recommended to work with rectangular rather than square matrices.

The various strategies of PCA use correlations between, or covariance of, columns in the data set. Correlation involves the normalization of data and giving equal statistical weight to all variables and destroying the original magnitude of the data. Covariance is unnormalized and weights the variables in proportion to their absolute value. The data can be considered to be distributed about the mean of column or about its origin. In the last case data are not adjusted; in the first, the data have no real zero and are moved to the center of the data.

Taking the logarithm makes the data more centrally distributed. Taking the square roots or reciprocals or dividing each column vector through by the first or largest element is in practical use in specialized applications.

Standardization of the data matrix gives unit variance to each column. All standarized variables then have the same length and thus the same influence on the PC model. PC loadings calculated from such a treated data matrix are eigenvectors of the correlation matrix.

Standarization of variables of small variability will cause scaling up. This means that such variables are overestimated and influence the final solution more. According to Wold (10) if the standard deviation of a variable is smaller than one quarter of the measurement error, it is better to leave the variable unscaled. With different types of variables, different pretreatment may be performed for each type. This could be done by dividing each variable by its standard deviation multiplied by the square root of the number of variables.

A very special and common problem is given by missing data. The "holes" in a data matrix must be filled, as, for instance, by substituting the mean value of that column in iterative processing with few iterations or by processing an uncomplete data matrix by an algorithm capable of dealing with missing data. The NIPALS algorithm is recommended (22).

D. Goodness-of-fit. Residuals

The amount of experimental data variability explained by the PC model may be expressed in different ways.

Eigenvalues λ of correlation (or covariance) matrix (centered matrix) placed in descending order allow the calculation of percent of total data variability explained by the PC model with 1,2... principal components from the expression

$$\frac{\text{eigenvalue}_1}{\sum_1^A \text{eigenvalues}} \cdot 100\%$$

The unexplained residuals $\varepsilon_{r,k}$ have the standard deviation SD_0

$$SD_0^2 = \frac{\sum_1^r \sum_1^k \varepsilon_{r,k}^2}{(k - A - 1)(k - A)} \tag{3}$$

which measure the "distance" between the PC model (on the levels $A = 1$, $A = 2$, $A = \ldots$) and the objects belonging to it. Standard deviations for columns and/or rows are also calculated, and used, in numerical form or as historgrams and allow us to follow the improvement of the PC model when adding further principal components.

Unexplained residuals are used to calculate the importance of each variable, by the modeling power ψ

$$\psi = 1 - \frac{SD_K}{SD_{K,0}} \tag{4}$$

in which $SD_K^2 = \sum_1^k \varepsilon_{r,k}^2/(k - A - 1)$.

Modeling power indicates the extent to which the variable k participates in the model. A value close to zero indicates the variable to be irrelevant and it may be delated, as contributing mainly noise.

E. "Abstract" Solution

The solution of the data matrix in terms of principal component scores and loadings l is called the "abstract" one. This type of solution is the best from a mathematical point of veiw and is relevant to the problem in hand. The abstract solution is not unique, however, because scores and loadings are undefined with respect to some constant K:

$$p_{1,k} \cdot l_{r,1} = (p_{1,k} \cdot K)\left(\frac{l_{r,1}}{K}\right) \tag{5}$$

Thus the PC model defines the final dimensionality of data; allows us to calculate the reconstituted data matrix with one, two, or more principal components; and inspect the structure of the data as projected into plane or space. The disadvantage of the abstract solution is that it has no chemical or physical meaning.

The advantage of the abstract solution lies in the fact that the data matrix D may be represented by a few plots. For example, orthogonal plots $P_{2,k}$ versus $P_{1,k}$, $P_{3,k}$ versus $P_{1,k}$, or appropriate plots of loadings show the most

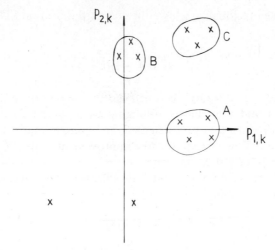

Figure 3 The plot of principal component $p_{2,k}$ versus $p_{1,k}$. For explanation, see text.

important patterns in **D**. Inspection of these plots will allow us to define similar objects or variables, to find outliers and so on. The artificial plot presented on Fig. 3 shows this.

The variables with large P_1 values and small P_2 values form a cluster A strongly associated with the first principal component, while variables with small P_1 and large P_2 form a cluster B associated with the second principal component. Variables in cluster C are strongly associated with both components and are less explanatory.

The other points may depict outlying variables not associated with ones belonging to class A – C.

The structure of points on plots depends on the nature of the problem and may vary from a disorder to quite well described clusters.

Figure 4 shows a plot of loadings representing various combinations of mixed solvent and temperatures for a solvolysis reaction (23). It is obvious that points of the same solvent composition lie on a line with positive slope and from left to right represent increasing temperature. Various solvent mixtures at equal temperatures occupy a line with a negative slope. The picture is very simple and regular. As the abstract solution is hard to explain in terms of chemistry or physics, some tools are needed to replace it by a real solution.

F. "Real" Solution

The real solution in PCA may be achieved by finding the value of the multiplicative constant K in Equation 5 or finding the transformation matrix

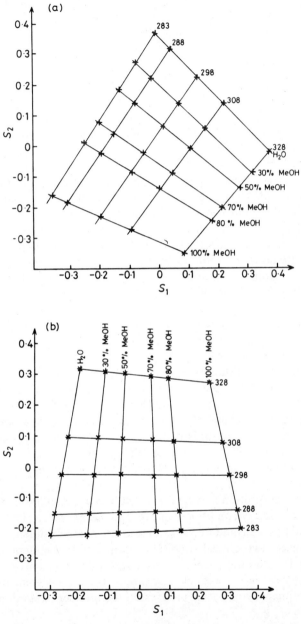

Figure 4 The plot of l_2 versus l_1 for solvolysis reaction (23): (*a*) "abstract" solution; (*b*) "real" solution.

T such that

$$P \cdot L = PTT^{-1}L \tag{6}$$

and putting

$$(PT)(T^{-1}L) = RS \tag{7}$$

where **R** and **S** are respectively the principal component matrix and the loading matrix in real solution.

A very simple and frequently useful way to find the real solution is rotation of axes to achieve a simpler structure of points. This may be explained using Fig. 4. Points tend to occupy almost parallel and perpendicular lines twisted from horizontal and vertical directions. Replacing abstract axes by new rotated real axes leads to calculation of new component and loading values; however, the distances between points remain unchanged. Now fewer corrdinates are necessary to describe the distribution of points on the plane. The first component after rotation accounts for a decreased percentage of total data variability, and the second one shows a gain, whereas the total remains unchanged.

More sophisticated versions of simple rotation are known as varimax, quartimax, and equimax (24). Very useful for solving various problems are quartimin, oblimax, or promax (25).

In all procedures distances between points representing principal components remain unchanged, but the points are projected onto new axes.

The results of rotation procedure themselves have the status of abstract solution; however, because of their greater simplicity they are easier to interpret. Such interpretation is possible through finding the relations (e.g., linear) between the PC and rotated PC models and some parameters describing the nature of the studied phenomenon known from the other experiments.

The abstract solution may be converted into a real solution by imposing critical and subsidiary conditions on principal components. This procedure has been used by Swain and Lupton (26) to create new substituent scales. With two components, two critical conditions (nature of substitutent effects) and four subsidiary conditions (definitions of zero and unity on the scales) are necessary to calculate scales having simple chemical significance. If the number of principal components is greater, the number of conditions grows rapidly.

A target transformation procedure proposed by Weiner (27) allows us to explain principal components one by one. A test vector prepared by the researcher (on the basis of knowledge of nature and complexity) and representing the idea to be target tested is input and a predicted vector having the best least-squares fit to the test vector is calculated. If the agreement between test and predicted vectors is satisfactory using chemical criteria, then the tested parameter is said to be identified. Additionally, the uniqueness test is performed to ascertain which objects exhibit the most typical and atypical properties.

The variant of target testing, called the *key-set*, uses a set of various columns in a matrix **D** to determine which subset gives the best model for the complete data matrix (28, 29).

G. Mathematical and Geometric Foundation of PCA

A data matrix **D** with r objects and k variables may be represented as r points in k-dimensional space. Each point is located at an Euclidean distance from the origin and has the properties of a vector. PCA in the geometric sense is a projection of that r points in k space down to a lower-dimensional subspace and calculation of new orthogonal coordinates (principal components). Figure 5 shows the reduction of dimensionality of three (a,b,c)-dimensional space. Experimental points in space tend to occupy the plane (f_1, f_2). The new direction f_1 represents the largest variability of experimental

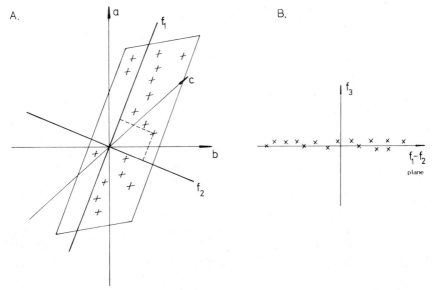

Figure 5 The experimental points (*) in three-dimensional space (axes a,b,c) are projected on plane (f_1, f_2) (*A*). Distribution of points along f_3 axis is neglegible (*B*).

data, and f_2 represents the next largest variability. The third dimension due to very small variability of experimental data is neglected. Then f_1 and f_2 are new, principal axes. Projection of experimental points on f_1 and f_2 gives their principal components. More exhaustive geometric interpretation of PCA is given in the literature (30).

The mathematical foundation of PCA is based on matrix calculation and properties. A data matrix **D** is transformed into a correlation matrix **R** or a covariance matrix **Z**. Each of them may be calculated for centered data (mean value of columns substracted) or against the origin. Both **R** and **Z** as square matrices have eigenvalues λ and

$$\mathbf{Z} = \mathbf{D}^T\mathbf{D} \tag{8}$$

eigenvectors e. Eigenvalues are computed and placed in descending order on the diagonal in a square matrix Λ. The value of $\lambda_{k,k} > 0$ indicates the rank of matrix and hence the number of principal components **A**. The basic structure of either **Z** or **R** matrix is

$$\mathbf{Z} = \mathbf{E}\Lambda\mathbf{E}^{-1} \tag{9}$$

Calculation of principal components is based on the relationships

$$\mathbf{P} = \mathbf{D}\mathbf{E} \tag{10}$$

and

$$\sum_1 P^2 = \mathbf{P}\mathbf{P}' = \mathbf{E}'\mathbf{D}'\mathbf{D}\mathbf{E} \tag{11}$$

Eigenvector elements are computed in such a way that variance of Equation 11 is maximized (if $E'E = 1$). This is valid (32) for e_1 being a unity eigenvector related to a maximum eigenvalue λ_1 in matrix **Z**; e_2 is related to λ_2 and thus to one. Putting Equations 8 and 9 into Equation 11, we get

$$\mathbf{P}\mathbf{P}' = \mathbf{E}'\mathbf{E}\Lambda\mathbf{E}^{-1}\mathbf{E} \tag{12}$$

and hence $\mathbf{E}'\mathbf{E} = \mathbf{I}$:

$$\mathbf{P}\mathbf{P}' = \mathbf{I}\Lambda\mathbf{I} = \Lambda \tag{13}$$

Principal components are thus calculated from matrix of eigenvalues. Equation 10 leads to the relation

$$\mathbf{D} = \mathbf{P}\mathbf{E}' \tag{14}$$

and in combination with Equation 2, we obtain

$$L = E' \qquad (15)$$

the loadings of principal components, as numbers in the range ± 1 and

$$\sum_{1}^{r} l_{1,r}^{2} = 1 \qquad (16)$$

In recent years the traditional methods of computation has been replaced by the NIPALS algorithm (32), and very recently the multimethod principal component analysis of data presented in three-dimensional (cube) form has been elaborated (33).

III. APPLICATION OF PCA IN ORGANIC CHEMISTRY

A. Substituent and Conformation Effects on ^{13}C NMR Spectra of Aromatic Compounds, Cyclohexanes, and Related Compounds

The growing applicability of ^{13}C NMR spectroscopy requires a better understanding of the shielding mechanism (34) in terms of experimental shifts caused by change of substituent and/or its conformation and the structure of molecular frame. The ^{13}C NMR substituent chemical shift (SCS) for a given nucleus has been approximated by a sum of diamagnetic contribution, paramagnetic contribution, and neighboring groups per nuclei contribution (35, 36). The main factors affecting paramagnetic contribution, which is dominating (37) in the case of ^{13}C NMR SCS, are charge polarization, variation of bond order, average energy of excitation, and other factors (38). Further application of ^{13}C NMR requires a better and consistent picture of the effects of changes in substituents and conformation on chemical shift (39). This approach has been begun by Dalling and Grant (40). The significance of a postulated relationship and the application of derived rules to unknown molecules of similar geometry should be based on rigorous standards of multiple regression analysis (15b, 41).

To obtain a reasonable correlation in NMR spectroscopy, the Hammett equation has often been replaced by two parameter equations with $\sigma_I - \sigma_R$ (42) or $\mathscr{F} - \mathscr{R}$ (43) scales. The significance of these treatments is in dispute because too much stress has been laid on the correlation coefficient as a criterion of the goodness of fit, and not enough on the number of independent mechanisms of the transmission of substituent effects.

1. Unsaturated and Aromatic Systems

Recently interpretation of SCS in aromatîxc compounds such as 4-substituted styrenes (44), 2-substituted indanes (45), chalcones (46), triazanes (47), and benzenes (48–51) by PCA shows that only one or two principal components are required.

For example, 99% of variability of ^{13}C NMR spectra of carbons C4–C7 in 15 derivatives of 2-substituted indanes (45) is described by the two-principal-components model. The modeling power of this model was higher than 0.92 for carbons C4 and C5 and about 0.73 for two others. The first component shows reasonably good correlation with mesomeric scales. Correlation ability increases after rotation of PC axes. The second component shows good correlation with the σ_I scale.

Two-component models are found by cross-validation to describe SCS of 54 chalcones and their thiophen and furan analogs (46). The data matrix consists of 54 objects × 11 chemical shifts. The plot of PC2 versus PC1 differentiates subseries—benzenes, thiophens, and furans—members of which occupy positions on three parallel lines. The similarity between the benzenoic carbons and carbonyl, and differentiation from ethylenic carbons, are evident from the plot of the two loadings. The results are promising for differentiations of carbons and compounds between similar series of compounds.

The results of calculation show two components are necessary in order to explain the data variability. Transformation of an "abstract" solution into a "real" one was done by simple rotation of orthogonal axes. New values of PC1 corresponds to σ_{para}, while the other distinguishes between three subsets of structures. Their nature was not explained by authors.

Two principal components were found to be necessary to model the ^{13}C NMR and ^{155}N NMR spectra of substituted triazines (47). However, both describe only 65% of data variability. The plot of two principal components allow us to divide substituents into classes of donors, acceptors, halogens, and alkyls. The phenyl substituent is an outlier.

The first principal component was explained as strongly related to charge density, for example, to the paramagnetic contribution, but the interpretation of the second principal component is not given.

A separate problem arising from this study deals with halogens. For all four halogen-containing derivatives, SCS are explained by only one principal component model with satisfactory precision. Their values are well correlated with data given by Wiberg et al. (50, 52). In two papers on aliphatic (50) and unsaturated as well as aromatic (52) halides, they found two large factors describing the predominating part of data SCS variability. The third factor was very small, but in many cases its importance was obvious. The first factor, a substituent parameter, "is essentially a constant for all the halogens and appears to be related to the polar effect of halogen vis-a-vis hydrogens" (50). The value of the second factor increases almost linearly on change from fluorine to iodine and is linearly dependent on a variety of non-NMR parameters (ionization potential, electron transition, bond polarizability (50, 52). The second factor was attributed to the "freeness" of valence electrons of halogen atoms.

In a subsequent paper (51) it was stressed that linearity among various properties of halogen-containing molecules is caused by polarizability. The polarizability of the halogens increase monotonically from F to I, while the C—X bond dipole moment is fairly constant (approximately $1.74 + 0.12D$).

The PCA of 82 monosubstituted benzenes (48) shows that 90% of compounds belong to one of the four clusters: alkyls, donors, acceptors, and halogens. Each of them is better described by separate one-parameter PC model. A similar grouping was found when σ_I was plotted against σ_R.

Exner and Budesinsky (53) investigated ^{13}C NMR SCS of meta- and para-substituted benzonitriles and found that CN carbon shifts correlate well with a dual-substituent parameter model in the para series but poorly in the meta. For a study of α carbon SCS in a broader set they selected from the literature eight series of α,β-unsaturated C=X groups adjacent to the phenyl ring. Fourteen meta and 14 para substituted derivatives in each series were available in the form of a 28 × 8 data matrix of Cα SCS. They found that one component is necessary to describe Cα shifts in meta compounds and two components for para. PCA models yield a significantly better fit than do various dual-substituents regression models.

The PC obtained for meta substituents is a linear function of σ_m with some exceptions, probably as a result of magnetic anisotropy of such substituents as I, Br, OMe, and NMe$_2$ (48). The first of two components in para-substituted compounds is even better explained by σ_m and the second, by σ_R. The deviations are rather small but sufficient to explan the better fit of the PCA model as compared to any dual substituent ones already tested.

Similar calculations were performed for Cβ SCS, showing that in meta-substituted compounds dual-substituent regresson involves one parameter in excess. Again the principal component was proportional to σ_m. The PCA model with two components is superior for para-substituted compounds. The first is explained by the σ_m substituent constant; however, the second is dissimilar to any kind of σ constant. Exner and Budesinsky explain the nature of this component by a very strong conjugation.

Principal components produced by PCA represent new scales of substituent constants, and their utility must be examined.

It was found in another factor analysis (49) that para SCS in 4-substituted phenols and 2-nitrophenols require two factors, whereas the ipso, ortho, and *meta* SCS required three factors. The explanation of factors for *para* SCS was successful with σ_F and σ_R° parameters (field and resonance). Factors yielded for other carbons correlate well with the appropriate *ipso*, *ortho*, and *meta* parameters determined by Reynolds (54).

It was stressed in numerous papers (45–48) that PCA allows us to detect incorrect shift data and false shift assignments.

2. Alicyclic Systems

The treatment of the various alicyclic systems of different structures and origins has been much less attempted by PCA. In a paper on 2-substituted norbornanes (55) (7-*exo* and 7-*endo*), three principal components were extracted. The first component is influenced mostly by carbon bearing substituent C2 and the second component is governed by C1 and C3, next to the substituted carbon. The third principal component is associated mainly with more remote carbons. Analysis of loading magnitudes for PC1 suggests its inductive character. Principal component PC3 discriminates between *exo* and *endo* forms, describing in itself only 5% of total data variability. Unsubstituted norbornane falls together with *exo* derivatives. Thus *endo* substituents appear to introduce an additional steric effect. Their loadings exclude an electronic nature for the third component. Only 0.9% of total data variability is accounted for by further principal component and is attributed mostly to carbons C4 and C5. The nature of the second principal component has not been explained.

The other papers and works deal with monosubstituted cyclohexanes (56), *t*-butyl-4-substituted cyclohexanes (57), and various steroids (58). In these studies conformation of substituents was known and fixed by low temperature (59a) or by rigid ring structure in steroids (60). The ^{13}C NMR substituent chemical shifts of 42 monosubstituted cyclohexanes (59), 26 monosubstituted 4-*t*-butylcyclohexanes (59a), and 41 steroid (60) (5α-cholestanes, 5-cholestenes, 5α-androstanes, 5α-pregnanes) substituted at various carbons were collected from the literature (59, 60).

Substituent chemical-shift data matrices for cyclohexanes contain four carbons; because of the symmetry of the molecule, carbon pairs C2–C6 and C3–C5 are identical. The number of carbon chemical shifts used for steroid molecules was 15 (from C1 to C14 and C19). Numbering of carbons and molecular formulas is given in Scheme 1. To avoid misleading numbering of carbons in various molecules, we prefer the notation *ipso* for carbon-bearing substituents and α, β, and γ for the others. The results of Table 1 show that three principal components explained almost all data variability. In all molecules, the $P_{1,k}$ describes more than 94% of total data variability, and the second adds from 2.2 to 5.5%. The small $P_{3,k}$ (0.1–0.5%) brings the total data variability

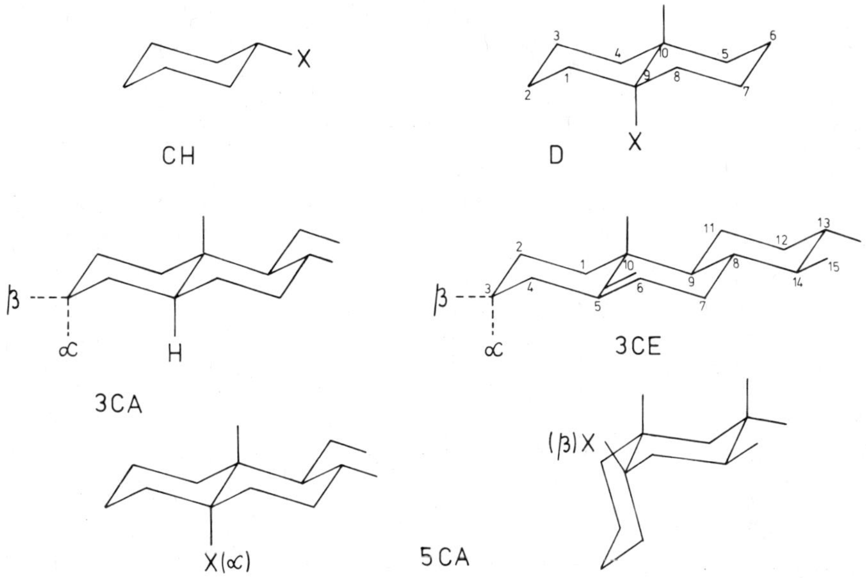

APPLICATION COMPONENT ANALYSIS IN ORGANIC CHEMISTRY 93

Scheme 1

explained to at least 99.9%, approximately. The unexplained residual variance for the PC model with one, two or three components is given in Table 2 for selected carbons. The calculations of SCS for C_{ipso} with two components (structures CH and D) or three three components (structures 3CA and 3CE) give results much better than does squared standard deviation of chemical shift measurement. The results for other carbons vary with the molecule, but

TABLE 1
Percent of Total Data Variability Explained by n Principal Components

Molecule Type[a]	$P_{1,k}$	$P_{2,k}$	$P_{3,k}$	$P_{4,k}$	N^b
CH(43)	97.41	2.23	0.36	—	4
D(6)	94.24	5.43	0.32	0.01[c]	6
3CA(20)	95.95	3.87	0.13	0.03	15
3CE(13)	94.84	4.40	0.69	0.05	15
5CA(8)	95.46	3.51	0.76	0.21	15
CH + 3CA(63)	96.16	3.36	0.47	0.01[c]	6

[a] For decoding abbreviations, see Scheme 1. In parantheses number of substituents X is given.
[b] Number of carbons.
[c] The sum of % equals 100.00.

TABLE 2
Residual Variance in PC1 and PC2 Models for Various Molecules

Type	Model	C_{ipso}	C_α	C_β	C_γ	$C_{\alpha'}$
CH	PC1	0.5822	15.7102	4.5937	4.0287	—
	PC2	0.0035	0.0030	1.5787	1.8789	—
	PC3	0.0000	0.0000	0.0000	0.0000	—
D	PC1	7.2583	20.4341[a]	3.7881	1.4920	—
	PC2	0.0001	0.1127	0.0642	1.4902	—
	PC3	0.0000$_2$	0.0099	0.0011	0.0003	—
3CA	PC1	2.8645	20.8371[b]	7.9192	1.0862	17.8053
	PC2	0.1526	0.1318	6.8819	0.7767	0.0601
	PC3	0.00041	0.0145	0.1606	0.0871	0.0191
3CE	PC1	4.2633	25.7276	5.1383	1.3247	25.1254
	PC2	0.1088	0.1337	4.7508	0.3393	0.0626
	PC3	0.0000$_2$	0.0232	0.2548	0.0135	0.0350

[a] $\alpha = C1, \beta = C2, \gamma = C3$.
[b] $\alpha = C2, \beta = C1, \gamma = C10, \alpha' = C4$.

in general the three-component model reproduces the data in satisfactory agreement with experiment.

Figures 6 and 7 show two projections of the three-dimensional space of the principal components. Figure 6 represents 3CA molecules and is similar to that for 3CE and CH, which was reported previously (56). The numerical values of $p_{1,k}$ are distributed over a 60-unit range and values of $P_{2,k}$, over a 17-unit range. The tendency toward linearity of negative slope is also achieved. All compounds fall into two clusters. The upper one (positive $P_{3,k}$) consists of axial substituents, and the lower one groups together equatorial substituents, in agreement with the norbornane compounds (55). Decalins show different patterns in Fig. 7 due to presence of only one type of conformer. The 5α- and 5β-substituted cholestanes (5CA) are also distinguishable from the $P_{2,k}$ versus $P_{1,k}$ plot.

It is a general tendency that magnitude of $P_{1,k}$ increases with higher electronegativity (61) of the substituent. For example, weakly electronegative substituents such as alkyl, iodine, or cyano groups have small $P_{1,k}$ values, whereas strongly electronegative substituents such as fluorine, nitro, or hydroxyl groups have large $P_{1,k}$. On the other hand, $P_{2,k}$ tends in a first approximation to behave oppositely. Closer inspection of the relationship between electronegativity and $P_{1,k}$ (or $P_{2,k}$) shows random scattering of points, with a straight line passing through halogens and NH_2 and OH groups. Other substituents for which electronegativity parameters (61) are available give large positive or negative deviation.

Further explanation of the nature of principal components may be possible after determining whether principal components derived for each

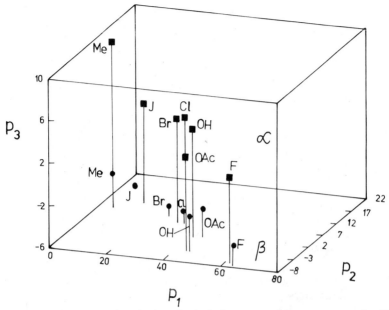

Figure 6 Three-dimensional projection of principal components p_3 versus p_2 and p_1 for 3-substituted cholestanes (3CA). The α substituents are designated by filled squares and β by circles.

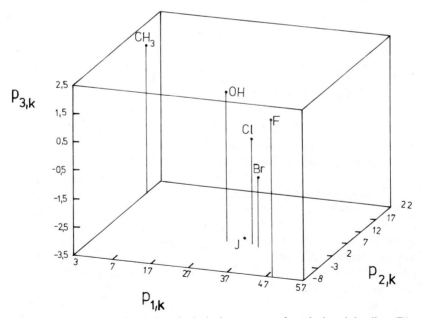

Figure 7 Three-dimensional plot of principal components for substituted decalines (D).

95

TABLE 3
Linear Regression of $p_{1,k}$ and $p_{2,k}$ for Various Steroids against Cyclohexane as Reference

Molecule	A	B	r	n
$p_{1,k} = A \cdot p_{1,k}^{CH} + B$				
D	0.75	10.05	0.63	6
3CE	1.06	2.99	0.997	13
3CA	1.09	2.43	0.998	14
5CA	0.39	16.77	0.09	8
$p_{2,k} = A \cdot p_{2,k}^{CH} + B$				
D	1.09	−2.08	0.85	6
3CE	1.57	−0.78	0.975	13
3CA	1.51	−0.69	0.982	14
5CA	0.001	3.53	0.708	8

class of compound studied are similar. This could be done by calculation of the regression parameters between $P_{1,k}$ or $P_{2,k}$. The cyclohexane data were chosen as explanatory variables, and the parameters of the linear equation are given in Table 3. It is obvious that a set of cyclohexane principal components may be used as a scale in a similarly model (15b) for further consideration.

Important information may be obtained from values of principal component loadings at particular carbon atoms and in various molecules studied. The PC loadings $l_{r,1}$ for C$ipso$ in all molecules are very large and positive. The values for other carbons are negative in CH. In other molecules $l_{r,1}$ are always negative at Cβ, Cβ', and Cγ; however, at Cα and Cα' they are positive but smaller than 0.1. This is likely to be the effect of the molecular frame (e.g., tertiary and quaternary carbons) and the total number of carbons in data matrices (see Equation 16). This means that a greater number of carbons causes smaller values of loadings.

Principal component loadings describe the intensity of transmission of effects explained by principal components. Thus this intensity decreases in the order shown in Scheme 2, C$ipso$ > Cα > (Cα') > Cγ > Cβ, and reflects polarization of σ bonds through substituents. The electronegative substituent attracts electrons from the skeleton and induces a partial positive charge at C$ipso$ (62), which diminishes with increasing distance from the place of substitution. Current experimental and theoretical results indicate that the electronegativity effect of the substituent beyond the second atom is generally negligible (63).

Principal component loadings $l_{r,2}$ are large and positive for Cα (and Cα'), while for other carbons they are negative (only in one case +0.0053). In

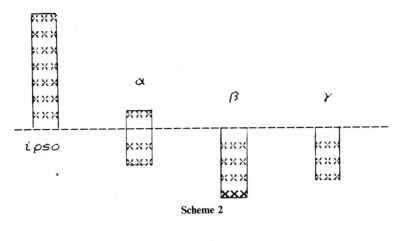

Scheme 2

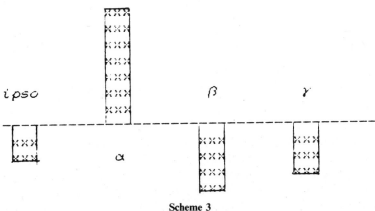

Scheme 3

cyclohexanes $l_{r,2}$ (see Scheme 3) is most negative at Cβ, but in more complex compounds *Cipso* is most negative. As a rule, Cγ loadings are less negative than Cβ loadings. This fact, combined with the conclusion that $P_{2,k}$ are linearly related to electronegativity (negative slope), means that the second component describes the inductive power of the substituent through-field effect (63, 64). Substituents acting through space induce a positive charge at Cα and perturb σ-orbitals. Negative charges are induced at both neighboring carbons (*Cipso*, Cβ) and decrease at the next carbons. The average pattern of $l_{r,2}$ is opposite for the first two carbons as compared to PCl. This may be caused by the nonproportional nature of the field and electronegativity effects (65–67).

Identification of the nature of the third principal component, which describes only a minor part of total data variability (0.13–0.76%), is not a

TABLE 4
Principal Component Loadings for Various Carbons in Different Compounds

Loading	Molecule	C_{ipso}	C_α	C_β	C_γ	$C\alpha'$	$C\beta'$
1	CH	.8488	−.1308	−.3950	−.3230	—	—
	D[a]	.8891	.0892	−.2021	−.1072	−.0293	—
	3CA[b]	.9302	.0944	−.1616	−.0840	.0877	−.1744
	3CE[b]	.9248	.0933	−.1711	−.0986	.0971	−.1767
	5CA[c]	.9115	.0719	−.1761	−.0907	.1141	−.1191
2	CH	−.1703	.8223	−.4253	−.2301	—	—
	D[a]	−.3806	.9533	−.2368	.0053	.1042	—
	3CA[b]	−.2432	.6720	−.1504	−.0821	.6234	−.1324
	3CE[b]	−.2578	.6399	−.0787	−.1255	.6332	−.1814
	5CA[c]	−.0300	.1267	−.1024	−.0999	.3640	.3731
3	CH	−.0195	−.1337	−.6093	.7430	—	—
	D[a]	−.0054	−.1619	−.1269	.6161	−.1207	—
	3CA[b]	−.0921	−.0808	−.6113	.1958	−.0527	−.6227
	3CE[b]	−.1061	−.1069	−.6820	.1836	−.0534	−.4689
	5CA[c]	−.2996	.6079	−.2549	−.0264	.4951	−.2211

[a] — ipso = C9, α = C1, β = C2, γ = C3.
[b] — ipso = C3, α = C2, β = C1, α' = C4, β' = C5.
[c] — ipso = C5, α = C4, β = C3, γ = C2, α' = C6, β' = C7.

simple and uniform task. The third component contributes to the overall shift effect by a moderate or weak shift of the β and γ carbons. Other carbons are almost unaffected. Of key importance for the contribution by a third mechanism is the sign and magnitude of the $P_{3,k}$, which is always negative for axial substituents and positive for equatorial ones. Thus $P_{3,k}$ designates upfield or downfield shift at Cβ and Cγ associated with the third mechanism. One can observe from Table 4 an accumulation of negative PC loadings at Cβ. The similar accumulation at Cγ is weakened by the positive value of $l_{r,3}$. Thus the product of multiplication—principal component times the loading—will be positive or negative, depending on the carbon atom in question and the conformation of substituent. Rationalization of this is complicated and may use various concepts such as CNDO and INDO calculations of charge density (59a), frontier orbital treatment (68), changes of torsional angles by axial substituents (60b) and others to support or reject an explanation.

A specific role of axial substituents has been proposed to explain Cβ shifts in *gauche* conformations by sterically induced charge polarization of the C—H bond (69) (the so called γ-*gauche* effect).

The importance of geometric changes for the third mechanism is evident is steroids. The relationship between $P_{3,k}$ in 3β-substituted steroids and equatorial monosubstituted cyclohexanes is very good:

$$P^{3\beta}_{3,k} = 0.9275 P^{eq}_{3,k} - 1.9544 \tag{17}$$

($r = .9846$, $s = 0.0770$, $n = 10$), while it fails for 3α-substituted steroid and axial monosubstituted cyclohexanes:

$$P^{3\alpha}_{3,k} = 2.0136 P^{ax}_{3,k} - 0.9753 \tag{18}$$

($r = .7688$, $s = 0.5699$, $n = 9$) as a result of strong changes in torsional angles (23b) and the presence of neighboring alkyl groups in steroids as compared to cyclohexanes.

The effect of the bulky t-butyl group in position 4 of substituted cyclohexanes on the geometry of the molecule is large. The angle $C_\alpha-C_\beta-C_\gamma$ increases by 2° and the angle X–C$ipso$–Cα slightly decreases (19) as compared to respective cyclohexanes lacking a t-butyl group.

The detection of specific long-range effects on the propagation of substituent effects by three independent paths is also possible from PCA calculation, as supported by earlier observations by Schneider et al. (60b).

The electronegativity effect described by the first principal component is a short-range effect. The efficiency of its transmission falls down very rapidly with the number of intervening bonds. Loadings $l_{r,1}$ decrease in sequence of carbons 3–2–1–10, 3–4–5–6, or 3–4–5–10, reaching a final limit of $-.078 \pm .005$ after the third carbon. As a consequence, the contribution of electronegativity to overall SCS at remote carbons (8, 9, 11, 12, 13, 14, and higher) is 0.2 ppm for 3α- and 3β-fluoro cholestanes and less for less electronegative substituents.

Much more irregular is the variation of loadings $l_{2,r}$ and $l_{3,r}$ with an increasing number of intervening bonds. The data for various paths are given in Table 5. This variation may result from a blend of distance, orientation between substituent and carbon atoms, and the number of secondary and tertiary carbons in the path. Remote carbons C9, C12, and C14 posses more negative $l_{2,r}$ values than do neighboring carbons, and their contributions to overall SCS are larger. The shielding effect caused by the second contribution (field effect) at remote carbons is larger in 3β-substituted cholestanes as compared to 3α isomers.

The numerical values of $l_{r,3}$ are negative for carbons close to the substituent (C1–C5) and positive for all others. The variation of the loadings for remote carbons is significant but independent of distance. The contribution of the third mechanism to SCS is very significant for carbons C1, C5, and C10; however, it is still significant for remote carbons. For example, the differences

TABLE 5
Loadings of the Second and Third Principal Components for Various Paths of Transmission

Path	$l_{2,r}$	$l_{3,r}$
A		
C3	−.243	−.092
C2	.672	−.081
C1	−.150	−.611
C10	−.082	.195
C9	−.094	.134
C11	−.066	.117
C12	−.076	.155
C13	−.064	.151
C14	−.077	.160
B		
C3	−.243	−.092
C4	.621	−.052
C5	−.137	−.622
C6	−.109	.129
C7	−.080	.154
C8	−.071	.169
C14	−.077	.160

TABLE 6
Experimental and Calculated SCS Values for 3β-Fluoro- and 3α-Methyl Cholestanes

Carbon	3β-Fluoro		3α-Methyl	
	Experimental	Calculated	Experimental	Calculated
1	−2.37	−1.99	−9.68	−8.32
2	6.28	6.33	5.71	5.82
3	65.85	65.84	0.52	0.51
4	6.17	6.15	5.68	5.57
5	−4.82	−3.16	−7.15	−8.38
6	−0.49	−0.44	−0.07	−0.54
7	−0.16	−0.10	0.03	−0.03
8	−0.10	−0.02	0.06	0.22
9	−0.55	−0.52	−0.03	−0.40
10	−0.78	−0.81	−0.62	0.31
11	0.42	0.41	−0.03	−0.20
12	−0.20	−0.10	0.00	0.02
13	−0.03	0.06	0.16	0.12
14	−0.22	−0.17	0.04	0.06
19	0.00	−0.47	−0.09	−0.28

in SCS as between 3β- and 3α-hydroxy cholestanes at C9 and C11 given by the third contribution are 0.94 and 0.84 ppm, respectively.

In view of our results, long-range effects observed by Schneider (60b) for certain carbons of rings B and C in steroids are associated with the second and the third mechanism in our model.

We can conclude that the principal component model with the three components

$$\delta_{SCS} = \bar{\delta}_{SCS} + P_{1,k}l_{r,1} + P_{2,k}l_{r,2} + P_{3,k}l_{r,3} + \varepsilon \qquad (19)$$

describes the effect of substituent on ^{13}C NMR substituent shift with acceptable accuracy. The three independent and additive modes of transmission of the substituent effects are more or less precisely defined. The Table 6 describes the experimental and calculated chemical shifts (in terms of the difference between substituted and unsubstituted cholestanes) for two molecules with strong and weak electronegative substituents, 3β-fluoro and 3α-methyl, respectively.

B. UV Spectra in Basicity Studies

A large area of PCA applications is the identification of mixture composition from a set of spectra. We use the sets of UV spectra measured at different acidity or basicity (from superacids and concentrated mineral strong acids, through dilute acidic anl basic solutions, to concentrated basic solutions) to calculate various acid–base equilibrium constants. Inside such a wide range of solvent acidity, an organic compound may take part in at least one dissociation process as donor or acceptor of proton. It is very probable that the neutral molecule and/or product absorb UV–visible light at a wavelength convenient for spectral measurement. It is necessary that $\Pi \to \Pi^*$ and $n \to \Pi^*$ transitions are allowed and their energy differs for both neutral and charged species. This is valid for more or less (by 1 unit) charged species.

The sensitivity of spectral transitions to solvent acidity or basicity causes the shift of absorption bands. A family of absorption curves resulting from measurements in solvents of various acidity (basicity) should intersect through an isobestic point if (1) the species under study obey the absorption laws ideally or (2) only two absorbing species take part in the equilibrium. Then the ionization ratio I can be calculated for the protonation reaction of base B

$$B + H^+ \rightleftharpoons BH^+ \qquad (20)$$

as

$$I = \frac{C_{BH^+}}{C_B} = \frac{A - A_B}{A_{BH^+} - A} \qquad (21)$$

where A is absorbance. It is assumed that A_B and A_{BH^+} are constant in all the acidity (basicity) ranges used. This assumption is, however, not always fulfilled by real solutions, and A_B and/or A_{BH^+} may vary across the range of acidities where both species contribute to the absorbance. Titration curve analysis was proposed only in the late 1960s (70), with factor analysis applied to spectra sets (71). PCA has been applied by Edward and Wong (72) on the basis of earlier works of an other author (6).

The family of curves reflecting equilibrium is described by PC model with one component:

$$A_{k,r} = \bar{A}_k + P_{1,k} l_{r,1} \tag{22}$$

The plot of $P_{1,k}$ versus k (wavelength) is a curve (Fig. 8) with all the necessary information concerning the shape of spectra, and $l_{r,l}$ is related to the concentration of B and BH$^+$ species.

Initially PCA was applied to find the number of independent, important factors, which were assumed to be connected with a number of species in a

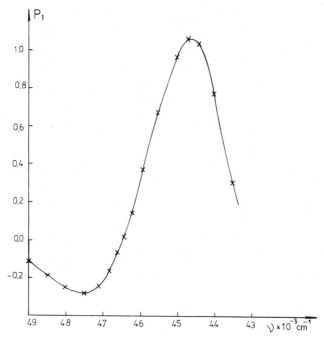

Figure 8 The plot of principal component $p_{1,k}$ versus wavelength k for mesityloxide in sulfuric acid

mixture, and for the reconstitution of data matrix connected with the equilibrium (72). PCA thus separates noise and other uncertainties. The TCA procedure (70) has also been linked with PCA in a one-step computer program (73) and explored in further works (74).

Ultraviolet absorbances for various fixed wavelengths form a row vector for given solvent and constitute a data matrix for r solvents. Initial data are centered about the mean in CVA (73), and then $P_{1,k}$ describes more than 93% of total data variability (72–74). It was assumed that the large first component can be associated with the spectral changes accompanying the establishment of equilibrium (71, 72). The reconstituted absorbance curves will be used in further calculation. The CVA–TCA procedure (73) selects columns in a reconstituted data matrix obeying the following criteria:

(i) $k = \max\limits_{k} P_{1,k}$ and/or $k = \min\limits_{k} P_{1,k}$

(ii) $k = \min (P_{2,k} - 0)$ (23)

(iii) $k = \max (A_{1,k} - A_{r,k})$

and calculates pK_{BH^+} automatically, assuming highest and lowest absorbance as fixed (A_B, A_{BH^+}) in Equation 21.

The identical final results may be achieved if spectra are centered about the origin; however, two principal components thus describe at least 93% of total data variability, each component describing one compound (11, 75):

$$A_{r,k} = P_{1,k}l_{r,1} + P_{2,k}l_{r,2} \qquad (24)$$

The curves of "abstract" $P_{1(2),k}$ versus k are shown in Fig. 9 for p-hydroxybenzoic acid, and the family of spectra might be reconstituted with Equation 24, giving a nice isobestic point. The ultimate objective of PCA is to convert an abstract solution into a "real" one. It was proposed (11, 75–77) that since absorbances in the calculated pure component UV spectra must be greater than or equal to zero, the following conditions hold:

1. A_B and $A_{BH^+} \geqslant 0$ at least one wavelength.
2. the absorbances are zero at tangential points of oblique axes of real components.

Projection of points in Fig. 10 on new, oblique axes allows us to calculate real components, proportional to real spectra of both B and BH^+ (Fig. 11).

The power of this procedure lies in the fact that even unmeasurable spectra can be calculated. For example, 3-cyano-5,5-dimethylcyclohex-2-en-1-

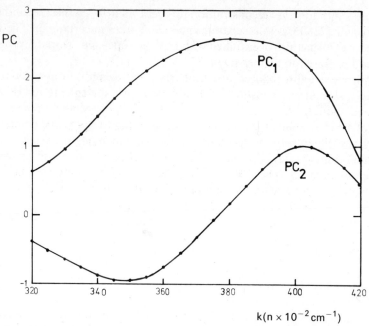

Figure 9 *p*-Hydroxybenzoic acid. "Abstract" principal components as a function of wavelength.

one is a very weak base, only partially protonated in pure sulfuric acid (78), and the spectrum of BH^+ is not measurable. The method described above, plus an additional constraint, allows us to calculate that spectrum and calculate pK_{BH^+}. That additional constraint deals with condition (2), which has been discarded by us. Not always in the measurement region, absorbance falls down to zero. Thus, the highest dissimilarity of real spectra condition has been imposed, and one or both real axes are not necessarily defined as tangential to the $P_{1,k}$ versus $P_{2,k}$ curve.

The transformation of real principal components shown in Fig. 11 into real spectra requires only multiplication by appropriate constants. Such constants are defined from a linear plot of loadings (Fig. 12). At points where $l_{r,2} = 0$ and $l_{r,1} = 0$, multipliers for the first and the second spectra are read, respectively.

Finally, the spectra of B and BH^+ are calculated, and results are shown by Fig. 13. One can observe a very high degree of similarity. However, the spectrum of BH^+ is more intense at the maximum. It is possible that 96% H_2SO_4 is not strong enough to provide full protonation of *p*-hydroxybenzoic acid. The corrected family of spectra, combined with calculated spectra of B and BH^+, gives a data set suitable for pK_{BH^+} calculation.

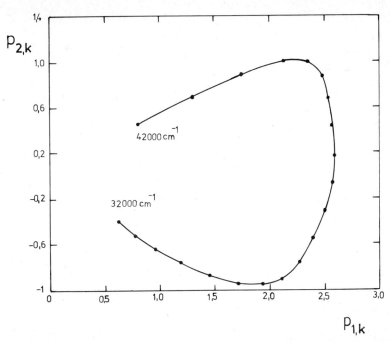

Figure 10 p-Hydroxybenzoic acid. The plot of $p_{2,k}$ versus $p_{1,k}$ as a function of wavelength from 32,000 to 42,000 cm^{-1} at 500-cm^{-1} intervals.

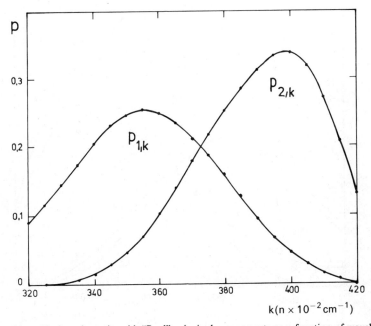

Figure 11 p-Hydroxybenzoic acid. "Real" principal components as a function of wavelength.

105

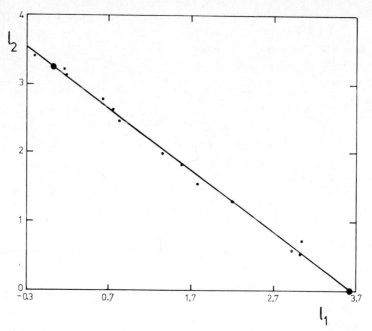

Figure 12 *p*-Hydroxybenzoic acid. The plot of loading for various concentrations of sulfuric acid (from 63 to 95%). Intersections between lines and with axes define multipliers.

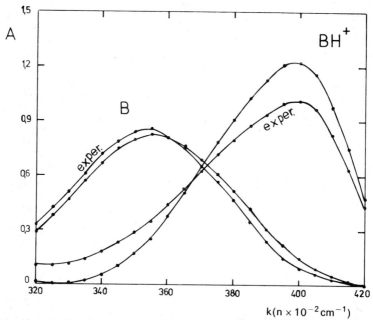

Figure 13 *p*-Hydroxybenzoic acid. Experimental and real spectra of B and BH$^+$ species.

In general, this procedure, might be applied to any set of curves describing discrete mixtures of two components of constant total concentration. If the total concentration is not constant or the thickness of the measured layer is not constant, each curve must be normalized by dividing it by the square root of the sum of the squared intensities (76).

C. Selected Spectroscopic and Chromatographic Problems

The general procedure described above for solving the specific problem of acid–base equilibria monitored by UV spectroscopy might be extended. In many chemical or physical processes (e.g., complex formation, tautomerism, conformation and other equilibria, phase changes, reactivity), two or more different species exist simultanously. Their composition at the moment is being modeled by various external factors (temperature, time, solvent polarity, acidity, preasure, etc.). The concentration changes of all components as a function of external factors were used to describe the process. Various analytical methods used to follow those changes produce results in a form of two-dimensional arrays (UV–visible, IR, Raman, NMR, MS, GC, HPLC, EPR, CD). which are convertible into absorbance or intensity data vector for each array, and into a data matrix for a family of arrays. The experimental data in the form of the data matrix are suitable for PCA processing and following will show same specific applications.

1. IR and Raman Spectroscopy

Most applications of PCA to certain spectral ranges of IR and Raman spectroscopy elucidate the number of species present under certain conditions. The factor analysis as a method complementary to band resolution has been applied in investigating of self-association of acetic acid (79). Four absorbing species in the carbonyl group range (1800–1700 cm^{-1}) were found, of which monomer and dimer were the most important. Also four absorbing species were found by analysis of CCl_3COOH in carbon tetrachloride (80). The spectra of particular components were calculated and species were defined as monomer, cyclic dimer, and linear polymers. In addition, the equilibrium constant monomer–dimer has been calculated at 25°C.

Three species were found and identified in solution of phenol in carbon tetrachloride (81) after PCA calculations applied to the OH stretching band at the temperatures in the range 25–60°C. Three components are assigned to monomer, dimer, and polymeric species (cyclic trimer), and their spectra were calculated. A comparison of results with achieved by computer program for band contour resolution (82) shows a good agreement. This was a first time that temperature dependence of an IR absorption band was studied by PCA (81). Calculated areas of the monomer and dimer bands and the enthalpy of dimerization $\Delta H° = -15.8$ kJ mole^{-3} have been obtained.

The PCA and band contour resolution methods applied to Raman spectra of acetic acid in aqueous solutions (83) supports earlier results on the number and type of species.

Both IR and Raman spectrophotometers give large numbers of data points, depending on the method of data aquisition and the spectral range used. Hundreds of points stored for dozens of samples create a problem of data storage and handling by personal computers. The data reduction is therefore generally necessary before PCA. Several specific methods of IR and Raman data pretreatment were offered. Partial least-squares regression (84), selection of wavelengths that were not highly correlated (85), and fast Fourier transform (FFT) (86) are used, among other methods. Thus FFT + PCA may be used for significant reduction of the large data matrix to achieve reasonable results (87).

In the following discussion we will consider other interesting external factors affecting organic compounds, such as temperature and pressure.

The Fourier transform IR spectra of short-chain hydrocarbon (undecane) and fatty acid ($C_{24}H_{49}COOH$) in the region 1240–1550 cm^{-1} were measured at various temperatures to explain phase transitions and phenomena of melting (88). On the basis of PCA results of spectra recorded in temperature ranges below 22.8°C (solid–solid transition) to 32°C (melting), undecane should be considered as a two-component mixture. Fatty acid in the range 25–83°C (melting) indicated three independent components with different IR spectra.

The effect of pressure (normal to 4 kbar) on polyethylene was studied by Raman spectroscopy (89). PCA of resulting spectra indicated the presence of two phases: crystalline and melt at lower pressure, while one more emerges at 4 kbar.

2. Gas and Liquid Chromatography

Both qualitative and quantitative analyses by means of chromatography rely on pure peaks for accurate results. The overlapped peaks will affect the quality of analysis. The peak overlapping might be obvious; if it is not, no evidence of contamination may be present (90). PCA is ideally suited to deal with such problems (77, 91, 92).

The following constraints were imposed to convert "abstract" solutions into "real" ones:

1. Negative elution profiles are prohibited.
2. Negative absorbances are prohibited (for cases when absorbance detection is applied).

To solve more complex mixtures, the constraints were formulated to select simplest elution profiles, defined as smallest area/norm ratios (93) or linear inequality in concentration domain (94).

In most methods no assumptions were made regarding the shape of elution profiles (91–94); however, modified Gaussian functions of profiles were used (95).

The efficiency of resolution of overlapped elution profiles is greater when multichannel detectors are applied at a sufficiently high rate. In the case of multichannel UV detector, the spectra of eluates are recorded in short time intervals, forming an absorbance matrix (94) very convenient for PCA. Similar performance may be achieved by PCA of repetitively scanned gas chromatography–mass spectrometry data (92, 96). The methods are sufficiently general, and identification of three and four components is possible.

Principal component analysis is also a very useful tool for studying various practical aspects and improving experiment design for future use. Hoovery and Soroka (97) analyzed retention indices for 7 liquid-phases and 49 straight- and branched-chain solutes. They found three factors as necessary for reproducing the data near experimental error. Target testing procedures identify such basic factors affecting variability of experimental data as carbon number, dipole moment and boiling point for solute, and dipole moment and molecular weight for solvent.

Walczak and coworkers, in a collection of papers (98–101), discussed various methodological and theoretical problems of reversed-phase HPLC. PCA of experimental data on 63 solutes (chalcones) with 23 available packings gave insight into variation of the selectivity with the column type (98). Principal components were identified as "hydrophobic" factors and "chemical–steric" factors influencing solute selectivity. The influence of packings is governed by the nature of organic ligand, the number of carbons, and the end-capping procedure, which was deduced from PC loadings. The energetics of retention and the similarity of retention mechanisms for 63 chalcones on 14 ODS packings has been studied (99). Principal components allow estimate of the Gibbs free energy for any phase pair. Selectivity of chalcones on a diol stationary phase (100) and ODS stationary phase (101) was investigated. Principal components and loadings enable one to study and improve experiment design for future application.

Three-dimensional PCA was applied (102) for two literature data sets on normal-phase HPLC (solutes × absorbents × eluents). It was shown that principal components fit retention data much better than does the commonly used Snyder equation (103).

3. Mass Spectrometry

The separation of mass spectra of mixtures has been a problem of vital importance in analytical chemistry for some time. The PCA method requires that the mixture have at least one mass peak, at which one component has a finite intensity and all other components have zero intensity (91). PCA was used to the artificial data sets representing two-component mixtures at five various concentrations and having nine masses. In addition, two real mixtures—cyclohexane–cyclohexene and cyclohexane–hexane—were tested with good results (91b). Target transformation of "abstract" PCA solution was used to determine whether a suspected compound was a component of mixture (91c). The test vector may be an incomplete mass spectrum. The poor agreement between the test vector and the predicted spectrum is evidence for the absence of the suspected substance.

A similar procedure was adopted for analyzing the three-component mixture (methylcyclohexane, toluene, and 4-methylpentan-2-one) (104) with a more reliable test for a number of real components.

The identification of components in mixtures by PCA of mass spectrometry data will be more effective when combined with gas chromatography or time-resolved pyrolysis mass spectrometry. Applications of the latter technique have been described in the current literature (105, 106) for biopolymer mixture, wood, and rubber copolymer. Total ion current obtained from time resolved mass analysis and mass spectra were used to calculate the number of species and their mass spectra using PCA and, finally, the variance diagram (VAARDIA) algorithm (107).

Principal component analysis has been used to interpret fragmentation patterns in mass spectrometry as well (108, 109).

D. Classification of Solvents

Physical and chemical differences between the numerous liquids and solvents, both inorganic and organic, causes many problems in useful classification. Reichardt, in his excellent book (110), uses five possible means of classification:

1. According to chemical constitution
2. Using physical constants.
3. In terms of acid–base behavior.
4. In terms of specific solute–solvent interactions.
5. Using multivariate statistical methods.

The main objective of physical organic chemistry is to investigate solvent effects on chemical and physical processes, specifically, reaction rates and equilibria, as well as spectral absorptions. Organic chemists usually explain solvent effects on such phenomena in terms of so-called solvent polarity. This term, despite a large effort, has not been precisely defined. One can understand the term "polarity" as either:

1. The permanent dipole moment of a solvent.
2. The dielectric constant of a solvent.
3. The sum of all molecular properties responsible for the interaction forces between solvent and solute molecule.

Definition 3 is closest to chemical intuition. It is not surprising that various physical (macroscopic) solvent parameters (e.g., dielectric constant, dipole moment, refraction index) and empirical (microscopic) parameters (e.g., Y, Z, $E_{T(30)}$, Π^*) fail to characterize all aspects of solvent polarity. Rather, they reflect the similarity between the model process and that under study. The chance that a solvent-dependent process will follow such a linear relationship depends strongly on the proper choice of the physical or empirical parameter as an independent variable.

Multiparameter equations were proposed to overcome these difficulties. The first of such equations has been applied by Katritzky and colleagues (111) and was followed by Koppel and Palm (112). The application of the multiparameter equation

$$P = P_0 + a_1 X_1 + a_2 X_2 + \cdots + b \tag{25}$$

requires a knowledge of the nature and complexity of the studied solvent–solute problem and sets of all appropriate solvent parameters X. To follow the formalism of multiparameter model (15b), one must be aware of the colinearity of various solvent parameters. Various empirical solvent polarity scales are linearly related (110), and this fact limits the reliability of the results achieved.

Awareness of these limitations turned researchers toward principal component analysis (PCA), factor analysis (FA), and characteristic vector analysis (CVA) as tools for explaining the data.

Factor analysis was used by Martin et al. (113) as early as in 1977. A data matrix built from 18 organic solvents and 18 parameters led to a solvent classification resembling the empirical classification by Parker (114). Analysis of a four-parameter set (Y, P, E, and B) for 51 solvents leads to a three-parameter equation explaining 94% of total data variability (115).

A set of 35 physicochemical constants and empirical parameters of solvent polarity known for 85 organic solvents has been used by Svoboda et al. (116). They found four factors explainable in terms of the Kirkwood function: $(\varepsilon - 1)/(2\varepsilon + 1)$, polarizability expressed as $(n^2 - 1)/(n^2 + 2)$, Lewis acidity, and Lewis basicity. This means that those parameters are required to describe in a quantitative way solvent effects on chemical reactions, equilibria, and absorption of electromagnetic radiation.

Carlson et al. (117) described a strategy of suitable solvent selection for organic synthesis based on PCA of eight common properties (m.p., b.p., ε, μ, n_D, $E_{T(30)}$, d, and log P) for 82 solvents. A two-component model describing 51% of total data variability has been established by cross-validation. The first component is attributed to polarity as explained mainly by ε, μ, and $E_{T(30)}$ and the second, to polcrizability. Both polarity and polarizability are almost orthogonal and thus do not contribute to each other.

A paper by Maria et al. (118) dealing with only 22 organic non-hydrogen-bond donor solvents and 5 (thermodynamic and spectroscopic) basicity scales led to two components that describe 93% of total data variability. The first principal component correlates with gas-phase affinity of solvent toward H^+ and K^+ and is attributed to a blend of electrostatic and charge-transfer (CT) phenomena. The character of the second component is electrostatic. A small third factor has been explained by steric hindrance to acid–base complexation.

A very interesting paper dealing with broader problems of intermolecular interaction in the liquid state has been published by Cramer (119). In that study, 114 organic solvents, organic liquids, and inorganic liquids described by six empirical parameters (aqueous solvation energy, partition coefficient, boiling point, molar refraction, volume, and vaporization enthalpy) yield two principal components, designated as B and C. Those two components describe 95.7% of total data variability. Decreasing the number of liquids (by increasing the number of experimental parameters) or increasing the number of liquids does not produce a significant change in the final results.

An interpretation of both principal components has been given. The value of B is a measure of some aspects of molecular bulk, being large for naphthalene or octanol and small for helium and hydrogen. The C parameter is a measure of molecular cohesiveness.

On this basis a hierarchically organized additive–constitutive model has been proposed that allows the prediction of 18 common physical properties with reasonable accuracy. The 749 predicted and actual values agree within a correlation coefficient of $r = .97$.

The series of papers by Chastrette and colleages (120) led to classification of 83 solvents on the basis of eight solvent parameters. These included macroscopic and microscopic properties and Lewis acidity–basicity parameters. Three components describing 82% of total data variability were generally explained.

The first of these properties might be interpreted as a polarizability index being strongly correlated to molar refraction, index of refraction, and highest occupied molecular orbital (HOMO) energy. The second represents polarity of solvent and is strongly related to the Kirkwood function, dipole moment, and boiling point. Finally, the third component—explained mainly by lowest unoccupied molecular orbital (LUMO) energy—describes Lewis acidity and electron affinity of solvent. Loadings of principal components for 83 solvents allow the classification of solvents by nonhierarchical methods into nine classes (120c) in good agreement with Parker's classification and chemical intuition. Only a few solvents ($\cong 10\%$) are not expected in the classes as found: CF_3COOH, $PhCH_2OH$, 1-octanol, CCl_4, n-hexane, cyclohexane, tetrahydrofurane, and 1,2-dimethoxyethane. The names of the classes are:

1. Aprotic dipolar.
2. Aprotic highly dipolar.
3. Aprotic highly dipolar and highly polarizable.
4. Aromatic apolar.
5. Aromatic relatively polar.
6. Electron pair donor.
7. Hydrogen bonding.
8. Hydrogen bonding strongly associated.
9. Miscellaneous.

Six physical and empirical parameters (b.p., ε, μ, n_D, $E_{T(30)}$, and δ_H) for a collection of 103 organic solvents were used for PCA and classification of solvents (121). Unfortunately, it was not possible to include a solvent Lewis basicity parameter because of a lack of data for many solvents. The mutual linear correlation between the six solvent parameters is poor (the best are δ_H and E_T ($r = 0.76$) and δ_H and ε ($r = .62$). Again three principal components describe approximately 88% of total data variability (46, 28, and 13%, respectively). An analysis of the plot of principal components $P_{2,k}$ versus $P_{1,k}$ as shown in Fig. 14 leads to the conclusion that the first component, explained mainly by ε, μ, and δ, is associated with solvent dipolarity. The second component, explained by n, is associated with polarizability. The magnitude of loading $l_{r,2}$ depends linearly on polarizability measured by the molar refraction for alcohols, nitriles, ketones, alkanes, and cycloalkanes. A classification of 103 solvents into 10 classes is presented in Scheme 4. Strange placements of solvents according to this classification are indicated by an asterisk. The solvent that is closest to the calculated class center within a class is underlined.

The reader may be confused when analyzing results of PCA applications to solvent problems. Because of the varied number of physical and empirical solvent parameters (which do not express all possible solute–solvent interactions) and various collections of solvents used in calculations, the results achieved are somewhat different. In spite of these limitations, the solvent classifications obtained by various authors are sufficiently similar (116, 117, 120, 121), especially these derived from largest data set (120, 121).

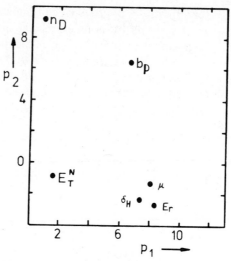

Figure 14 Plot of principal components for six solvent parameters (121).

The problem of solvent classification is still open for discussion. The increasing number of solvents and parameters made available by experimental measurements will improve future PCA results and classification.

E. Separation of Inductive and Resonance Substituent Effects and Related Phenomena

The Hammett substituent constant $\sigma_{m,p}$ measures the inductive and resonance effects of a substituent on the reaction center in benzene derivatives (13). It is fulfilled by reactions in which the blend of both contributions is almost constant. The through-conjugation effect modifies this blend and leads to the so-called sliding scale of substituent constants, with σ^- or σ^+ at the extreme.

A separation of $\sigma_{m,p}$ substituent constants into inductive and resonance constants attracted attention from the early days of LFER, and various approaches were proposed and reviewed (41). Detailed classical analysis of this problem is beyond the scope of this contribution, which is devoted to various PCA approaches in this field.

The first who tried to solve the problem were Swain and Lupton (26, 43). They identified two factors, termed *inductive* and *resonance*, from the literature sets of various substituent constants for a large number of substituents. The most controversial part of their work was the second critical condition imposed to achieve a "real" solution. It states that the $(CH_3)_3N^+$ substituent is

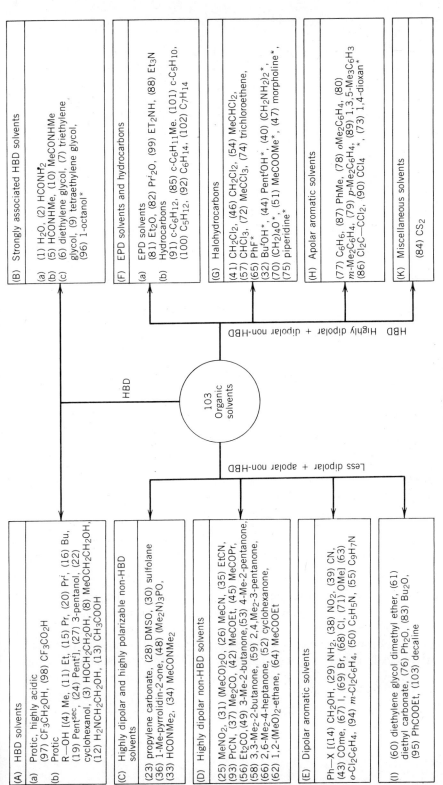

Scheme 4

never more electron donating or more electron attracting than H by resonance. This condition was criticized by many leading authors (122).

Three substituent parameters were extracted by PCA of 76 reaction series and 17 substituents (123), including those not following the σ_I–σ_R Taft model. No arbitrary critical conditions were used. "Abstract" solutions were rotated into "real" solutions and three sets of substituent parameters were derived: σ_I, σ_R, and σ_E. The nature of the first two is self-explanatory, while σ_E represents a mixture of unknown composition of electrical effects. The resulting three-parameter equation has been claimed (124) as an "optimum linear free-energy relationship" for the prediction of missing data for substituent effects.

The disjoint principal component model (Soft independent modeling of class analogy) (SIMCA) was used by Wold and his colleagues (125) on literature data sets for the reactivity of aromatic compounds. The σ_I scale thus reported corresponds as closely as possible to the σ_I scale of inductive effect operating in aromatic reactivities.

Despite long-term studies and the various treatments used, the problem of inductive and mesomeric effects of substituents on reactivity, equilibria, and other properties has not been finally solved. Interpretation of experimental results by various multivariate methods, including PCA and theoretical calculations (63, 67), will for certain explain these complex phenomena.

F. Acidity of Nonideal Solvents

Acidity of the solvent controls various processes such as acid–base catalyzed reactions or acid–base equilibria. Such processes may occur in a very broad range of acidity, from concentrated strong mineral acids to concentrated alkaline solutions, with the ordinary pH region in the middle. The term "nonideal solvent" for extremes of the acidity range is frequently used. To describe the acidity of the medium beyond the ordinary pH range, Hammett (126) developed the acidity function, by H_0:

$$H_0 = -\log C_{H^+} \frac{\gamma_H + \gamma_B}{\gamma_{BH^+}} \tag{26}$$

assuming the activity coefficient term to be a linear function in a series of solutions of overlapping acidity, and for two compounds B and A, the term log $(\gamma_B \gamma_{AH^+})/(\gamma_{BH^+} \gamma_A)$ (cancellation assumption) is equal to zero. On this basis the practical indicator method by overlapping technique was imposed (126) and widely used. It becomes apparent that H_0 was not unique but only one of many possible acidity functions (127, 128). The failure of the H_0 acidity function has been explained by various factors; solvation (hydration) is the most important (127). During the last five decades more than 400 various acidity functions have been created for mineral and organic acids in water and organic solvents (129). The interest in strongly basic media is less.

Such an enormous number of acidity functions creates many problems in their practical use and presents difficulties in comparison of results. Thus, some work was done to replace the collections of acidity functions per solvent system by one universal acidity function. The proton activity scale log a_{H^+} (130) or τ "solvation coefficient" (131) are examples. Several more fruitful treatments were proposed following LFER formalism and overcoming the cancellation assumption. Among them is the Bunnett–Olsen method (132) and the "excess" acidity method (133, 134). Recently Edward and coworkers (135) suggested that an aquadratic term (e.g., excess acidity x^2) must be included in equations explaining basicity or kinetics in strongly acidic or basic solvents.

The last sentence of an acidity function review (129) stated that "we can certainly continue to expect new insights into equilibria and reactions in nonideal media, resulting from development and use of acidity functions." One can find in the literature at least two papers serving as a proof. Scorrano et al. (136) stated that both pK and solvation coefficient m^* in the LFER Equation 27:

$$pK_c = m^*X + pK_\alpha \qquad (27)$$

are necessary for definition on the strength of the base in a medium of known acidity X.

The other paper (137) attempts a more direct approach to the acidity function problem on the basis of PCA. More than 170 various acidity functions for 15 inorganic and organic strong acids were used. Literature data on various acidity functions were organized in two matrices as functions of concentration expressed as percent or molarity. The data matrices were not complete along the concentration axis. For example, most data for sulfuric acid were available up to 97% but for hydrochloric acid, up to only 36%. Because of the requirements of the PCA method used, the initial incomplete matrices were divided into a few submatrices taking concentration mode as the criterion. This was a compromise between unnecessary omission of existing data and/or extrapolation to nonexisting data. Six submatrices were created for various percentage ranges from zero to 40, 55, 70, 80, 90, and 95, and five for various ranges of molarity from zero to 2.5, 4.0, 5.5, 7.0, and 10.0. One principal component explains more than 98% of the variance in all submatrices, except two of lowest molar concentration. The two-component PC model improves the explanation to more than 99.5%, with one exception (0–4 M range).

The P_1 and P_2 values are the only significant variables. The curve P_1 versus concentration of acids in the range 0–95%, which is presented in Fig. 15, is in general similar to curve for H_0 acidity function and the newest X acidity

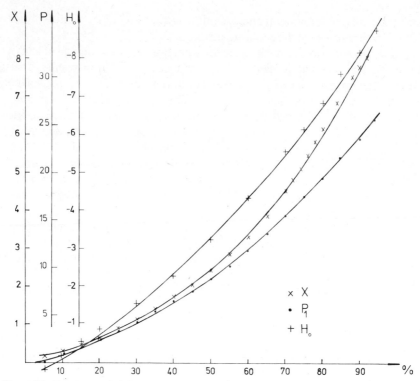

Figure 15 Parameters $P_1(12)$, $H_0(1)$, and $X(111)$ as functions of acid concentration (0–96%); H_0 and X are for sulfuric acid.

function data set (136). The P_1 values for various submatrices are linearly related to each other with correlation coefficient $r > .995$ and standard deviation $SD < 0.14$ (0.5% of P_1 range). The P_1 for concentration range 0–95% is linearly related to various acidity function ($r > .97$).

The second principal component is not very large in all submatrices and can be neglected if the second term value in the PC model does not exceed the experimental error (0.05 units of acidity function). The P_2 value is associated with experimental error and failure of the overlapping procedure. Ionization ratio as a function of concentration is not always perfectly linear and parallel, thus introducing cumulative errors of uncertain magnitude (138).

The explanation of P_1 is obvious. It is a universal acidity measure independent of the nature of the acid but dependent on its concentration. This dependence consists of two parts: linear at low concentration (0–35%

or 0–5 M) and curved

$$P_1 = aC^b \qquad (28)$$

at higher concentration. As a consequence, there appears to be no more than one type of variable that gives significant and independent difference in the acidity function data set.

The second parameter derived from PCA, loading $l_{r,1}$, reflects the nature of the acid system and the relation between various acidity functions. For example, in the H_0 and H_R acidity functions in sulfuric acid reported for the temperature ranges 25–90°C, the loading values are linearly related (with a very small curvature) to the reciprocal of temperature. Different acidity functions have identical loadings (e.g., $H_0^3, H_0', H_0'''; H_I, H_T; C_H, \Delta, H_{GF}$) and consequently the same nature.

A linear relation of $l_{r,1}$ for various acids with $l_{r,1}$ for sulfuric acid as reference suggests that the nature of the acidity functions of identical types is independent of the acid and that activity coefficient terms are proportional. The slope (proportionality factor) larger than one means that the acid is stronger than sulfuric (e.g., $HClO_4$ and HNO_3). The slope smaller than one is characteristic for weaker acids (e.g., H_3PO_4 toluene p-sulfonic acid, HCl, HBr).

A very common acidic system is sulfuric acid, used in the investigation of weak base equilibrium constants pK_{BH^+} and reaction rates. Examination of complicated kinetic problems and elucidation of mechanisms require detailed knowledge of species present in sulfuric acid–water mixtures. Attention was devoted to concentration profiles of species vs. mole fraction of sulfuric acid (139). The further work on resolution of the Raman spectra of aqueous sulfuric acid mixtures using PCA (140, 141) yielded the relative amounts of the three components and allows the identification of the chemical species. These are: undissociated H_2SO_4, combination of free SO_4^{2-} and HSO_4^- (not ion paired), and combination of ion-paired SO_4^{2-} and HSO_4^-. Ion paired species exist in dynamic equilibrium with the remaining species. The ratio between sulfate and bisulfate remains relatively unchanged with increasing concentration of sulfuric acid; however, the position of the equilibrium between free and ion-paired species would be moderately dependent because of the decreasing water concentration:

$$mH_2O \cdot HSO_4^- + nH_2O \cdot H_3O^+ \rightleftharpoons HSO_4^- \cdot H_3O^+ + (m+n-1)H_2O \qquad (29)$$

PCA studies are very useful in the analysis of the Raman spectra and providing a concentration profile of individual species present in the complex mixture H_2O–H_2SO_4; however, higher spectral resolution will be necessary for better distinction between possible species and their role as nucleophilic agents (142) (e.g., HSO_4^- aqueous and/or HSO_4^- ion-paired).

G. Reactivity of Organic Molecules

The application of PCA for elucidating the mechanism of reaction or the structure of the transition state are scare. Albano and Wold, in a very interesting paper, give a classification of reactants in a solvolytic reaction paralleling charge delocalization in the transition state (143).

The solvolysis reaction has attracted continuous attention following Winstein's proposal for nonclassical carbonium ions being intermediates in certain reactions (144). The positive charge of nonclassical ions would be delocalized by a single bond or multiple bond except in allylic position, whereas in classical carbonium ion it is delocalized by resonance. Studies of solvolytic reactions of molecules (the norbornyl one is very popular) under different experimental conditions have been used in attempting to obtain information on the transition states.

Albano and Wold (143) compared substrates that undergo solvolysis with known mechanisms and substrates with undefined mechanisms to estimate a PCA model and then projected the principal components for the substrates in question on a formerly developed model. They used processes with a delocalized charge in the transition state (e.g., benzyl tosylate) and those lacking such charge delocalization (e.g., methyl tosylate) as well as many other in between, and tried to elucidate degree of charge delocalization in the transition state. The data matrix consists of the reaction rates for 26 substituted benzenesulfonates, toluene-p-sulfonates, and p-bromobenzenesulfonates at 14 variables (seven solvents and two temperatures). Among the substrates five classes of leaving groups were distinguished:

1. Eight primary alkyls and eight secondary alkyls.
2. Two known to have resonance delocalization (benzyl and allyl).
3. Three suspected to have nonclassical delocalization (exo-2-norbornyl and cyclopropylmethyl).
4. Two not suspected to have delocalization (endo-2-norbornyl).
5. Three that were difficult to classify (1-adamantyl and cyclooctyl).

The PC model with two components were extracted from the data matrix and is presented schematically in Fig. 16 as the plot of principal components. Substrates of class form two clusters. The left-hand one with negative values of p_1 collects all primary alkyls and phenylethyl group, in which delocalization of positive charge is not possible. The righthand cluster collects all secondary alkyls and cycloalkyls. The nine remaining substituents are distributed as follows: benzyl and allyl (12) occupy positions below the right-hand cluster; *exo*-2-norbornyl and cyclopropylmethyl, as well as 1-adamantyl, are more to the right than this cluster; and *endo*-2-norbornyl are inside the right-hand cluster.

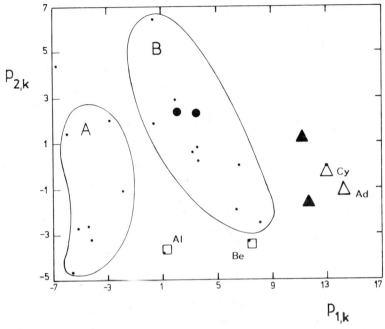

Figure 16 Plot of p_2 versus p_1 for solvolysis kinetic data. Filled squares refer to primary compounds (cluster A) and secondary compounds (cluster B). Filled circles refer to *endo*-2-noroboranes, filled triangles refer *exo*-2-norboranes. Allyl (Al) and benzyl (Be) are denoted by open squares and cyclopentylmethyl (Cp) and adamantyl (Ad) by open triangles.

The difference between secondary alkyls and *exo*-2-norbornyl compounds is similar to that between primary alkyls and benzyl and allyl groups, and is caused by different charge delocalization in the transition state. The charge delocalization increases with magnitude of p_1 and is greatest for *exo*-2-norbornanes as a result of nonclassical behavior.

The second principal component seems to be related to the size of alkyl group and/or inductive effect and polarizability of substituent.

Factors governing the solvolysis rate of β-dinitroalkylamines of general formula $R^1C_6H_4NHCH_2C(NO_2)_2C_6H_4R^2$ were investigated by PCA (23). Experimental rates of S_N1 reaction (145) were influenced by four independent variables: six different substituents R^1, seven different substituents R^2 (total number of compounds 42); five different temperatures, and six solvent mixtures from pure water to pure methyl alcohol. All kinetic data—1260 rates—have been arranged in data matrix of format 42×30 and processed using a characteristic vector analysis (CVA) program. In contrary to the 16-variable equation proposed by Pivovarov (145). CVA yielded the two-principal-component model. The first component describes 97.18% of total data variability, and two components describe 99.25% with variance of unexplained residue in range 0.001 to 0.035, thus less than the variance of the experimental rate.

The magnitudes of p_1 and p_2 are very closely related to a Hammett substituent constants and reflect the effect of substituents on reactivity. Explanation of loadings is more complicated. The plot of l_2 versus l_1 was presented in Fig. 3 and indicated a complex pattern. The points form several straight lines representing different solvent mixtures and different temperatures. The simple rotation of axes was used to simplify the pattern. The rotated loadings l_1 for water do not overlap with values for other solvent systems, and l_2 for various temperatures do not overlap, either.

The loading of the first principal component represents the solvent composition and for a given temperature is linearly related to Y scale of solvent polarity. The loading of the second principal component describes the effect of temperature and for a particular solvent system is linearly related to $1/T$.

It has been shown (146) that the qualitative concepts that an organic chemist uses to rationalize observations and to explain reaction mechanisms can be translated into quantitative concepts. This new idea contrasts with the widely used LFER approach, in which a molecule is arbitrarily separated into skeleton, substituent, and reaction site. The new idea prefers to treat molecule as a whole, although some effects can be located on specific atoms or bonds. The basic set of information widely used by organic chemists consists of quantitative, semiquantitative, and qualitative data—for example, whether bonds are reactive or not inductive effect, resonance effect, bond dissociation energy, steric effects, solvent effect, partial atomic charges, various spectral data, rates of reactions, equilibrium constants, and so on. Thus the intuitive models of organic chemistry and mathematical models of chemometry can be combined for better understanding of organic reactivity.

In an organic molecule, the atoms and bonds might be characterized by several readily calculated numbers or by various binary (1 or 0, "yes" or "no") states. These data can be used by various data processing methods, including PCA.

In order to evaluate and explain chemical reactivity, available measures of physical and/or chemical effects can be used as coordinates in a multidimensional space. Each chemical species is represented by a point in such space and can be treated as a vector. The chemical consequences in three-dimensional space (bond polarizability, charge difference, and electronegativity) spanned for BrJ and HF are self-explanatory (146). The same method may be used for multiatom molecule (2-cyclopentanone carboxylic acid) in a space spanned by the resonance effect, bond dissociation energy, and charge flow (146). A three-dimensional picture, combined with knowledge of possible bond-breaking ability, may be used to separate such bonds from unbreakable ones. The distance from the origin favors OH bond in the carboxylic group as the most reactive.

The expansion of this three-dimensional model was used to study the reactivity of 29 aliphatic molecules with typical functional groups and structure patterns (146). Each of 385 bonds, characterized by bond dissociation energy, bond polarizability, σ-charge difference, charge flow, σ-electronegativity, and resonance effect has two choices for shifting the charges on heterolysis. In this study 116 bonds out of 770 (2 × 385) were selected to construct a 116 bonds × 6 properties data matrix. PCA applied to this data set indicated three components as describing 85% data variability. The first component explains effects connected with σ-electron distribution; and the second, with bond dissociation energy, bond polarizability, and resonance. The following cluster analysis grouped bonds of similar types in a common clusters and indicates interesting chemical relationships.

Further applications of this method for prediction of the course and products of complex organic reactions have been reported (147).

ACKNOWLEDGMENTS

The author express his gratitude to the the Polish Academy of Sciences for grant 01.12.10.20, in the frame of which a large part of Sections III.A and III.B was prepared. Thanks also go to Dr. John Shorter (Hull) for his comments.

REFERENCES

1. K. Pearson. *Phil. Mag.*, 2, (6) 559 (1901).
2. H. Hotelling, *J. Educ. Psych.*, 24, 417, 498 (1933).
3. L. L. Thurstone, *Multiple Factor Analysis*, Chicago University Press, 1947.
4. G. Golub and C. VanLoan, *Matrix Computations*, The John Hopkins University Press, Oxford, 1983.
5. K. Karhunen, *Ann. Acad. Sci. Fennicae Ser. A*, 137 (1947); M. Loeve, *Process Stochastiques et Mouvement Brovuien*, Herman, Paris, 1948.
6. J. L. Simonds. *J. Opt. Sci. Am.*, 53, 968 (1963).
7. T. W. Andersson, *Introduction to Multivariate Analysis*, Wiley, New York, 1958; M. G. Kendall, *A Course in Multivariate Analysis*, Griffin, London, 1965.
8. S. J. Press, *Applied Multivariate Analysis*, Holt, Rinehardt and Winston, New York, 1973; P. M. Mather, *Computational Methods of Multivariate Analysis in Physical Geography*, Wiley, New York, 1976.
9. (a). T. W. Anderson, *An Introduction to Multivariate Statistical Analysis*, 2nd ed., Wiley, New York, 1984; (b) W. R. Dillon and M. Goldstein, *Multivariate Analysis: Methods and*

Application, Wiley, New York, 1984; (c) L. Lebart, A. Morineau, and K. M. Warwick, *Multivariate Descriptive Analysis: Correspondance Analysis and Related Techniques*, Wiley, New York, 1984; (d) P. P. Mager, *Multivariate Chemometrics in QSAR*, Wiley, New York, 1987; (e) K. Takeuchi, H. Yanai, and B. N. Mukherjee, *The Foundation of Multivariate Analysis; A Unified Approach by Means of Projection onto Linear Subspace*, Wiley, New York, 1982; (f) J. L. A. van Rijckevorsel and J. de Leeuw, Eds., *Component and Correspondance Analysis*, Wiley, New York, 1988; (g) B. Flury, *Common Principal Components and Related Multivariate Models*, Wiley, New York, 1988.

10. S. Wold, K. Erbensen, and P. Geladi, *Chemometr. Intel. Lab. Syst.*, 2, 37 (1987).
11. E. R. Malinowski and D. G. Howery, *Factor Analysis in Chemistry*, Wiley New York, 1980.
12. L. S. Ramos, K. R. Beebe, W. P. Carey, E. Sauchez, B. C. Erikson, B. E. Wilson, L. E. Wangen, and B. R. Kowalski, *Anal. Chem.*, 58, 294R (1986).
13. L. P. Hammett, *Physical Organic Chemistry*, 2nd ed., McGraw-Hill, New York, 1970.
14. (a) C. Hansch and A. Leo, *Substituent Constants for Correlation Analysis in Chemistry and Biology*, Wiley, New York, 1979; (b) Y. C. Martin, *Quantitative Drug Design—A Critical Introduction*, Marcel Dekker, New York, 1978; (c) E. J. Ariêns, Ed., *Drug Design*, Vols. I–XI, Academic Press, New York, 1971–1983; (d) F. O. Menger and U. V. Venkataram, *J. Am. Chem. Soc.*, 108, 2980 (1986).
15. (a) J. Shorter and N. B. Chapman, *Correlation Analysis in Chemistry, Recent Advances*, Plenum Press, New York, 1978; (b) T. M. Krygowski and R. I. Zalewski, *Match Communication in Mathematical Chemistry*, 24, 119 (1989).
16. E. R. Malinowski, *Anal. Chem.*, 49, 606 (1977).
17. R. A. Herman, J. H. Screevens, K. R. Jennings, and M. J. Farncombe, *Chemometri Intel. Lab. Syst.*, 1, 167 (1987).
18. H. M. Cartwright and H. A. Farley, *J. Chemometri.*, 2, 111 (1988); H. M. Cartwright, *J. Chemometri.*, 1, 111 (1987).
19. S. Wold, *Technometrics*, 20, 397 (1978).
20. H. Eastment and W. Krzanowski, *Technometrics*, 24, 73, (1982).
21. S. Wold, *Chemometr. Intel. Lab. Syst.*, 2, 37 (1987).
22. H. Wold, *Research Papers in Statistics*, Wiley, New York, 1966, pp. 411–444.
23. R. I. Zalewski and Z. Geltz, *J. Chem. Soc. Perkin II*, 1885 (1984).
24. C. B. Crawford and G. A. Fergusson, *Psychometrika*, 35, 321 (1970); P. Horst, *Factor Analysis of Data Matrices*, Holt, Rinehardt and Winston, New York, 1965.
25. J. B. Carroll, *Psychometrika*, 18, 23 (1953); Saunders, *Psychometrika*, 2%, 199 (1960); A. E. Hendrickson and P. O. White, *Brit. J. Statist. Psychol.*, 17, 65 (1964).
26. C. G. Swain and E. C. Lupton, *J. Am. Chem. Soc.*, 90, 432 (1968); C. G. Swain, *J. Org. Chem.*, 49, 2005 (1984).
27. P. Weiner, *Can. J. Chem.*, 50, 448 (1972).
28. Ü. Haldna, this volume.
29. A. Ebber, Ü. Haldna, and A. Murshak, *Org. React.*, 23, 40 (1986).
30. J. C. Davis, *Statistics and Data Analysis in Geology*, Wiley, New York, 1973.
31. D. F. Morrison, *Multivariate Statistical Methods*, 2nd ed., McGraw-Hill, New York, 1976.
32. S. Wold, C. Albano, W. J. Dunn III, U. Edlund, K. Esbensen, P. Geladi, S. Hellberg, E. Johansson, and M. Sjöström, "Multivariete Data Analysis in Chemistry," in B. R. Kowalski, Ed., *Proceedings of NATO Advanced Study Institute of Chemometrics*, Cosenza, Italy, 1983, pp. 17–97.

33. S. Wold, P. Geladi, K. Esbensen, and J. Öhman, *J. Chemometr.*, *1*, 41 (1987); C. L. de Ligny and J. Lohmöller, "Variables and PLS Parameter Estimation," paper presented at European Meeting of the Psychometrics Society, Groningen, Holland, 1980.
34. C. L. Nelson and E. A. Williams, *Proogr. Phys. Org. Chem.*, *12*, 229 (1976).
35. A. Saika and C. P. Slichter, *J. Chem. Phys.*, *22*, 26 (1954).
36. J. B. Stothers, *Carbon-13 Nuclear Magnetic Resonance Spectroscopy*, Academic Press, New York, 1972.
37. M. Karplus and J. A. Pople, *J. Chem. Phys.*, *38*, 2803 (1963).
38. D. M. Grant et al., *J. Am. Chem. Soc.*, *93*, 1880, 1887 (1971); *92*, 2186 (1970).
39. G. E. Maciel, *Top. Carbon-13 NMR Spectra*, *1*, 1 (1974).
40. D. K. Dalling and D. M. Grant, *J. Am. Chem. Soc.*, *96*, 1827 (1974).
41. J. Shorter, *Correlation Analysis of Organic Reactivity*, Research Studies Press, Wiley, Chichester, U.K., 1982.
42. R. W. Taft, *J. Am. Chem. Soc.*, *80*, 2436 (1958).
43. C. G. Swain, S. H. Unger, N. R. Rosenquist, and M. S. Swain, *J. Am. Chem. Soc.*, *105*, 492 (1983).
44. U. Edlund and S. Wold, *J. Magn. Reson.*, *37*, 183 (1980).
45. B. Eliasson and U. Edlund, *J. Chem. Soc. Perkin. II*, 403 (1981).
46. G. Musumarra, S. Wold, and S. Gronowitz, *Org. Magn. Reson*, *17*, 118 (1981).
47. W. J. Dunn III, C. Lins, G. Kumar, T. Monimara, S. Grigoras, U. Edlund, and S. Wold, *Org. Magn. Reson.*, *21*, 450 (1983).
48. D. Johnels et al., *J. Chem. Soc. Perkin II*, 863 (1983).
49. H. M. Hutton, K. R. Kuntz, J. D. Bozek, and B. J. Blackburn *Can. J. Chem.*, *65*, 1316 (1987).
50. K. B. Wiberg, W. E. Pratt, and W. F. Bailey, *J. Org. Chem.*, *45*, 4936 (1980).
51. W. F. Bailey and K. B. Wiberg, *J. Org. Chem.*, *46*, 4225 (1981).
52. W. F. Bailey, E. A. Cioff, and K. B. Wiberg, *J. Org. Chem.*, *46*, 4219 (1981).
53. O. Exner and M. Budesinsky, private communication.
54. W. F. Reynolds, A. Gomes, A. Maron, D. W. McIntyre, A. Tanin, and I. A. Reat, *Can. J. Chem.*, *61*, 2376 (1983).
55. J. Arunachalam and S. Gangadharan, *Anal. Chim. Acta*, *157*, 245 (1984).
56. R. I. Zalewski, *J. Chem. Soc. Perkin II*, 495 (1986).
57. Z. Urbaniak and R. I. Zalewski, *Polish J. Chem.*, *61*, 773, (1987).
58. R. I. Zalewski and Z. Urbaniak, *Proceedings of XVIII Seminar on Application of NMR*, Krakow, Poland, 1986.
59. (a) H. J. Schneider and V. Hoppen, *J. Org. Chem.*, *43*, 3866 (1978); (b) O. A. Subbotin and W. M. Sergeyev, *J. Chem. Soc. Chem. Comm.*, 141 (1976). *J. Am. Chem. Soc.*, *97*, 1080 (1975); (c) D. K. Dalling and D. M. Grant, *J. Am. Chem. Soc.*, *89*, 6612 (1967); (d) Janecke and H. Werner, *J. prakt. Chem.*, *322*, 247 (1980); (e) Le Roy, F. Johnson, and W. C. Jankowski, *Carbon-13 NMR Spectra. A Collection of Assigned. Coded and Indexed Spectra*. Wiley, New York, 1972.
60. (a) W. J. Blunt and J. B. Stothers, *Org. Magn. Reson.*, *9*, 439 (1977); (b) H. J. Schneider and W. Gschwendter, *J. Org. Chem.*, *47*, 4216 (1982).
61. J. E. Huheey, *J. Phys. Chem.*, *69*, 3284 (1965); *70*, 2086 (1966).
62. E. T. McBee, I. Serfaty, and T. Hidgins, *J. Am. Chem. Soc.*, *93*, 5711 (1971).

63. R. W. Taft and R. D. Topsom, *Progr. Phys. Org. Chem.*, 17, 1 (1987).
64. F. H. Westheimer and J. G. Kirkwood, *J. Chem. Phys.*, 6, 513 (1938).
65. W. F. Reynolds, *J. Chem. Soc. Perkin II*, 985 (1980).
66. W. F. Reynolds, *Progr. Phys. Org. Chem.*, 14, 165 (1983).
67. R. D. Topsom, *Progr. Phys. Org. Chem.*, 16, 125 (1987).
68. N. S. Zefirow, *Tetrahedron Lett.*, 1077 (1975).
69. H. J. Schneider and E. P. Weigand, *J. Am. Chem. Soc.*, 99, 8362 (1977).
70. R. I. Zalewski and G. E. Dunn, *Can. J. Chem.*, 47, 2263 (1969).
71. R. L. Reeves, *J. Am. Chem. Soc.*, 88, 2240 (1966).
72. J. T. Edward and S. C. Wong, *J. Am. Chem. Soc.*, 99, 4249 (1977).
73. R. I. Zalewski, *J. Chem. Soc.*, 1637 (1979).
74. Z. Geltz, H. Kokocinska, R. I. Zalewski, and T. M. Krygowski, *J. Chem. Soc. Perkin II*, 1069 (1983); M. R. Sharif and R. I. Zalewski, *Bull. Polon. Acad. Sci. Ser. Sci. Chem.*, 29, 385 (1982).
75. W. H. Lawtone and E. A. Sylvestre, *Technometrics*, 13, 617 (1971).
76. P. C. Gillette, J. B. Lando, and J. L. Koenig, *Anal. Chem.*, 15, 630 (1983).
77. D. Macnaughtan, Jr., L. B. Rogers, and G. Wernimont, *Anal. Chem.*, 44, 1421 (1972).
78. R. I. Zalewski and S. Geribaldi, *J. Chem. Soc. Perkin II*, 113 (1988).
79. J. F. Bulmer and H. F. Shurwell, *J. Phys. Chem.*, 77, 256 (1973).
80. J. F. Bulmer and H. F. Shurwell, *Can. J. Chem.*, 53, 1251 (1975).
81. B. U. Putelenz and H. F. Shurwell, *Can. J. Chem.*, 58, 353 (1980).
82. R. N. Jones, Ed., *Computer Programs for Infrared Spectrophotometry*, National Research Council of Canada, Ottawa, 1978.
83. D. Jesse and H. F. Shurwell, *J. Phys. Chem.*, 91, 496 (1987).
84. H. Martens and S. A. Jensen, *Progress in Cereal Chemistry and Technology*. Vol. 5a, Elsevier, Amsterdam, 1983, p. 607.
85. P. Robert and D. Berrand, *Sciences des Aliments*, 5, 501 (1985).
86. W. F. Mc Clure, A. Hamid, F. G. Giesbrecht, and W. W. Weeks, *Appl. Spectrosc.* 38, 322 (1984).
87. M. F. Devaux, D. Bertrand, P. Robert and J. L. Morat, *J. Chemometr.*, 1, 103 (1987).
88. G. Ramana Rao and G. Zerbi, *Appl. Spectrosc.*, 38, 795 (1984).
89. L. B. Shih and R. G. Priest, *Appl. Spectrosc.*, 38, 687 (1984).
90. J. C. Giddings and J. M. Davis, *Anal. Chem.*, 57, 2166 (1985).
91. (a) G. L. Ritter, S. R. Lowry, and T. L. Isenhour, *Anal. Chem.*, 48, 591 (1976); (b) F. J. Knorr and J. H. Futrell, *Anal. Chem.*, 51, 1236 (1979); (c) E. R. Malinowski and M. Melue, *Anal. Chem.*, 49, 284 (1977).
92. H. B. Woodruff, P. C. Tway, and L. J. Cline Lowe, *Anal. Chem.*, 53, 81 (1981).
93. B. Vanderginste, R. Essers, T. Bosman, J. Reijnen, and G. Kateman, *Anal. Chem.*, 57, 971 (1985); B. Vanderginste, W. Derks, and G. Kateman, *Anal. Chim. Acta*, 173, 253 (1985).
94. P. J. Gemperline, *Anal. Chem.*, 58, 2656 (1986).
95. J. M. Harris, M. L. McConnell, and S. D. Frans, *Anal. Chem.*, 57, 1552 (1985).
96. J. Halket, *J. Chromatogr.*, 175, 229 (1979); M. Chien, *Anal. Chem.*, 57, 348 (1085).
97. D. G. Hoovery and J. M. Soroka, *Anal. Chem.*, 58, 3091 (1986).

98. B. Walczak, L. Morin-Allory, M. Lafosse, M. Dreux, and J. R. Chretien, *J. Chromatogr., 395*, 183 (1987).
99. J. R. Chretien, B. Walczak, L. Morin-Allory, M. Dreux, and M. Lafosse, *J. Chromatogr., 371*, 253 (1986).
100. B. Walczak, L. Morin-Allory, J. R. Chretien, M. Lafosse, and M. Dreux, *Chemometr Intel. Lab. Syst., 1*, 79 (1986).
101. B. Walczak, J. R. Chretien, M. Dreux, L. Morin-Allory, and M. Lafosse, *Chemometr. Intel. Lab. Syst., 2*, 177 (1987).
102. C. L. De Ligny, M. C. Spanjer, J. C. Van Houwelingen, and H. M. Weesie, *J. Chromatogr., 301*, 311 (1984).
103. L. R. Snyder, *Principles of Absorption Chromatography*, Marcel Dekker, New York, 1968.
104. R. Angus Hearman, J. H. Screewens, K. R. Jennings, and M. J. Farncombe, *Chemometr. Intell. Lab. Syst. 1*, 167 (1987).
105. W. Windig, E. Jakob, J. M. Richards, and H. L. C. Menzelaar, *Anal. Chem., 59*, 317 (1987).
106. W. Windig, T. Chakrzwarty, J. M. Richards, and H. L. C. Menzelaar, *Anal. Chim. Acta., 191*, 205 (1986).
107. W. Windig and H. L. C. Menzelaar, *Anal. Chem., 56*, 2297 (1984).
108. D. R. Burgard, S. P. Perone, and J. L. Wiebers, *Anal. Chem., 48*, 1444 (1977).
109. R. W. Rozett and E. M. Petersen, *Anal. Chem., 47*, 2286 (1975).
110. C. Reichardt, *Solvents and Solvent Effects in Organic Chemistry*, VCH Varlagsgesellshaft, Weinheim, 1988.
111. F. W. Fowler, A. R. Katritzky, and R. J. D. Rutherford, *J. Chem. Soc. Part B*, 460 (1971).
112. I. A. Koppel and V. A. Palm, *The Influence of Solvent on Organic Reactivity*, in *Advances in Linear Free Energy Relationships*, N. B. Chapman and J. Shorter, Eds., Plenum Press, London, 1972, Chap. 5, p. 203.
113. M. Bohle, W. Kollecker, and D. Martin, *Z. Chem., 17*, 161 (1977).
114. A. J. Parker, *Quart. Rev. (London), 16*, 163 (1962); *Chem. Rev., 69*, 1 (1969); *Pure Appl. Chem., 25*, 345 (1971).
115. J. Elguero and A. Fruchier, *Annales de Quimica, Ser. C, 79*, 72 (1983).
116. P. Svoboda, O. Rytela, and M. Vecera, *Collect. Czech Chem. Commun., 48*, 3287 (1983).
117. R. Carlson, T. Lundstedt, and C. Albano. *Acta Chem. Scand., Part B, 39*, 79 (1985).
118. P. C. Maria, J.-F. Gal, J. de Franceschi, and E. Fargin, *J. Am. Chem. Soc., 109*, 483 (1987).
119. R. D. Cramer III, *J. Am. Chem. Soc., 102*, 1837, 1849 (1980).
120. (a) M. Chastrette, *Tetrahedron, 35*, 144 (1979); (b) M. Chastrette and J. Carretto, *Tetrahedron, 38*, 1615 (1982); (c) M. Chastrette, M. Rajzamann, M. Chanon, and K. F. Purcell, *J. Am. Chem. Soc., 107*, 1 (1985).
121. R. I. Zalewski, H. Kokocińska, and Ch. Reichardt, *J. Phys. Org. Chem., 2*, (1988).
122. W. F. Reynolds and R. D. Topsom, *J. Org. Chem., 49*, 1989 (1984); M. Charton, *J. Org. Chem., 49*, 1997 (1984); A. J. Hoefnagel, W. Oosterbeck, and B. W. Wepster, *J. Org. Chem., 48*, 1993 (1984).
123. G. H. E. Nieuwdrup and C. L. de Ligny, *J. Chem. Soc. Perkin II*, 537 (1979).
124. C. L. de Ligny and H. C. Van Houvelingen, *J. Chem. Soc., Perkin*, 559 (1987).
125. M. Sjöstrom and S. Wold, *Chemica Scripta, 6*, 114 (1974); S. Alunni, S. Clementi, U. Edlund, D. Johnels, S. Halberg, M. Sjöstrom, and S. Wold, *Acta Chem. Scand., Part B, 37*, 47 (1983).

126. L. P. Hammett and A. J. Deyrup, *J. Am. Chem. Soc.*, *54*, 2721 (1932).
127. M. M. Arnett and G. W. Mach, *J. Am. Chem. Soc.*, *88*, 1177 (1966).
128. R. Boyd, *J. Phys. Chem.*, *67*, 737 (1963).
129. R. A. Cox and K. Yates, *Can. J. Chem.*, *61*, 2225 (1983).
130. K. Yates and R. A. McClelland- *Progr. Phys. Org. Chem.*, *11*, 323 (1974).
131. P. Dominique and J. M. Caspentier, *J. Chem. Res.* (S), 58 (1979).
132. J. F. Bunnett and F. P. Olsen, *Can. J. Chem.*, *44*, 1899, 1917 (1966).
133. R. A. Cox and K. Yates, *J. Am. Chem. Soc.*, *100*, 3864 (1978).
134. N. C. Marziano, P. G. Traverso, and R. C. Passerini, *J. Chem. Soc. Perkin II*, 306 (1977).
135. J. T. Edward, M. Sjöstrom, and S. Wold, *Can. J. Chem.*, *59*, 2350 (1981).
136. A. Bagno, G. Scorrano, and R. A. More O' Ferrall, *Rev. Chem. Intermed.*, *7*, 13 (1987).
137. R. I. Zalewski, A. Y. Sarkice, and Z. Geltz, *J. Chem. Soc., Perkin II*, 1059 (1983).
138. T. A. Modro, K. Yates, and J. Janata, *J. Am. Chem. Soc.*, *97*, 1492 (1975).
139. R. A. Cox, *J. Am. Chem. Soc.*, *96*, 1059 (1974).
140. R. A. Cox, U. L. Haldna, K. L. Idler, and K. Yates, *Can J. Chem.*, *59*, 2591 (1981).
141. E. R. Malinowski, R. A. Cox, and U. L. Haldna, *Anal. Chem.*, *56*, 778 (1984).
142. N. B. Librovich and V. D. Maiorov, *Izw. Akad. Nauk. SSSR Ser. Khim.*, 684 (1977).
143. C. Albano and S. Wold, *J. Chem. Soc. Perkin II*, 1447 (1980).
144. S. Winstein and D. S. Trifan, *J. Am. Chem. Soc.*, *71*, 2953 (1949); S. Winstein, *Bull. Soc. chim. France*, *C55*, (1951).
145. S. A. Pivovarov, V. P. Selianov, and B. V. Gidaspov, *Org. Reactiv.* (Tartu), *12*, 305 (1975).
146. J. Gasteiger, H. Saller, and P. Loew, *Anal. Chim. Acta 191*, 111 (1986).
147. J. Gasteiger, M. S. Huthings, B. Christoph, L. Gann, C. Hiller, P. Loew, M. Marsili, H. Saller, and K. Yuki, *Topics Curr. Chem.*, *137*, 19 (1987).

Physicochemical Preconditions of Linear Free-Energy Relationships

BY OTTO EXNER

Institute of Organic Chemistry and Biochemistry
Czechoslovak Academy of Sciences
Prague, Czechoslovakia

CONTENTS

I. Statement of the Problem 129
II. Survey of Thermodynamic Quantities 130
III. Energy Terms in Isodesmic and Nonisodesmic Reactions 133
IV. Extrapolation to Zero Temperature 137
V. Correlations between Various Thermodynamic Quantities 141
VI. Linear Relationships for Reaction Enthalpy 144
VII. Importance of Polar Effects 153
VIII. Correlations between Solvation Enthalpies and Entropies 155
IX. Conclusions 158
References 160

I. STATEMENT OF THE PROBLEM

Empirical relationships have been found for many numerical quantities in chemistry (1), but they are most elaborated for Gibbs energies of reaction ($\Delta G°$) or activation ($\Delta G^{\#}$), which were obtained experimentally mostly as logarithms of the equilibrium or rate constants. The Hammett equation is a classical example (1–3). With the previously common term "free energy" instead of "Gibbs energy", the phrase "linear free-energy relationships" (LFER) has been widely used. In many textbooks the existence of LFER is accepted almost as an axiom (4), and sometimes this term is used in a broader sense even for relationships that are not restricted to Gibbs energies. However, $\Delta G°$ is a relatively complex quantity in terms of thermodynamics and can be successively partitioned into several terms (1, 2, 5, 6). In addition, it is strongly dependent on temperature (7). Therefore, the following question arises: Is it

[†]Presented in part at the IVth Conference on Correlation Analysis in Organic Chemistry, Hammett Memorial Symposium, Poznań, Poland, 1988.

possible that relationships between the more complex quantities ($\Delta G°$) are, in fact, simpler than those between its more fundamental components ($\Delta G°, \Delta S°, \Delta H_0°, \Delta E_p$), or do we have the LFER only because $\Delta G°$ is the most readily accessible quantity?

II. SURVEY OF THERMODYNAMIC QUANTITIES

The question posed above was recognized earlier by Hammett (2), who divided the Gibbs energy into several terms and tried to estimate their relative importance. Using a more modern nomenclature we can divide first $\Delta G°$ into the enthalpy ($\Delta H°$) and entropy ($\Delta S°$) terms (Equation 1), and further, $\Delta H°$ into the enthalpy at absolute zero ($\Delta H_0°$) and the heat content function $[\Delta(H_T° - H_0°)]$ (Equation 2). In the last step, $\Delta H_0°$ can be further separated— this time merely theoretically, using spectroscopic data—into the residual or zero-point energy (ΔE_r), representing the rest of the vibrational energy present at absolute zero, and the potential energy (ΔE_p) (Equation 3). There is a rather commonly accepted opinion (2, 5) that this potential energy is the proper quantity to be related to any chemical theory (not involving temperature), particularly to the common language of chemical formulas. Equations 1–3 concern isolated molecules, that is, the ideal-gas state. Since most experiments are carried out in solution, it is necessary to add the solvation enthalpy (ΔH_S) and entropy (ΔS_S) (Equation 4):

$$\Delta G° = \Delta H° - T\Delta S° \qquad (1)$$

$$\Delta G° = \Delta H_0° + \Delta(H_T° - H_0°) - T\Delta S° \qquad (2)$$

$$\Delta G° = \Delta E_p + \Delta E_r + \Delta(H_T° - H_0°) - T\Delta S° \qquad (3)$$

$$\Delta G°_{sol} = \Delta E_p + \Delta E_r + \Delta(H_T° - H_0°) - T\Delta S° + \Delta S_S - T\Delta S_S \qquad (4)$$

Quite similar equations can, in principle, be written even for Gibbs energies of activation, but a practical separation cannot be achieved (5). In the following we shall deal only with equilibria and $\Delta G°$. Hammett (2) used a somewhat different terminology in which the entropy and heat content function were replaced by one term, the kinetic energy, involving the partition functions. His most important conclusion was that $\Delta H°$ is not a simpler quantity than $\Delta G°$; hence separation according to Equation 1 does not help. Only in one particular case the named quantities may acquire a simpler physical meaning. In that case one may assume that in certain, rather similar reactions the partition functions can be so similar that the values of both the kinetic and residual energy are practically equal. In this case the two reactions can be subtracted and either the relative Gibbs energy ($\delta \Delta G°$) or enthalpy ($\delta \Delta H°$) can be substituted for the unknown potential energy ($\delta \Delta E_p$) since the remaining terms cancel:

$$\delta \Delta G° = \delta \Delta H° = \delta \Delta E_p \tag{5}$$

For the same reason the entropy term is equal in the two reactions and cancels by their subtraction:

$$\delta \Delta S° = 0 \tag{6}$$

The validity of Equation 6 in a particular case may be easily tested experimentally (7). Once it is proved, it should also justify the assumption in Equation 5. In Equations 5 and 6 the operator δ was introduced (6) to denote differences between two related reactions. These may be, for instance, the reactions of a substituted and an unsubstituted compound with the same reagent, or the same reaction in an arbitrary solvent and in the standard solvent, and so on. The letter δ is used in contradistinction to Δ, which relates to the difference between reactants and products. When we are concerned only with equilibria, the operator δ is not quite necessary since $\delta \Delta G°$ also represents $\Delta G°$ of another equilibrium reaction, a so-called isodesmic reaction (8). For instance, if the two similar reactions are ionizations of benzoic and 4-chlorobenzoic acid, $\delta \Delta G°$ corresponds to an isodesmic reaction in which a proton is transferred from the 4-chlorobenzoic acid to the anion of benzoic acid. Further examples will be given later. The term "isodesmic" means that on either side of the reaction equation there is the same number of molecules, the same number of charges, and the same number of bonds between particular atoms (bonds C—C, C—H, etc.). The principle of isodesmic reactions has been broadly exploited (1, 5). Their equilibrium constants and other thermodynamic quantities can be evaluated even in cases when the reaction itself cannot be realized under suitable conditions.

Of course, Equation 6 need not be valid for all isodesmic reactions, but its validity must be proved in every particular case by experiment. Actually, it is relatively seldom fulfilled. Remarkably, Hammett did not continue work on this problem in the second edition of his book (3), but less severe conditions were searched for by others. It was argued (6) that $\delta \Delta G°$ and $\delta \Delta H°$ need not be equal as in Equation 5, but only proportional to each other. In this case $\delta \Delta H°$ and $\delta \Delta S°$ are proportional (Equation 7); this is so-called isokinetic relationship. (The term "isokinetic" is often used even for equilibria.) Then either $\delta \Delta H°$ or $\delta \Delta G°$ can serve as an estimate of relative values of $\delta \Delta E_p$ within related reactions series, but actual values of $\delta \Delta E_p$ cannot be calculated. Experimental verification of Equation 7 is connected with some rather weighty statistical problems (7); hence many claims of its existence (6) are unwarranted:

$$\delta \Delta H° = \beta \delta \Delta S° \tag{7}$$

Several other kinds of partitioning $\Delta G°$ have been advanced (9–12), particularly the theory of Hepler (9, 10), which has gained some popularity. It applies only to ionic equilibria, essentially in water. The total Gibbs energy is divided into enthalpic and entropic terms and each into an internal and an external contribution (Equation 8). The external terms can be identified with the solvation terms ΔH_S and $T \Delta S_S$ of Equation 4. With several assumptions, it has been concluded (9) that $\delta \Delta G°$ may approach $\delta \Delta H_{int}$ very closely while the other terms compensate. This is at variance with the enthalpies of dissociation in the gas phase (13). Also, the inherent provisions are open to criticism; in particular, an assumed isokinetic relationship between the quantities $\delta \Delta H_{ext}$ and $\delta \Delta S_{ext}$ has been challenged (14). The gas-phase data available now make it possible to check in some detail both the original theory (9) and its more sophisticated form (10):

$$\delta \Delta G° = \delta \Delta H_{int} - T\delta \Delta S_{int} + \delta \Delta H_{ext} - T\delta \Delta S_{ext} \tag{8}$$

In this chapter several attempts will be reported to proceed a step further in understanding the relative importance of various thermodynamic quantities and the relations between them. All investigations will be purely empirical in character, that is, based only on selected series of reactions. This selection is controlled partly by the availability of experimental data and partly by our intention to study well-behaved series, mostly those known from the Hammett equation and similar relationships. The first task is to estimate quantitatively the relative importance of individual thermodynamic quantities—say, the terms of Equation 4—in concrete examples. Further, the temperature dependence should be followed to see the possibilities of extrapolation to absolute zero. Another possible approach is to search for linear relationships between thermodynamic quantities other than Gibbs energies, say, enthalpies or potential energies. Ultimately, the theory expressed by Equation 8 will require a comparison between gas-phase and solution data. Many of these investigations will be prevented by the lack of good experimental data, but in the author's opinion this is quite a typical situation in the correlation analysis. The supply of necessary data is no worse now than was the supply of Gibbs energy data at the time when Hammett discovered his equation.

III. ENERGY TERMS IN ISODESMIC AND NONISODESMIC REACTIONS

Let us consider halogenation of substituted benzenes (Reactions A and B, below) as a typical example of reaction series. They do not obey the original Hammett equation but obey its extended version as introduced by Brown (15). The values of $\Delta G°$ and $\Delta H°$ at temperatures between 298 and 1000 K can be obtained by summing the tabular values (16, 17) of the enthalpies of formation or Gibbs energies of formation, respectively, for products and reactants. The values below 298 K up to absolute zero, the residual energies ΔE_r, and potential energies ΔE_p were obtained by us (18, 19) by statistical and mechanical calculations based on published spectral data:

$$XC_6H_5 + F_2 = XC_6H_4F + HF \tag{A}$$

$$XC_6H_5 + Cl_2 = XC_6H_4Cl + HCl \tag{B}$$

$$XC_6H_4 + CH_4 = XC_6H_4CH_3 + H_2 \tag{C}$$

$$XC_6H_5 + C_6H_5F = XC_6H_4F + C_6H_6 \tag{D}$$

$$XC_6H_5 + C_6H_5Cl = XC_6H_4Cl + C_6H_6 \tag{E}$$

$$XC_6H_5 + C_6H_5CH_3 = XC_6H_4CH_3 + C_6H_6 \tag{F}$$

$$C_6H_4X_2\text{-}(1,2) = C_6H_4X_2\text{-}(1,3) \tag{G}$$

$$C_6H_4X_2\text{-}(1,2) = C_6H_4X_2\text{-}(1,4) \tag{H}$$

$$C_6H_4X_2\text{-}(1,3) = C_6H_4X_2\text{-}(1,4) \tag{J}$$

The results of these reactions, listed in Table 1, allow the following conclusions. The potential-energy term dominates as expected, and this statement can be safely extended to any reaction series controlled by sufficiently strong polar effects. It is also confirmed by a comparison with Reaction C, which is to be denoted as methylation of substituted benzenes with methane as methylating agent. Let us consider ΔE_p as a "signal" responsible for the observed regularities, and the contributions from ΔE_z, $T\Delta(H°_{298} - H°_0)$, and $-T\Delta S°_{298}$ as "noise." Then the signal-to-noise ratio would be > 300, ~ 250, and $100\text{–}350$ for the three named factors, respectively, in the case of the fluorination reactions (Reaction A). This ratio

TABLE 1

Thermodynamic Data for Nuclear Substitution of Benzene Derivatives[a]

Reaction and Substituent X	ΔE_p	ΔE_r	$\Delta H_0°$	$\Delta H_{298}°$	$\Delta G_{298}°$ [b]	$\Delta H_{1000}°$	$\Delta G_{1000}°$ [b]	$\Pi\sigma$ [c]
				Nonisodesmic				
(A) H	−112.55	−0.35	−112.90	−112.48	(−111.33)	−111.65	−114.45 (−109.51)	$\frac{1}{12}$
(A) 2-F	−107.53	−0.26	−107.79	−107.33	(−106.18)	−106.61	−105.65 (−104.27)	$\frac{1}{2}$
(A) 3-F	−111.21	−0.25	−111.46	−111.03	(−109.84)	−110.26	−109.27 (−107.89)	$\frac{1}{2}$
(A) 4-F	−110.53	−0.30	−110.83	−110.37	(−109.21)	−109.58	−107.35 (−107.35)	1
(A) 4-CH$_3$	−112.60	0.04	−112.56	−112.13	(−110.97)	−111.55	−110.33 (−108.95)	$\frac{1}{2}$
(B) H	−27.63	−2.42	−30.05	−29.49	(−28.64)	−28.50	−32.53 (−27.59)	$\frac{1}{12}$
(B) 2-Cl	−25.73	−2.13	−27.86	−27.29	(−26.30)	−26.40	−26.22 (−25.81)	$\frac{1}{2}$
(B) 3-Cl	−25.50	−2.22	−27.72	−28.13	(−29.25)	−27.18	−27.50 (−27.09)	$\frac{1}{2}$
(B) 4-Cl	−27.28	−2.26	−29.54	−28.95	(−28.01)	−28.00	−26.75 (−26.75)	1
(C) H	13.34	−3.93	9.41	10.02	(11.73)	12.82	8.61 (13.55)	$\frac{1}{12}$
(C) 2-CH$_3$	14.08	−4.40	9.68	10.48	(12.54)	13.39	13.69 (15.07)	$\frac{1}{12}$
(C) 3-CH$_3$	13.87	−4.65	9.22	10.06	(11.79)	13.01	12.15 (13.53)	$\frac{1}{2}$
(C) 4-CH$_3$	13.67	−4.19	9.48	10.23	(11.94)	13.08	13.70 (13.70)	1
(C) 4-F	12.84	−3.09	9.75	10.37	(12.09)	12.92	12.73 (14.11)	$\frac{1}{2}$
				Isodesmic				
(D) 2-F	5.02	0.09	5.11	5.15	(5.15)	5.04	8.80 (5.24)	6
(D) 3-F	1.34	0.10	1.44	1.45	(1.49)	1.39	5.18 (1.62)	6
(D) 4-F	2.02	0.05	2.07	2.11	(2.12)	2.07	7.10 (2.16)	12
(D) 4-CH$_3$	−0.05	0.39	0.34	0.35	(0.36)	0.10	4.12 (0.56)	6
(E) 2-Cl	1.90	0.29	2.19	2.20	(2.34)	2.10	6.31 (2.75)	6
(E) 3-Cl	1.13	0.20	1.33	1.36	(1.39)	1.32	5.03 (1.47)	6
(E) 4-Cl	0.35	0.16	0.51	0.54	(0.62)	0.51	5.78 (0.84)	12

(F)	2-CH$_3$	0.74	−0.47	0.27	0.46	1.87	(0.81)	0.57	5.08	(1.52)	6
(F)	3-CH$_3$	0.53	−0.72	−0.19	0.04	1.12	(0.06)	0.19	3.54	(−0.02)	6
(F)	4-CH$_3$	0.33	−0.26	0.07	0.21	1.68	(0.21)	0.26	5.09	(0.15)	12
(F)	4-F	−0.50	0.84	0.34	0.35	1.42	(0.36)	0.10	4.12	(0.56)	6
(G)	F	−3.68	0.01	−3.67	−3.70	−3.66	(−3.66)	−3.65	−3.62	(−3.62)	1
(H)	F	−3.00	−0.04	−3.04	−3.04	−2.62	(−3.03)	−2.97	−1.70	(−3.08)	2
(J)	F	0.68	−0.05	0.63	0.66	1.04	(0.63)	0.68	1.92	(0.54)	2
(G)	Cl	−0.76	−0.09	−0.86	−0.84	−0.95	(−0.95)	−0.78	−1.27	(−1.27)	1
(H)	Cl	−1.55	−0.13	−1.68	−1.66	−1.31	(−1.72)	−1.59	−0.53	(−1.91)	2
(J)	Cl	−0.78	−0.04	−0.82	−0.82	−0.36	(−0.77)	−0.81	−0.75	(−0.63)	2
(G)	CH$_3$	−0.21	−0.25	−0.46	−0.42	−0.75	(−0.75)	−0.38	−1.54	(−1.54)	1
(H)	CH$_3$	−0.41	0.21	−0.20	−0.25	−0.19	(−0.60)	0.31	0.01	(−1.37)	2
(J)	CH$_3$	−0.20	0.46	0.26	0.17	0.56	(0.15)	0.07	1.55	(0.17)	2

[a] In kcal mole^{-1}, ideal-gas state (18, 19).
[b] In parentheses—corrected for symmetry factors, i.e., the term $RT\Sigma \ln \sigma$ subtracted.
[c] Products of the external symmetry numbers, reaction products in the numerator, reactants in the denominator.

allows us to neglect these three contributions within the common accuracy of the correlation analysis. However, the signal-to-noise ratio drops dramatically for Reaction C, not controlled by polar effects, and attains the values ~ 4, ~ 20, and 8–50, respectively. At the same time these figures become more sensitive to experimental errors and to possible misassignment of spectral signals (19). This casts some doubt on the admissibility of alkyl substituents in the Hammett equation and similar relationships. We shall return to this question later. Note only that the difference between the two reactions would be much less dramatic at the temperature of 1000 K instead of 298 K, only at the lower temperature the term $T\Delta S°$ could be neglected in either case.

The latter entropic term should be reduced by calculating separately that part of the entropy that is connected only with the symmetry of the molecules of reactants and products. It is given by Equation 9, where the summation extends over all molecules, with products being taken with the positive and reactants with the negative sign. Of the symmetry numbers, σ and n, only the external symmetry numbers σ are of importance for the reactions under consideration. The symmetry corrections are well known in the correlation analysis (1), but because of the low symmetry of most reactants, they are applied mostly only to dissociation of polybasic acids:

$$S_{sym} = -R\sum \ln(\sigma n) \qquad (9)$$

In Table 1, the external symmetry numbers for individual reactions are given in the last column, and the symmetry-corrected values of $\Delta G°$ are given in parentheses for the two temperatures, 298 and 1000 K. In some cases the correction does not seem to represent any progress; the corrected values of $\Delta G°$ are often less close to the enthalpies $\Delta H°$ than are the original uncorrected $\Delta G°$.

Some of the conclusions mentioned above change quite fundamentally when one proceeds to isodesmic reactions. There are two possibilities of combining the reactants and products. Reactions D and E may be considered as halogenations with fluorobenzene and chlorobenzene, respectively, acting as halogenating agents. Reaction F, in the same terms, may be called "methylation with toluene." In the reverse direction these reactions may be called "disproportionation" (20) or in any direction, simply "redistribution" (21). Another type of isodesmic reaction, particularly simple, are the isomerization reactions, Reactions G, H, and J. The results attained from the two types of reaction are very similar. Again the potential energy is the dominating term in the case of halogen-substituted compounds, but the contributions from $\delta\Delta E_r$, $\delta\Delta(H_T° - H_0°)$, and $-T\delta\Delta S°$ gain relatively more importance. The "signal-to-noise" ratio would be 15–50, 50–100, and ~ 40, respectively, in the case of the halogenation reactions, and similar but still more variable values are obtained for the isomerizations. In the case of methyl derivatives,

TABLE 2
Enthalpies of Solvation for Isodesmic Reactions of Chlorobenzenes[a]

Reaction		ΔH_s° in the Solvent	
		CCl_4	Nitromethane
Chlorination	(E) X = 2-Cl	0.05	0.05
	(E) X = 3-Cl	0.00	0.19
	(E) X = 4-Cl	0.22	0.45
Isomerization	(G) X = Cl	−0.05	0.14
	(H) X = Cl	0.17	0.40
	(J) X = Cl	0.22	0.26

[a] In kcal mole^{-1}, 298 K, reference 22.

this ratio is reduced to such an extent that $\delta \Delta E_p$ is sometimes no longer the deciding term; in particular, $\delta \Delta E_r$ becomes important. On the other hand, the isodesmic reactions evidence very clearly the importance of symmetry correction. In the case of halogen derivatives, this correction renders $\delta \Delta H°$ and $\delta \Delta G°$ almost equal and reveals that the main part of entropy at 298 K originates in symmetry. This is less evident for methyl derivatives because of the greater relative error. One can thus suggest that the easiest interpretable data are obtained for isodesmic reactions with sufficiently strong polar effects.

The last two terms in Equation 4, which are related to solvation, are available for only few equilibria, although they can be determined relatively easily. Particularly the enthalpies of solvation, ΔH_S, can be obtained from the experimental enthalpies of solution and enthalpies of vaporization of all the compounds involved. Some values are listed in Table 2. The values of H_{solv} for an individual compound are of the order of 10 kcal mole^{-1}; the main contribution comes from the enthalpy of vaporization (22). For a chemical reaction, however, these values largely cancel and the net effect sometimes only slightly exceeds the experimental error, which may exceed 0.1 kcal mole^{-1} in unfavorable cases. The data in Table 2 do not show much regularity, and attempts to predict them in a simple way, such as in terms of the reaction-field theory, have not been successful (22). The main result is that the solvent effects are of importance only in nitromethane, and in tetrachloromethane they are almost negligible, even in the case of polar solutes. In our opinion the effects on equilibria, observed or predicted in the gas phase, can be essentially transferred to solutions in less polar solvents.

IV. EXTRAPOLATION TO ZERO TEMPERATURE

Some of the previous conclusions can be illustrated more objectively in the graphical representation of temperature dependence. Some of the results shown in Figs. 1–3 reveal the possibilities of obtaining ΔH_0° and/or ΔE_p from measurements at common temperatures. Figure 1 concerns nonisodesmic fluorination. Both ΔG° and ΔH° depend on the temperature very steeply, and at first glance, it would appear to be impossible to extrapolate these curves to absolute zero if only their course above 298 K were known. It would also be difficult to estimate ΔH° from a single value of either ΔH_{298}° or ΔG_{298}°. Even if

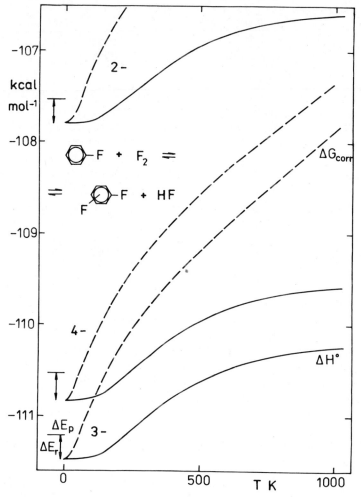

Figure 1 Relation between thermodynamic parameters and their temperature dependence for the fluorination of fluorobenzene with F_2: ΔH°—full lines; ΔG_{corr}°—broken lines; ΔE_r—arrows; ΔE_p—short given lines. Reproduced with permission from O. Exner and Z. Slanina, *Thermochim. Acta*, in press.

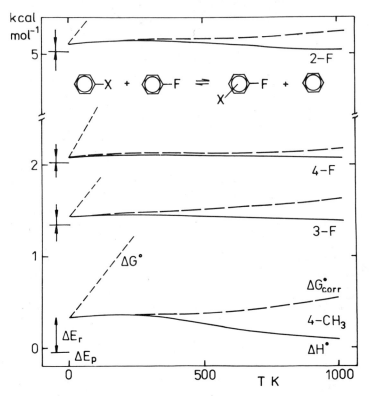

Figure 2 Relation between thermodynamic parameters for isodesmic fluorination with fluorobenzene as agent. Symbols as in Fig. 1, $\Delta G°$ uncorrected—dotted lines.

the course of the curves up to 0 K were known, it would still be impossible to estimate ΔE_r and to obtain ΔE_p. The position is considerably improved with the use of isodesmic reactions (Fig. 2). The curves are quite flat, and the section between 0 and 298 K on these curves can be foreseen with an acceptable degree of uncertainty. Figure 2 also shows very clearly the importance of symmetry corrections to $\Delta G°$; the uncorrected values are quite different from $\Delta H°$ and would not allow any extrapolation (dotted lines).

As in all cases observed, the picture is changed in the case of methyl derivatives (Fig. 3), where not only the values of $\Delta G°$ and $\Delta H°$ are smaller but the temperature dependence is greater than in the case of halogen derivatives, and no relation to $\Delta H_0°$ can be seen. In all the cases there remains still the residual energy, which is not negligible even for polar effects (Fig. 2) and can hardly be predicted from thermodynamic quantities. In the case of methyl derivatives the contribution from ΔE_r is even sufficient to reverse the order of individual positional isomers. In addition, this value is sensitive to experi-

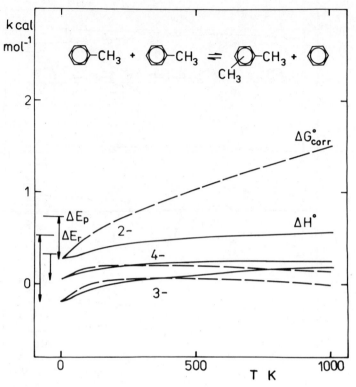

Figure 3 Relation between thermodynamic parameters for an isodesmic methylation with toluene as agent. Symbols as in Figure 1, reproduced from Z. Slanina, Collect. Czech. Chem. Commun., 55 (1990), in press.

mental errors and assignment in the spectra. For these reasons ΔE_r appears to be a somewhat unreliable quantity and its inherent uncertainty is transferred to ΔE_p.

At this point some remarks concerning the reliability of all thermodynamic quantities seem appropriate. Let us consider Fig. 2 as an arbitrary example of this. Although the individual quantities may be determined by various procedures and in different orders (16, 17), it is typically the combustion calorimetry that serves as starting point. If the reaction involves four compounds, we need four experimental enthalpies of combustion and, in addition, four enthalpies of vaporization (sublimation) to obtain $\Delta H°_{298}$: any error in these eight quantities is transferred into the final value. In the tables the errors are estimated rather cautiously, for instance, to less than 2 kcal mole^{-1} in the data of average quality and less than 0.5 for a good quality (17). The value of $\Delta H°_{298}$ thus yields a point on the $\Delta H°$ curve on which the whole graph is anchored (Fig. 2).

The next task is to determine $\Delta S°_{298}$, measuring the distance from the $\Delta G°$ curve, and further to extend the two curves to either side. Typically, both the entropy and heat content functions are obtained from statistical mechanics, using the translational, rotational, and vibrational partition functions; the latter is based on measuring and assigning the vibrational spectra. Of the two sides of the curves, those toward the low temperatures can be determined much more reliably. At lower temperatures the contribution of vibrational energy decreases and the translational and rotational energies are known with much greater accuracy. The extrapolation up to absolute zero is aided by the Third Law of Thermodynamics, requiring zero slopes for both the $\Delta H°$ and $\Delta G°$ curves (Fig. 2). In the other direction—toward high temperatures—the uncertainty increases with temperature as the vibrational energy becomes more important. The main sources of uncertainty are incomplete assignment of spectral lines and imperfections of the models of anharmonic oscillator and rigid rotor; these imperfections are of importance particularly at higher temperatures. Similar problems arise in calculating ΔE_r, which depends entirely on the vibrational energy since there is no translation and rotation at absolute zero. In some cases low-temperature heat capacity measurements can be used in addition to, or partly instead of, statistical mechanics. In this case it is again the low-temperature part of the functions that is further improved. Summarizing, there are three main sources of uncertainty: imperfections of theoretical models affect mainly $\Delta S°$ at high temperatures, incomplete assignment of vibrational spectra affects $\Delta S°$ and also ΔE_r, and the combustion calorimetry controls the relative position of all the values with respect to another compound.

For these reason we suggest (18) abandoning potential energy as a theoretically required quantity. In addition to the problems mentioned above regarding its accuracy, there are still theoretically objections that it is a combination of two theories and does not actually exist in the real world. The enthalpy at absolute zero, $\Delta H°_0$, is the quantity of choice. Although in principle unattainable, it is the first theoretical quantity with an evident physical meaning and can be rather well approached by smooth extrapolation. It is also a quantity that should be obtained by quite rigorous quantum-chemical calculations not based on the Born–Oppenheimer approximation.

V. CORRELATIONS BETWEEN VARIOUS THERMODYNAMIC QUANTITIES

In principle, it would be quite easy to decide on empirical grounds which thermodynamic quantity is most suitable for theoretical interpretations. It would be necessary to choose only a well-behaved reaction series, such as the *meta*- and *para*-substituted benzene derivatives, and process successively the values of $\Delta H°_{298}$, $\Delta H°_0$, and ΔE_p in the same way as the $\Delta G°_{298}$ values have been processed in the Hammett equation. We can postulate that the best quantity is the one that yields the most accurate relationships. The necessary mathematical algorithm is the principal component analysis (PCA), which has already been applied in the case of the Hammett equation (23). However, the data actually available are very restricted. Among our reaction series we find as the most suitable the fluorination of substituted benzenes, that is, either Reaction A or Reaction D; for the purpose of PCA it is irrelevant whether the reaction is written in the isodesmic form. This series involves only five reactions, which do not belong to the range of validity of the Hammett equation but to Brown's extension (15). In addition, there is one *ortho* derivative present, which should violate empirical relationships of all known types. Nevertheless, there is very good correlation between all thermodynamic quantities.

For the reasons outlined above we chose $\Delta H°_0$ as standard and plotted against it successively the values of ΔE_p, $\Delta H°_{298}$, $\Delta G°_{298}$, $\Delta H°_{1000}$, and $\Delta G°_{1000}$. The graph is not shown because of its simplicity; five straight lines are obtained with the slopes not statistically different from unity, and the quality of the fit cannot be seen from the graphical representation. A statistical treatment is, in fact, of little value with five points; nevertheless, we can obtain very good linearity for $\Delta H°_{298}$ and $\Delta H°_{1000}$ with standard deviations 0.012 and 0.11, respectively (in kcal mole^{-1}). For $\Delta G°_{298}$ and $\Delta G°_{1000}$ the standard deviations are of the same order (0.021 and 0.10, respectively), provided the symmetry correction has been applied, without it, a complete scatter would arise (standard deviations 0.58 and 1.97). This is the most significant proof of the need for this correction.

The above figures also reveal a significant difference in accuracy between the measurements at 298 and 1000 K. Since most equilibria (and kinetics) were investigated at ambient or slightly elevated temperatures, one can guess that this fact is also partly responsible for the success of LFER. Remarkable is also the correlation of ΔE_p with $\Delta H°_0$, which is of relatively low precision (standard deviation 0.17). This confirms our reserved judgment of potential energy as a theoretical reference quantity. The values of $T\Delta S°$, $\Delta(H_T - H_0)$, and ΔE_r show almost no relation to $\Delta H°_0$ since their small values disappear in the noise. If we attempt a similar correlations with the methylation reaction (Reaction C or F), we obtain a result similar to results from previous investigations on this series—there is no correlation at all.

An approximate proportionality of all thermodynamic quantities within a series of isodesmic reactions seems to be a relatively frequent phenomenon. Its preconditions are that the reactions involved are sufficiently similar on one hand, but are controlled by relatively strong polar effects on the other hand. The polar effects simply outweigh the other effects, which are less understood and more difficult for modeling. This postulate was, in principle, confirmed with several reaction series (e.g., those listed in Table 1 and on others); however, most of these series include either too few reactions or reactions too similar to each other. It also turned out that the proportionality holds better at low temperatures, say, up to 500 K, as in Fig. 2 and 3.

The parallelism of all thermodynamic quantities includes an important corollary, the much discussed isokinetic relationship (Equation 7). This relationship has often been claimed where it is not fulfilled and vice versa, as a result of erroneous statistical treatment (7); the latter is particularly important in cases where enthalpy and entropy have been obtained together from the temperature dependence of rate or equilibrium constants. Using correct statistics, it was found (14) that the isokinetic relationship occurs very commonly within a narrow range of temperatures; its range of validity is certainly broader than that of the Hammett or Taft equations. If both the Hammett equation and the isokinetic relationship are valid in a given reaction series, the latter is always fulfilled with a higher degree of accuracy (24). In our reaction series, $\Delta H°$ and $\Delta S°$ have been obtained independently; in this case Equation 7 can be tested by simply plotting these two values against each other (25). Such plots are quite uninteresting since the values of $\delta \Delta S°$ are almost negligible; the reaction series may be considered as isoentropic. This finding is an apparent contradiction to Fig. 2, where the entropic terms is represented by the distance between the two curves for $\Delta H°$ and $\Delta G°$ and is not negligible. The difference is in scaling. While the term $T \Delta S°$ at high temperatures may appear important for a single reaction (say, the reaction at the bottom in Fig. 2), it may be unimportant in the whole graph with the usual scaling: 1 kcal mole^{-1} on the $\Delta H°$ axis, approximately as long as 1 cal K^{-1} mole^{-1} on the $\Delta S°$ axis.

For a more accurate comparison, we have plotted one example in the fashion usual in kinetics (7): in the plot log K versus reciprocal temperature (Fig. 4). We obtain a family of van't Hoff straight lines; if they intersect in one point, the isokinetic relationship is valid. The figure reveals that the reaction series is isoentropic with a high degree of accuracy since the lines intersect at $T^{-1} = 0$. According to conventional statistics (7), specifically, standard deviation SD 0.010 log units, $\psi = 0.0094$, the relationship is one of the most accurate ones. We obtained the same results for several isodesmic reactions in the gas-phase, although again for methyl derivatives, the precision is much less than that for halogen derivatives. One can assume that isoentropic series are common in the gas-phase and that an isokinetic relationship (Equation 7) with a finite value of β comes into existence only with the effect of the solvent, which makes the relationship less precise and shifts the position of the point of intersection.

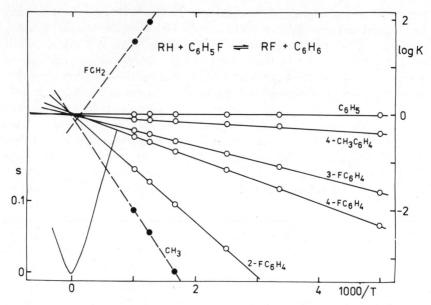

Figure 4 Isokinetic relationship for isodesmic fluorination with florobenzene. Shown are the van't Hoff lines in the coordinates log K versus the reciprocal temperature and the dependence of the residual standard deviation on the assumed isokinetic temperature (the curve at the bottom).

Figure 4 also confirms that the isokinetic relationship may be valid within a broader range than usual for free-energy relationships. When an aliphatic compound is included ($R = CH_3$), the precision is not significantly impaired; only the line for $R = FCH_2$ deviates distinctly from the common point of intersection. In this way a reaction series may be defined, that is, the range of validity of a relationship delimited, referring only to the data on this particular series. Usually the range of validity is determined according to the deviation from a Hammett plot, or more directly from an immediate comparison of two reaction series (26).

VI. LINEAR RELATIONSHIPS FOR REACTION ENTHALPY

Of the quantities considered, only ΔH°_{298} are at disposal in a sufficient number to attempt a similar correlation as they are known for ΔG° (i.e., for log K). With respect to the tabulated data (16, 17, 27), we have selected two reaction series. The first one corresponds to the range of validity of the Hammett equation: *meta*- and *para*-substituted benzene derivatives, with a reaction taking place in the side chain. It consists of only four reactions

(Reactions K–N) and is clearly insufficient; at best, it is suitable only for the first orientation. The second series is of the Hammett–Brown type, since a substitution reaction proceeds immediately on the benzene ring. It is represented generally by reaction P, or more generally by Reaction Q. All the reactions are written in the isodesmic form, but the fictive reagent used is irrelevant for the correlation since the pertinent values cancel within the series. Some of these reactions—particularly in their nonisodesmic form with a more realistic reagent—could have a chemical meaning and could be even realized under certain conditions, the other probably not. Even this is irrelevant for calculating $\Delta H°$ or $\Delta G°$ and for the correlations.

$$XC_6H_4CH_3 + C_6H_5COOH = XC_6H_4COOH + C_6H_5CH_3 \qquad (K)$$

$$XC_6H_4C_2H_5 + C_6H_5COOH = XC_6H_4COOH + C_6H_5C_2H_5 \qquad (L)$$

$$XC_6H_4CH_3 + C_6H_5C_2H_5 = XC_6H_4C_2H_5 + C_6H_5CH_3 \qquad (M)$$

$$XC_6H_4NO_2 + C_6H_5NH_2 = XC_6H_4NH_2 + C_6H_5NO_2 \qquad (N)$$

$$XC_6H_5 + C_6H_5Y = XC_6H_4Y + C_6H_6 \qquad (P)$$

$$XC_6H_4Y + C_6H_5Z = XC_6H_4Z + C_6H_5Y \qquad (Q)$$

We have calculated $\Delta H°_{298}$ of these reactions from the tabulated enthalpies of formation (16, 17, 27) of the products and reactants. The values obtained were processed by PCA similarly as applied to $\Delta G°$ and $\Delta G^{\#}$ in the case of the Hammett equation (23). In principle, the experimental $\Delta H°_{ij}$ for the ith reaction and the jth substituent must be reproduced by the empirical Equation 10. The parameters α_i, β_i, γ_i belong to the ith reactions, the parameters b_j, c_j characterizes the substituent and its position (meta or para), ε_{ij} is the error. All the parameters are determined simultaneously by the least-squares condition (Equation 11). Some computational details were mentioned (28). In our examples in Table 3, Equation 10 had to be restricted to one term only ($\gamma_i = 0$).

$$(\Delta H°_{298})_{ij} = \alpha_i + \beta_i b_j + \gamma_i c_j + \cdots + \varepsilon_{ij} \qquad (10)$$

$$\sum \varepsilon_{ij}^2 = \min \qquad (11)$$

Some statistics of the correlations are listed in Table 3. They deserve two comments. First, the input data are of unequal quality and the possible experimental error is mostly unknown (17). Therefore, it appears necessary to identify outliers and to exclude them from the correlation on a posteriori

TABLE 3
Principal Component Analysis of Reaction Enthalpies for Reactions of Benzene Derivatives

Type of Correlation	Hammett			Hammett–Brown				
Thermodynamic Quantity	$\Delta H°_{298}$			$\Delta H°_{298}$	$\Delta H°_{298}$	$\Delta H°_{298}$	$\Delta H°_{298}$	$\Delta G°_{298}$
Series[a]	1	2	3	4	5	6	7	8
Number of reactions	4	4	3	8	8	8	7	5
Number of substituents	9	8	7	15	13	10	11	9
Number of known values[b]	27	25	21	81	67	54	53	31
Filling of the data matrix (%)	75	78	100	68	64	68	69	69
Degrees of freedom[c]	<u>12</u>	<u>10</u>	<u>10</u>	52	40	30	30	<u>14</u>
SD, kcal mole^{-1}	1.2	0.9	0.4	1.6	1.0	0.9	0.7	1.4
SD, log units	0.9	0.7	0.3	1.2	0.7	0.6	0.5	1.1
Variance explained[d] (%)	93.3	95.7	99.0	64.1	86.5	90.1	97.1	82.1
$\psi^{d,e}$	0.38	0.30	0.14	0.74	0.47	0.42	0.22	0.64

[a] Series 1: Reactions K–N, substituents H, 3-CH$_3$, 4-CH$_3$, 3-C$_2$H$_5$, 4-C$_2$H$_5$, 3-COOH, 4-COOH, 4-NH$_2$, 4-NO$_2$. Series 2: the same without 4-NO$_2$. Series 3: without Reaction N and substituent 4-NH$_2$. Series 4: Reaction P with Y = CH$_3$, C$_2$H$_5$, COOH, NH$_2$, NO$_2$, OH, F, Cl; substituent X = H, 3-CH$_3$, 4-CH$_3$, 3-C$_2$H$_5$, 4-C$_2$H$_5$, 3-COOH, 4-COOH, 3-NH$_2$, 4-NH$_2$, 4-NO$_2$, 3-OH, 4-OH, 3-Cl, 4-Cl. Series 5: as the preceding one without the substituents 3-CH$_3$ and 3-Cl and without five outliers (the combinations CH$_3$–3-NH$_2$, NH$_2$–4-CH$_3$, NH$_2$–4-NH$_2$, OH–4-OH, OH–4-Cl. Series 6: as the preceding one without the substituents 3-OH, 4-OH, and 4-NO$_2$. Series 7: Reaction Q with substituent Y = COOH; Z = H, CH$_3$, C$_2$H$_5$, NH$_2$, NO$_2$, F, Cl; substituent X = H, 3-CH$_3$, 4-CH$_3$, 3-C$_2$H$_5$, 4-C$_2$H$_5$, 3-COOH, 4-COOH, 3-NH$_2$, 4-NH$_2$, 4-F, 4-Cl. Series 8: Reaction P with substituent Y = CH$_3$, NH$_2$, OH, F, Cl; substituents X = H, 4-CH$_3$, 3-NH$_2$, 4-NH$_2$, 3-OH, 4-OH, 3-Cl, 4-Cl, 4-F.

[b] Data taken from references 16, 17, and 27.

[c] The underlined figures mean that there are less degrees of freedom than the number of parameters; hence the available data are insufficient for any conclusion.

[d] These statistics are related to $\Delta H°_{298}$ or $\Delta G°_{298}$ of the isodesmic reaction, that is, to the substituent effect as the primary variable.

[e] According to an empirical proposal $\psi < 0.2$ means a fair correlation, and $\psi < 0.5$ is the utmost acceptable limit (29).

grounds. This "polishing" of data is more extensive and more important than usual in PCA. Second, the input data, $\Delta H°_{298}$, are not primary experimental results but are constructed from several enthalpies of formation $\Delta H°_f$; each $\Delta H°_f$ can be involved in several $\Delta H°_{298}$. Hence the actual number of degrees of freedom is not known if we use Equation 10, which corresponds to the usual, unsymmetric form of the Hammett equation. Alternatively, it would be possible to correlate the original values of $\Delta H°_f$ directly using a symmetric equation (28), but the close similarity to the Hammett equation would be lost.

The correlation of side-chain reactions clearly suffers from the lack of independent data. Of the four reaction series, Reaction M depends on

Reaction K and L, and the fourth reaction, Reaction N, was finally excluded. Nevertheless, all the correlations appear relatively imprecise in spite of polishing of the input data. This is particularly evident when the standard deviations in kcal mole^{-1} are recalculated to decimal logarithmic scale (Table 3), comparable to log K used in the Hammett equation. It is also true that the substituent effects themselves are larger than usual in the Hammett correlations, but even the statistic ψ (29), expressing the ratio of the two standard deviations, is unsatisfactory. A deciding factor for the final conclusion is the numerical instability during the calculations and changes of the substituent constants when the data set is restricted. The original selected set (Table 3, column 1) was dominated by the enormous effect of the 4-NO$_2$ substituent. When this was eliminated (column 2), the reaction constant of Reaction N was excessive. Only when this reaction was omitted was a consistent choice obtained (column 3), but it contains only three, not independent, reactions (Reactions K–L), and there are too few degrees of freedom. The substituent effects are controlled by the carboxyl group, but it would be useless to discuss the substituent constants.

TABLE 4
Components (Substituent Constants) from PCA for Reactions of Aromatic Compounds

Substituent	Series[a]			
	4[b]	5[b]	6[b]	7[c]
H	0	0	0	0
3-CH$_3$	−0.18	—	—	−0.49
4-CH$_3$	−0.28	−0.33	−0.23	−0.62
3-C$_2$H$_5$	0.22	0.26	−0.01	−0.05
4-C$_2$H$_5$	−0.78	−0.70	−0.42	−1.03
3-COOH	−2.58	−2.58	−2.57	−2.61
4-COOH	−4.37	−4.27	−4.50	−4.79
3-NH$_2$	0.22	0.29	−0.06	−0.29
4-NH$_2$	−0.36	−0.18	−0.62	−0.51
4-NO$_2$	6.27	3.29	—	—
3-OH	−7.59	−6.95	—	—
4-OH	−7.44	−6.16	—	—
4-F	0.27	0.28	0.08	−0.46
3-Cl	−0.99	—	—	—
4-Cl	−1.00	−1.14	−1.20	−1.71

[a] The numbers of series correspond to those in Table 3.
[b] The substituents constants were normalized by anchoring the loading for the standard reaction (Reaction P with Y = COOH) to the value of 10 kcal mole^{-1}.
[c] The standard reaction was Reaction Q with Y = COOH and Z = H, the loading equal to −10 kcal mole^{-1}.

More definite results can be obtained from the second series, which concerns substitution on the benzene nucleus. The original set (Table 3, column 4) consisted of eight reactions of the Reaction P type, which in the nonisodesmic form have a clear chemical meaning (halogenation, nitration, alkylation, carboxylation, or decarboxylation). The mathematical treatment revealed several individual outliers, which were eliminated. Then a consistent picture was obtained (column 5), controlled by strong positive effect of the 4-NO_2 substituent and strong negative effects of the OH group in either position (Table 4). These effects agree with the lowered values of $\Delta H°$ observed in the presence of one acceptor and one donor group in the case of nitroanilines (21) and for several substituted anilines, this time even in the liquid-phase data (30). The generally positive deviations in the presence of two acceptors or two

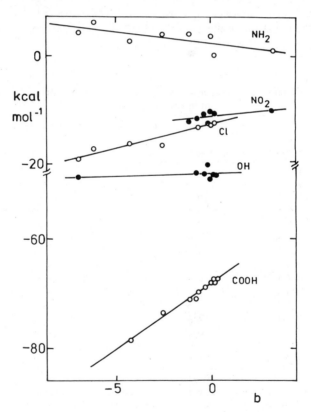

Figure 5 Principal component analysis of $\Delta H°_{298}$ for aromatic substitution reaction series (Reaction P). Part of the data is shown with Y = NH_2, NO_2, Cl, OH, COOH. Experimental reaction enthalpies are plotted against the components b of the substituent X.

donors (21) would also be consistent. What is not understandable, however, are the negative constants for the COOH substituent in either position, which should indicate a donor character. These features are also evident from a graphical representation (Fig. 5). As in a conventional Hammett plot, the experimental values of ΔH°_{298} have been plotted against the "substituent constants," specifically, the components of Table 4. For the purpose of this graphical representation all the reactions have been rewritten into simple nonisodesmic forms, namely, chlorination with Cl_2, nitration with HNO_3, oxidation with H_2O_2, amination with NH_3, and carboxylation with CO_2. Of course, this transformation does not affect the fit but allows us to separate the individual lines by shifting them along the y-axis. Figure 5 reveals first a large

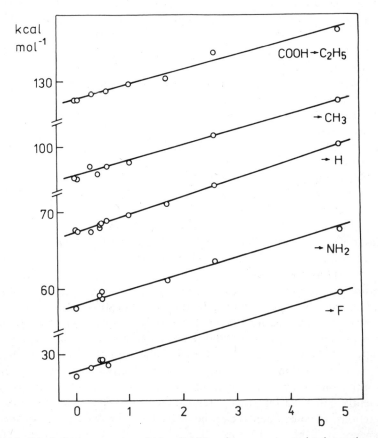

Figure 6 Principal component analysis of ΔH°_{298} for a more restricted reaction series (Reaction Q with Y = COOH). A part of the data is shown for Z = C_2H_5, CH_3, H, NH_2, F; see Fig. 5.

scatter, in particular with respect to the low values of the slopes (corresponding to Hammett ρ). Deciding for the slopes of many regression lines are the extreme points for substituents OH and 4-NO_2, but even in the best-behaved reaction (carboxylation) these extreme points are lacking. For these reasons we proceeded a step further in reducing the experimental data and eliminated these extreme substituents (column 6 in Table 3). The overall precision has only slightly improved, and the substituent parameters have not changed dramatically (Table 4). Instead of the eliminated substituents, the overall picture is now controlled by the strong effects of the substituent COOH in either position. Therefore, we made the last attempt by replacing the basic reaction (Reaction P; substitution on the benzene nucleus) by the reaction (Reaction Q) with Y = COOH, specifically, replacement of the carboxyl group by another group. Some reactions of this series have a clear chemical meaning, at least in one direction (decarboxylation, Hofmann degradation, Hunsdiecker reaction, oxidations). This series gave relatively good results (see Table 3, column 7 and Fig. 6). For the purpose of graphical representation the reactions have been again rewritten into nonisodesmic forms: thermal decarboxylation, oxidation of toluene or ethylbenzene with O_2, decarboxylation, halodecarboxylation with F_2 or Cl_2, and similar. Although the fit in Fig. 6 seems to be rather good, there is an imperfection in that all the lines have practically the same slope: we have to do almost with an additive behavior rather than with a Hammett-type equation. Furthermore, the substituent COOH in both positions is very important, almost controlling the entire correlation. All the pertinent points depend on the enthalpies of formation of isophthalic and terephthalic acids (27), which appear to be quite anomalous. According to the previously mentioned regularities (21, 30), a disubstituted benzene derivative with two equal substituents should always be destabilized, that is, show a higher enthalpy of formation as expected from the mono derivative. This is fulfilled for most of the substituents involved in our correlations: F, Cl, OH, NH_2, NO_2, and even CH_3 and C_2H_5, although sometimes the effects are within the limits of experimental error (alkyls, meta-OH, para-NH_2, para-NO_2). Of the other substituents the same effect is observed, for example, for CN and I. The only exception is for COOH since benzenedicarboxylic acids are stabilized by an interaction of substituents very strongly (5.8 and 10.9 kcal mole^{-1} for meta and para positions, respectively). Note, however, that in no case could the fit be effectively improved by taking more than one term from Equation 10. Most series would simply not be sufficiently extensive for such a refinement. Further insight into possible interactions of substituents can be traced in the values of substituent parameters in Table 4. This table gives several sets of values obtained from different sets of reactions as defined in Table 3, but the differences between them are not of importance for a general survey. According to the rule just discussed (21), one would expect values for acceptors and donors to be of opposite signs. This would be in accord with the largest positive and largest negative values for NO_2 and OH, respectively, but is violated strongly by COOH and in addition by several small, not quite conclusive values (4-F and possibly 3-C_2H_5). Hence the signs already exclude any parallelism between these values and any kind of constants σ, and even closer inspection has not revealed any correlation of this kind. There is not even any similarity to the constants φ, which have been derived only for enthalpies but from compounds with a direct interaction of the two substituents (31). The available φ values (32) are all positive and cannot be interpreted in a straightforward way.

Summarizing, we can say that there is a fundamental difference between the correlation of reaction enthalpies and convential LFER: the former are much less precise and are not simply controlled by substituent polar effects. One possible conclusion could be that simpler relationships exist between Gibbs energies than between enthalpies. We tried to test this hypothesis by treating similarly the Gibbs energies of the same reactions that were used in enthalpy correlations. However, the numerical values of ΔG°_{298} are more scarce and the data matrix had to be rather drastically reduced (Table 3, column 8). A PCA treatment gave still worse results than for ΔH°_{298}, with fewer degrees of freedom. Although the result is not convincing as far as the value of the standard deviation is concerned, it is, in our opinion, sufficient for a conclusion that correlations of ΔG°_{298} are no better than those of ΔH°_{298}. Note also that a common method of estimating unknown values of ΔH°_f is an additive scheme (16, 17). This scheme would predict $\Delta H^\circ_{298} = 0$ for any isodesmic reaction and thus represents a lower degree of approximation. In this section we are searching just for the deviations from the additivity principle.

These relatively high requirements are combined with experimental errors that may be relatively greater for thermochemical data than for ionization constants or rate constants. These errors were discussed in Section IV. Although errors up to 2 kcal mole^{-1} are admitted (17), thus larger than the standard deviations in Table 3, we are of the opinion that the scatter is not due exclusively to experimental error. One reason is that elimination of the deviating points has only restricted effect, but the main argument is based on the values of substituent parameters (Table 4). We suggest as the most probable explanation that the equilibria in Reactions K–Q are not controlled prevailingly by polar effects, and for this reason the substituent effects are not simply transferable from one reaction to the other. The experimental background of the Hammett equation and its extensions differ not so much in using $\Delta G°$ instead of $\Delta H°$, but the character of the reaction is essential. These are practically restricted to ionization equilibria and rate processes. In the former case there is a charge in either a product or reactant; in the latter the transition state is almost always strongly polar. It follows that the Hammett substituent constants should describe the behavior of substituents under the influence of polar effect, that is, the ability to delocalize a charge. In some special constants (σ^+, σ^-) this ability is differentiated according to whether a positive or a negative charge is to be delocalized (1).

Correlations of equilibria between noncharged particles are rather few. For instance, in a classical review (33) there were seven examples, two involving strongly polar molecules, while the most extensive collection of constants (32) included only prototropic equilibria and hydrogen-bonded complexes. From a purely mathematical point of view, if a correlation reaction holds for the rate constants of forward and reverse reactions, it must also hold for the equilibrium constant (1): the question is with how much accuracy. It is sometimes assumed that for a reverse reaction the sign of ρ should by reversed (34). Then the absolute value of ρ should be greater for an equilibrium than for kinetics (34) and the relative accuracy higher. However, some reactions can approach the symmetric reaction for which the rates are equal from either side and ρ_{eq} equals zero (1). According to the data in Table 4 (footnotes), the ρ constants of our reactions would be relatively large: of the order of 10, thus relatively large even among the reaction proceeding immediately on the benzene nucleus (1, 15).

Some indirect evidence of the substituent effects in a neutral molecule can be obtained from ^{13}C chemical shifts. In *meta*- and *para*-substituted toluenes **1** (35) or ethylbenzenes (structure **2** below) (36), the substituent chemical shifts are rather small and irregular but are slightly related to Hammett σ constants. On the other hand, in substituted methyl benzoates (**3**) the shifts correlate excellently with common constants σ_m and σ_p on the same regression line (37). This is probably due to the presence of the electron-attracting carbonyl group, which makes the situation more similar to the standard Hammett compounds:

1 X—C₆H₄—CH₃
2 X—C₆H₄—CH₂CH₃
3 X—C₆H₄—C(=O)OCH₃

We conclude that equilibria between noncharged molecules would certainly deserve more extensive investigation, but from the information available one can guess that they are not just a suitable field for correlation analysis. In any case, correlations of reaction enthalpies are no better than those of Gibbs energies. Of the other quantities, one could still consider the quantum-chemically calculated energies, which could be used instead of ΔE_p in an attempted correlation. The most extensive data are on the STO-3G level (38) and their accuracy is not sufficient for this purpose; the values obtained for isodesmic reactions are too small.

VII. IMPORTANCE OF POLAR EFFECTS

In the foregoing discussion we encountered several times the polar effects as the deciding factor. This means that interactions between a sufficiently polar substituent and a polar reaction center can control the observable thermodynamic quantities and cause their similarity in different situation, while interactions of weakly polar substituents escape among experimental errors and other little known effects; see particularly Section V and compare Figs. 2 and 3. Therefore, a question may be posed as to whether even the success of well-tried relationships such as the Hammett equation is conditioned only by such "strong" substituents (halogens, CN, CF$_3$, but mainly NO$_2$), whereas for the "weak" substituents the equation could actually be invalid. The validity has been tested mostly for the correlation coefficient and/or standard deviation in one series, and these values are little affected by substituents with small σ constants. We tried to carry out an independent test for a single substituent by rewritting the Hammett equation into the form

$$\log k_p - \log k^\circ = \left(\frac{\sigma_p}{\sigma_m}\right)(\log k_m - \log k^\circ) \tag{12}$$

The symbol k stays for either an equilibrium constant or a rate constant. Equation 12 allows us to test the validity of the Hammett equation for each substituent separately by plotting ($\log k_p - \log k^\circ$) against ($\log k_m - \log k^\circ$) for the same reaction. A similar test, called *extended selectivity treatment* (15), was based on plotting ($\log k - \log k^\circ$) against the reaction constant ρ. This variant is apparently more efficient since it allows us to test separately each substituent in either position. It is, however, less convincing since ρ is not an experimental quantity and the two variables are not independent.

We preferred a direct test according to Equation 12. The results for the substituents NO$_2$ and CH$_3$ are shown in Figs. 7 and 8, respectively. In either case the experimental data consisted about equally of rate constants and dissociation constants and there was no need to distinguish them in the graphs. The range of validity was restricted to reactions not requiring either σ^+ or σ^- constants. The excellent correlation for the substituent NO$_2$ is not surprising since it was already obtained even for several acceptor substituents

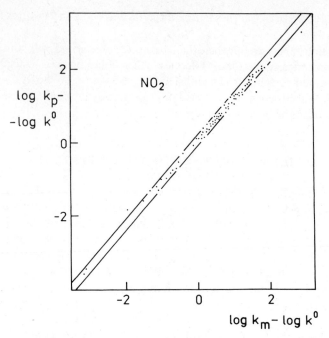

Figure 7 Testing the Hammett equation in the form of Equation 12 only for the substituent NO_2. The straight lines have the theoretical slope σ_p/σ_m, and their distance is 0.2 log units.

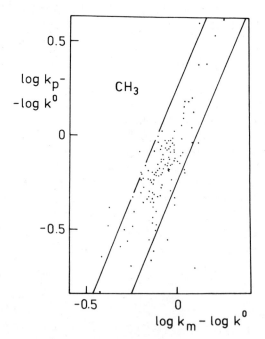

Figure 8 The same plot as in Fig. 7 for the substituent CH_3.

with a common straight line (39) and used as a proof (39,40) that all these substituents are negligibly conjugated with the benzene nucleus. The correlation for the substituent CH_3 is at first sight much worse but the scatter is only relatively greater—that is, compared to the much weaker substituent effect, not absolutely. This was shown by constructing straight lines, not by regression, but with the expected slopes (σ_p/σ_m), and drawing two parallel lines in the same distance in both Figs. 7 and 8. The result is that the deviations are approximately equal in the case of the two substituents, and that they exceed distinctly the assumed experimental error. The methyl group (and other weakly polar substituents) do not violate the Hammett equation and can be included in its range of validity but would not be sufficient by themselves. A correlation by this equation must, in any individual case, be based on some strongly polar substituents. This is well known and is recommended in textbooks (1, 3–6). Even so, the result from Fig. 8 seems to be somewhat more favorable than those from Section V or Fig. 3. A possible explanation would be that some unspecified effects on Gibbs energies can partly compensate in Fig. 8 when they are operative in both the *meta* and *para* positions. Hence Fig. 8 need not be the most sensitive test, and weakly polar substituents may impair the validity of LFER to a greater degree than this test predicts.

VIII. CORRELATIONS BETWEEN SOLVATION ENTHALPIES AND ENTROPIES

Of all the energy terms, the solvation enthalpy and solvation entropy, the last two terms of Equation 4, are probably the most difficult to calculate or to correlate, either with another quantity or among themselves. The most developed approach, the reaction field theory (41), is in principle not empirical, although it uses some empirical estimates, and will not be criticised here in detail. Note only that it has been applied mostly to conformational equilibria in nonpolar or little polar solvents and that it fails (22) for isodesmic reactions, as do the reactions in Table 2.

The second most popular approach is that of Hepler (9, 10), advanced for ionization equilibria measured mainly in water. As mentioned in the introduction, the total Gibbs energy of an isodesmic reaction is formally partitioned into internal and external enthalpies and entropies (Equation 8). The internal terms represent pure substituent effects in the gas phase and external terms, the change of substituent effects due to the solvent. The following three approximations have now been suggested (9):

1. The internal entropy term can be neglected:

$$\delta \Delta S_{\text{int}} = 0 \qquad (13)$$

2. The external terms are connected by an isokinetic relationship that is primary and more precise than a possible relationships between gross $\delta\Delta H°$ and $\delta\Delta S°$:

$$\delta\Delta H_{ext} = \beta_{ext}\delta\Delta S_{ext} \qquad (14)$$

3. The constant β_{ext} in the preceding equation, the "external" isokinetic temperature, is in many series near to the experimental temperature:

$$\beta_{ext} \simeq T_{exp} \qquad (15)$$

If Equations 13–15 are approximately valid, the observed $\delta\Delta G°$ in solution equals approximately $\delta\Delta H_{int}$. This should explain the success of the Hammett equation and similar LFER: the experimental Gibbs energy in solution should

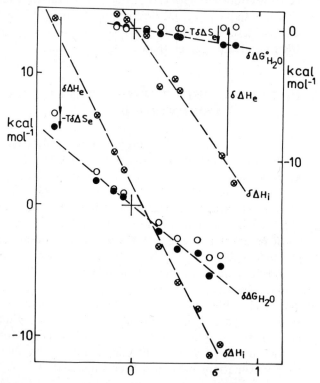

Figure 9 Hammett plot of the thermodynamic quantities involved in Equation 8: $\otimes\ \delta\Delta H_{int}$, $\bigcirc\ \delta\Delta H°_{H2O}$, $\bullet\ \delta\Delta G°_{H2O}$; at the top, substituted benzoic acids, at the bottom, substituted pyridines. Data from reference 13, 42, and 43.

be very near to the theoretically much simpler quantity, the enthalpy in the gas phase.

With the data now available, these above assumptions can be successively tested. Equation 13 seems to be a very reasonable approximation for both ionic ionization equilibria (13,42,43) and equilibria between noncharged particles, investigated in Section V (of course, with a symmetry correction). Equation 14 is very doubtful (14, 42). The reaction series thoroughly investigated are relatively few (42–44), and the external entropy is the least important term of the three remaining in Equation 8. In Fig. 9 various thermodynamic quantitites are plotted against the Hammett σ, but the fit and kind of σ constants are not important in the present context. For instance, the regression line passing the origin—at the bottom—reveals the polarizability effect of all substituents (45, 46). Important are the relative values of individual quantities: while $T\delta\Delta S_{ext}$ is small and irregular (sometimes almost hidden in experimental errors), $\delta\Delta H_{int}$ and $\delta\Delta H_{ext}$ are opposite in sign and comparable in absolute value. If one wanted to establish an isokinetic relationship of the two external parameters, it would be both very imprecise and nearly isoentropic, with a very high value of β_{ext}. Therefore, it is also clear that the third condition (Equation 15) is very far from reality.

Just from general considerations, one cannot understand how the isokinetic temperature can be found always at the experimental temperature, since no theory can anticipate at what temperature a given experiment will be carried out. Furthermore, it has been shown on a statistical analysis of a large number of examples (14) that the isokinetic temperature never occurs within the interval experimentally investigated. It is, rather, to be viewed as a product of mathematical extrapolation without any immediate physical meaning. [Some very rare exceptions were claimed (47) in kinetics of complex reactions, not guaranteeing that the rate constants are simple.] If the premises do not hold, it is no wonder that the final conclusion does not hold either. In the two examples in Fig. 9 $\delta\Delta G°$ in water is not approximately equal to $\delta\Delta H_{int}$ but several times smaller, although the trend of the values is maintained. A simple refutation of the whole theory consists also in its application to an additional solvent. Since there is no reason why the assumptions in Equations 13–15 should apply only to water solution, the theory would predict that $\delta\Delta G°$ are equal in all solvents.

In a more sophisticated form (10) of this theory, Equation 14 is extended by an additional term:

$$\delta\Delta H_{ext} = \beta\delta\Delta S_{ext} + \gamma\delta\Delta H_{int} \qquad (16)$$

This equation can be tested by multiple linear regression. Since $\delta\Delta H_{ext}$ and $\delta\Delta S_{ext}$ are strongly dependent on each other through the process of their calculations (7), we carried out several regressions with the quantities $\delta\Delta G°_{H_2O}$, $\delta\Delta H°_{H_2O}$, and $\Delta G°_g$. In the series of substituted pyridines the result can be anticipated from Fig. 9. A fair correlation of $\delta\Delta G°_g (=\delta\Delta H_{int})$ and $\delta\Delta G°_{H_2O}$ is not improved by adding the third term; the multiple correlation coefficient equals the correlation coefficient of the simple regression. The second example concerned the data on aliphatic amines (42), thus exceeding the range of validity of the Hammett equation. There was no correlation whatsoever: the multiple correlation coefficient, partial correlation coefficients, and correlation coefficients of simple regressions were all below 0.7.

Instead of Equation 16, a simplified version was advanced (42):

$$\delta\Delta H_{ext} = h\delta\Delta H_{int} \qquad (17)$$

which, in combination with Equation 14, postulates simply that all thermodynamic quantitites are proportional to each other, in both the gas phase and solution. This is certainly an attractive approximation, but it does not say how precisely it is fulfilled in a concrete example. From Fig. 9 one can see that Equation 17 holds as a rough approximation, while Equation 16 is actually invalid, but the entropic term is too small to destroy the relationships between the others. This modest conclusion seems to be rather general for solvent effects. One can anticipate that relatively simple relationships in the gas phase are impaired by irregular solvent effects, enthalpic and entropic, but these effects are seldom strong enough to destroy the regularity completely. For instance, the gas acidities of certain rigid molecules (excluding steric interaction) are well described by three terms: induction, resonance, and polarizability (46). In solution the third term is canceled but a number of solvation terms come into existence (45) that can hardly be quantitatively estimated. Nevertheless, the main character of the correlations is maintained in solution, for instance, the values of many substituent constants. As mentioned earlier, the solvent effects have been investigated mostly on ionization reactions in water. The solvent effects of less polar solvents on the equilibria of noncharged species are possibly still more difficult to interpret. The main reason is that they are too small (Table 2).

IX. CONCLUSIONS

The following conclusions are not always original. Some of them have already been formulated several times, more or less explicitly, although it would be difficult to give exact citations. On the other hand, some of the conclusions are not well supported and are to be viewed merely as postulates or suggestions. We attempted to list them in order beginning with those that have been proved soundly:

1. When dealing with equilibria, the reaction should be formulated as isodesmic, and in the case of reaction rates relative values should be used. This is inherent in the Hammett equation (2). A very convincing proof is comparison of Figs. 1 and 2, many similar examples could be constructed from available data.

2. In the correlations of Gibbs energy, entropy, and equilibrium or rate constants the symmetry correction should always be applied; that is, the part of entropy due only to symmetric properties of the molecules should be subtracted. A good proof is Fig. 2, and also the correlations presented in Section V. Many examples are known from correlations with σ constants where some points deviate, unless the correction has been applied.

3. There exists a concept of reaction series. Its definition is somewhat fuzzy, but its validity cannot be denied. The usual proof is based on a plot versus σ constants where some points deviate; more exactly, one can directly compare reactivities in two series and delimit the range of validity of a given relationship (26). Sometimes the reaction series can be defined on only one series, comparing two different properties: Fig. 4 is an example.

4. Of the various thermodynamic quantities, ΔH_0° would be the theoretically preferred quantity. The potential energy ΔE_p is, from a theoretical point of view, a compromise, and in practice it is extracted from the thermodynamic quantities with a considerable degree of uncertainty (see Figs. 2 and 3).

5. Either ΔH_T° or ΔG_T° at finite temperatures may be used for correlations, with approximately the same precision, provided the temperature is not too elevated, generally up to 500 K. Below this temperature the entropy term $-T\Delta S^\circ$ can be neglected (for isodesmic reactions in the gas phase) and Equation 6 can be taken as valid (see Fig. 2).

6. In solution the simple isoentropic relationship (Equation 6) is usually impaired by solvent effects and transformed into a more general isokinetic relationship (Equation 7). But there is no general isokinetic relationship between the solvation enthalpy and the solvation entropy.

7. For a correlation between two reaction series a common dominating mechanism of substituent effects appears to be necessary. Most reliable are polar effects that are able to dominate all others and define a scale of substituent constants. From this point of view, for instance, the Hammett equation hlolds only for strongly polar substituents; nonpolar substituents such as CH_3 are unable to disturb the overall validity by their weak effects, but taken by themselves they do not obey the equation.

8. The well-defined polar effects require not only a polar substituent but also a polar reaction centre; most important are either ionized groups in

ionization equilibria or strongly polar transition states in kinetics. The common substituent constants thus at least express partly the ability of substituents to delocalize a charge.

REFERENCES

1. O. Exner, *Correlation Analysis of Chemical Data*, Plenum Press, New York, 1988.
2. L. P. Hammett, *Physical Organic Chemistry*, McGraw-Hill, New York, 1940.
3. L. P. Hammett, *Physical Organic Chemistry*, 2nd ed., McGraw-Hill, New York, 1970.
4. V. A. Palm, *Osnovy Kolichestvennoi Teorii Organicheskikh Reaktsii*, Izdatelstvo Khimiya, Leningrad, 1967 and 1977.
5. R. W. Taft, in *Steric Effects in Organic Chemistry*, M. S. Newman, Ed., Wiley, New York, 1956, p. 556.
6. J. E. Leffler and E. Grunwald, *Rates and Equilibria of Organic Reactions*, Wiley, New York, 1963.
7. O. Exner, *Progr. Phys. Org. Chem.*, 10, 411 (1973).
8. W. J. Hehre, R. Ditchfield, L. Radom, and J. A. Pople, *J. Am. Chem. Soc.*, 92, 4796 (1970).
9. L. G. Hepler, *J. Am. Chem. Soc.*, 85, 3089 (1963).
10. L. G. Hepler, *Can. J. Chem.*, 49, 2803 (1971).
11. D. J. G. Ives and P. D. Marsden, *J. Chem. Soc.*, 649 (1965).
12. E. Nieboer and W. A. E. McBryde, *Can. J. Chem.*, 48, 2565 (1970).
13. T. B. McMahon and P. Kebarle, *J. Am. Chem. Soc.*, 99, 2222 (1977).
14. O. Exner, *Collect. Czech. Chem. Commun.*, 40, 2762 (1975).
15. L. M. Stock and H. C. Brown, *Adv. Phys. Org. Chem.*, 1, 35 (1963).
16. D. R. Stull, E. F. Westrum, and G. C. Sinke, *The Chemical Thermodynamics of Organic Compounds*, Wiley, New York, 1969.
17. M. Bureš, R. Holub, J. Leitner, and P. Voňka, *Sb. Vys. Sk. Chem.-Technol. Praze, Fys. Chem.*, No. 8, 5 (1987).
18. O. Exner and Z. Slanina, *Thermochim. Acta*, in press.
19. Z. Slanina, *Collect. Czech. Chem. Commun.*, 55 (1990), in press.
20. Z. Friedl, *Can. J. Chem.* 62, 2337 (1984).
21. L. Nuñez, L. Barral, S. Gavilanes Largo, and G. Pilcher, *J. Chem. Thermodyn.*, 18, 575 (1986).
22. Z. Friedl, J. Biroš, and O. Exner, *J. Chem. Soc. Perkin II*, in press.
23. S. Wold and M. Sjöström, *Chem. Scr.*, 2, 49 (1972).
24. O. Exner, *Collect. Czech. Chem. Commun.*, 39, 515 (1974).
25. O. Exner, *Collect. Czech. Chem. Commun.*, 38, 799 (1973).
26. O. Exner, *Collect. Czech. Chem. Commun.*, 40, 2781 (1975).
27. J. D. Cox and G. Pilcher, *Thermochemistry of Organic and Organometallic Compounds*, Academic Press, London, 1970.
28. O. Exner, *Collect. Czech. Chem. Commun.*, 41, 1516 (1976).
29. O. Exner, *Collect. Czech. Chem. Commun.*, 31, 3222 (1966).
30. N. D. Lebedeva, N. M. Gutner, and V. L. Ryadnenko, *Zh. Fiz. Khim.*, 45, 999 (1971).

31. B. I. Istomin and V. A. Palm, *Reakts. Sposobnost. Org. Soedin.* (*Tartu*), 8, 845 (1971).
32. V. A. Palm, Ed., *Tables of Rate and Equilibrium Constants of Heterolytic Organic Reactions*, Vol. 5 (II), VINITI, Moscow, 1979.
33. H. H. Jaffé, *Chem. Rev.*, 53, 191 (1953).
34. A. E. Shilov and H. S. Venkataraman, *J. Chem. Soc.*, 4993 (1960).
35. N. Inamoto, S. Masuda, K. Tokumaru, K. Tori, M. Yoshida, and Y. Yoshimura, *Tetrahedron Lett.*, 3707 (1976).
36. W. Adcock, W. Kitching, V. Alberts, G. Wickham, P. Barron, and D. Doddrell, *Org. Magn. Reson.*, 10, 47 (1977).
37. M. Buděšínský and O. Exner, *Magn. Reson. Chem.*, 27, 585 (1989).
38. A. Pross and L. Radom, *Progr. Phys. Org. Chem.*, 13, 1 (1981).
39. O. Exner, *Collect. Czech. Chem. Commun.*, 31, 65 (1966).
40. O. Exner, *Org. Reactiv.* (*Tartu*), 21, 3 (1984).
41. R. J. Abraham and E. Bretschneider, in *Internal Rotation of Molecules*, W. J. Orville Thomas, Ed., Wiley, London, 1974, p. 481.
42. D. H. Aue, H. M. Webb, and M. T. Bowers, *J. Am. Chem. soc.*, 98, 318 (1976).
43. E. M. Arnett, B. Chawla, L. Bell, M. Taagepera, W. J. Hehre, and R. W. Taft, *J. Am. Chem. Soc.*, 99, 5729 (1977).
44. T. Matsui, H. C. Ko, and L. G. Hepler, *Can. J. Chem.*, 52, 2906 (1974).
45. R. W. Taft, *Progr. Phys. Org. Chem.*, 14, 247 (1983).
46. R. W. Taft and R. D. Topsom, *Progr. Phys. Org. Chem.*, 16, 1 (1987).
47. A. B. Dekel'baum, B. V. Passet, and G. F. Fedorov, *Reakts. Sposobnost Org. Soedin.* (*Tartu*), 10, 637 (1973).

The Quantitative Description of Amino Acid, Peptide, and Protein Properties and Bioactivities

BY MARVIN CHARTON

Chemistry Department
School of Liberal Arts and Sciences
Pratt Institute
Brooklyn, New York

CONTENTS

- I. Introduction 164
- II. The Intermolecular Force Equation 165
  - A. Intermolecular Forces 165
    1. Van der Waals Interactions 166
    2. Ion–Dipole and Ion–Induced Dipole Interactions 168
    3. Hydrogen Bonding 169
    4. Charge Transfer 169
  - B. Steric Effects 170
    1. Introduction 170
    2. Branching Equations 172
  - C. The IMF Equation 174
  - D. Problems in the Application of the IMF Model 175
  - E. Method 176
- III. Amino Acids 177
  - A. Hydrophobicities 177
    1. Introduction 177
    2. Volume and Surface Area Parameters 188
  - B. Solution Properties 192
  - C. Chromatographic Properties 194
    1. Thin-layer Chromatography 194
    2. Other Chromatographic Methods 204
  - D. Other Amino Acid Properties 204
  - E. Amino Acid Bioactivities 209
    1. Bioactivity Model 209
    2. Hydrolytic Enzymes 211
    3. Isoleucyl-tRNA Synthetase 219
    4. Amino Acid Transport 220
    5. Other Amino Acid Bioactivities 221
  - F. Extension of the Model to Variant Amino Acids 221

IV. Peptides 228
   A. Introduction 228
   B. Peptide Properties 229
   C. Peptide Bioactivity, an Overview 237
      1. X^n Substitution 237
      2. X^C Substitution 248
      3. Z^oZ Substitution 248
      4. X^P Substitution 249
      5. Multiply Substituted Peptides 250
   D. Peptide Antibiotic Bioactivities 251
      1. Introduction 251
      2. Inhibition of Ristocetin Binding 259
V. Proteins 265
   A. Protein Conformation 265
   B. Protein Bioactivities 267
VI. Conclusions 269
   References 270
Appendix 1. Amino Acid Side-Chain Parameter Tables 275
   List of Nonstandard Abbreviations 275
   Table A. Intermolecular Force Parameters 275
   Table B. Steric Parameters 278
Appendix 2. The Zeta and Omega Methods 281
   A. The Zeta Method 281
   B. The Omega Method 283

I. INTRODUCTION

Over the last fifty years, a large body of data has accumulated on the chemical properties, conformation, and biological activities of amino acids, peptides, and proteins. It seemed of interest to develop a quantitative correlation analysis model for the effect of side-chain structure on these quantities. Such a model would be expected to be useful to biological, medicinal, agricalatural, environmental, and bioorganic chemists.

The properties of interest all involve the transfer of a substrate from an initial to a final phase, a process which involves a change in intermolecular forces (Fig. 1). What is required then for modeling those quantities is an equation that parameterizes these intermolecular forces. Thus,

$$\Delta G_{inf} = (imf)_f - (imf)_i \qquad (1)$$

where imf indicates intermolecular forces, and i and f represent the initial and final states. The ΔG in Equation 1 may be written as a sum of contributions from the various types of intermolecular forces. We shall show later that a steric term is also required.

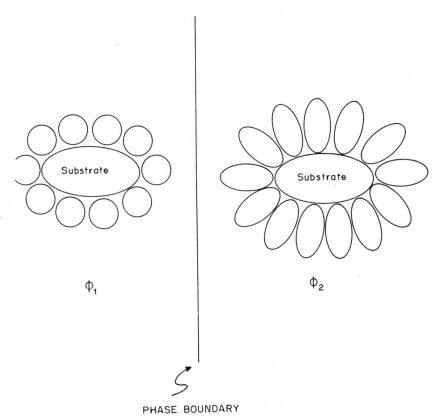

Figure 1A Stylized illustration of a solvated substrate in each of two phases separated by a phase boundary.

$$\Delta G = \Delta G_{dd} + \Delta G_{di} + \Delta G_{ii} + \Delta G_{Id} + \Delta G_{Ii} + \Delta G_{hb} + \Delta G_{ct} + \Delta G_S \quad (2)$$

$$= \sum_{j=1}^{m} \Delta G_j \quad (3)$$

II. THE INTERMOLECULAR FORCE EQUATION

A. Intermolecular Forces

The intermolecular forces of interest are reported in Table 1. They result from the interaction between a substrate and the medium which surrounds it. We now consider their parameterization.

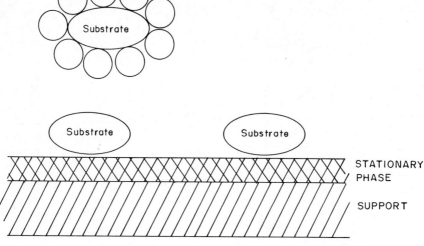

Figure 1B Stylized illustration of a solved substrate in a mobile phase in equilibrium with substrate interacting with a stationary phase in column or thin-layer chromatography.

TABLE 1
Intermolecular Forces

Interaction (Abbreviation)	Parameters
Dipole–dipole (dd)	$\sigma_l, \sigma_d, \sigma_e$
Dipole–induced dipole (di)	$\sigma_l, \sigma_d, \sigma_e, \alpha$
Induced dipole–induced dipole (ii)	α
Hydrogen bonding (hb)	$\sigma_l, \sigma_d, \sigma_e, n_H, n_n$
Charge transfer (ct)	$\sigma_l, \sigma_d, \sigma_e, n_A, n_D$
Ion–dipole (Id)	$\sigma_l, \sigma_d, \sigma_e, i$
Ion–induced dipole (Ii)	i, α

Parameters are described and their sources given in Section 2A. Tables of parameter values for amino acid side chains are reported in Appendix 1.

1. Van der Waals Interactions

Classical physical chemistry states that energies of the dipole–dipole (dd), dipole–induced dipole (di), and induced dipole–induced dipole interaction (ii) are given (1) by the relationships:

$$E_{dd} = -2\mu_a^2 * \mu_b^2 / 3k_B T r^6 D^2 \qquad (4)$$

$$E_{di} = \alpha_a \mu_b^2 / r^6 D^2 - \alpha_b \mu_a^2 / r^6 D^2 \qquad (5)$$

$$E_{ii} = -3(h\nu_a h\nu_b)(\alpha_a \alpha_b)/2(h\nu_a + h\nu_b)r^6 \qquad (6)$$

where k_B is the Boltzmann constant, T is the absolute temperature, r is the distance between the interacting species, D is the bulk dielectric constant, ν_a and ν_b are those frequencies of the oscillating electrical dipoles responsible for the induced dipoles, α_a and α_b are the polarizabilities of the interacting species, and mu_a and mu_b are their dipole moments. In the data sets of interest in this work, one of these species is the amino acid side chain X. As within the data set only X is allowed to vary we may rewrite Equations 5 and 6 in the form

$$E_{dd} = \mu_X^2 k_{dd} \qquad (7)$$

$$E_{di} = -\alpha_X k'_{di} - \mu_K^2 k''_{di} \qquad (8)$$

A similar relationship may be written for Equation 6 as follows:

$$E_{ii} - -(\alpha_X k_{ii}) \qquad (9)$$

In order to describe these interactions for amino acid side chains, we need to parameterize the dipole moment and the polarizability. Dipole moments have been shown by a number of authors to be a function of electrical effect substituent constants (2–4). The polarizability of a substituent has been shown to be a function of its group molar refractivity, MR_X. We should therefore be able to parameterize van der Waals forces by a two-parameter equation such as

$$Q_X = \rho\sigma_X + A*MR_X + h \qquad (10)$$

This relationship was tested for a very simple case by correlating the ratio (a/b) where a and b are the constants of the van der Waals equation of state with Equation 10. The results obtained for systems of the type $H(CH_2)_n x$ were quite good (5), validating the parameterization of Equation 12 where the electrical effect parameter is σ_l. Note that σ_l is identical to σ_I (6). We have shown that there are several other parameters which are very highly collinear in MR. They include the group molar volume, the group van der Waals volume, and the group parachor. Obviously, any of these parameters might equally well be used to represent polarizability. We have chosen MR because of the ease of estimation of values for new side chains of interest.

In keeping with other correlation analysis parameters, we have defined a polarizability parameter, α, from group molar refractivities which has a value of zero when X = H (7). Thus,

$$\alpha \equiv (MR_X - MR_H)/100 \quad (11)$$

where MR_X and MR_H are the group molar refractivities of X and H, respectively. Division by 100 is intended to make the coefficient of this variable roughly comparable to those of the other variables in the correlation equation. Then we may write

$$\Delta G_{dd} = L_{dd}\sigma_{lX} + h_{dd} \quad (12)$$

$$\Delta G_{di} = L_{di}\sigma_{lX} + A_{di}\alpha_X + h_{di} \quad (13)$$

$$\Delta G_{ii} = A_{ii}\alpha_X + h_{ii} \quad (14)$$

2. Ion–Dipole and Ion–Induced Dipole Interactions

The Id and Ii interactions are given by the expressions

$$E_{Id} = q^2\mu^2/3k_B T r^4 D^2 \quad (15)$$

$$E_{Ii} = \alpha q^2/2r^4 D^2 \quad (16)$$

where q is the charge on the ion.

The effect of an ionic side chain can be represented by the parameter i which takes the value 1 for a charged side chain and 0 for a neutral side chain. We have found that the results are not improved by using 1 for positive and -1 for negative side chains. For Id and Ii interactions where the ions are in the medium, the side-chain interactions are a function of σ_{lX} and α. It remains only to determine whether or not the side chain of an amino acid is charged. We define ionizable basic side chains as those with a pKa > 7 and ionizable acidic side chains as those with pKa < 7. It follows then that

$$\Delta G_{Id} = I_{Id}i + h_{Id} \quad (17)$$

$$\Delta G_{Ii} = I_{Ii}i + h_{Ii} \quad (18)$$

In the parameterization of the intermolecular forces, we must take into account the fact that the entire substrate is involved in intermolecular interactions with the medium.

Consider a substrate with the structure XGY in which X is a variable substituent, Y is some consant functional group, and G is a skeletal group to which X and Y are bonded. When physical properties or chemical reactivities are studied, the phenomenon measured occurs at Y which is then called the active site. In these cases, only the effect of X on Y need be considered in the parameterization. When chemical properties or bioactivities are studied, the interaction between X and the medium must of course be parameterized. In the case of hydrogen bonding media, we must also consider the effect of X on Y if Y is capable of hydrogen bonding. This is also true of G if it is capable of hydrogen bonding.

The same situation exists in the case of charge transfer. We must take into account the effect of X and G and on Y if they are capable of charge transfer.

3. Hydrogen Bonding

There are three interactions to model. They are X-medium, the effect of X on the Y-medium interaction, and the effect of X on the G-medium interaction. In modeling the X-medium interactions, we assume that in 55.5 M water hydrogen bonding is maximal. The extent of hydrogen bonding to groups X with NH and/or OH bonds is assumed to be proportional to n_H, the number of such bonds in the group. Similarly, the extent of hydrogen to groups X with lone pairs on N or O atoms is assumed to be proportional to the number of these on the X group. Then the X-medium interaction, X_H, is given by

$$\Delta G_{hbX} = H_1 n_H + H_2 n_n + h_{hbX} \tag{19}$$

For amino acids, peptides, and proteins, the effect of X on hydrogen bonding in the G-medium, and Y-medium interaction is given by

$$\Delta G_{hbG} = L_{hbG} \sigma_{lX} + h_{hbG} \tag{20}$$

$$\Delta G_{hbY} = L_{hbY} \sigma_{lX} + h_{hbY} \tag{21}$$

as only the localized electrical effect need be considered. Then combining Equations 19, 20, and 21 gives

$$\Delta G_{bh} = H_1 n_H + H_2 n_n + L_{hb} \sigma_{lX} + L_{hbY} \sigma_{lX} + h_{hb} \tag{22}$$

4. Charge Transfer

The most difficult interaction to parameterize for the amino acid sidechain is charge-transfer. If the side chain interacts with a surrounding medium, as is the case for amino acid and peptide properties, the magnitude of the interaction depends in part on the capacity of the medium to act as a charge-transfer donor or acceptor. Solvents such as water and alcohols are generally not charge-transfer acceptors and are very poor charge-transfer donors. If the side-chain is to interact with other side-chains as is the case with peptide and protein conformation, or with side-chains in a receptor site as is the case with bioactivity, then we must characterize the extent to which side-chains can donate or accept an electron. The capacity of a side-chain to act as an electron donor could be parameterized by the first ionization potential of a suitable model compound such as HX or MeX. Then we can define a charge-transfer donor parameter ε_D by the expression

$$\varepsilon_{DX} = IP_X - IP_H \tag{23}$$

where IP represents the first ionization potential of the model compound. A similar definition of a charge-transfer acceptor parameter could be made using the electron affinities of the model compounds. Thus

$$\varepsilon_{AX} = EA_X - EA_H \tag{24}$$

As values of IP and EA for model compounds are not available for all the side chains of interest, we have had to make use of a simpler approach. We have defined the parameter n_D as 1 for those side-chains for which HX or MeX acts as a charge-transfer donor and 0 for those side-chains for which they do not. Similarly, we have defined the parameter n_A as 1 for those side chains for which HX or MeX acts as a charge-transfer acceptor and 0 for those side-chains for which they do not. The effects of X on charge-transfer to G or Y are given in the general case by the electrical effect parameters σ_l, σ_d, and σ_e of the triparametric LDR equation (6, 8, 9). Again, in the case of amino acids, peptides, and proteins, the σ_1 constant alone suffices.

$$\Delta G_{ctX} = B_A n_A + B_D n_D + h_{ctX} \tag{25}$$

$$\Delta G_{ctG} = L_{ctG}\sigma_{1X} + h_{ctG} \tag{26}$$

Then
$$\Delta G_{ctY} = L_{ctY}\sigma_{1X} + h_{ctG} \tag{27}$$

$$\Delta G_{ct} = B_A n_A + B_D n_D + L_{ctG}\sigma_{1X} + L_{ctY}\sigma_{1X} + h_{ct} \tag{28}$$

B. Steric Effects

1. Introduction

Inspection of the water-solvated amino acid molecule (Fig. 2) shows that a large side-chain may interfere with the solvation of the amino and carboxyl

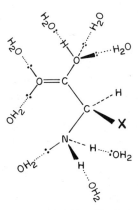

Figure 2 Possible hydration sites other than the side chain in an α-amino acid. Clearly the extent of hydration is dependent on steric effects exerted by the side chain X.

groups present in all α-amino acids. It follows then that there should be a steric effect on the intermolecular forces. To account for it, we must include steric parameterization of the side chain in the IMF equation. Thus,

$$\Delta G_S = S\psi + h \tag{29}$$

There are two possible ways to do this. We may opt for the use of a single steric parameter such as v which is based on van der Waals radii (10, 11). It is defined as

$$v = r_{VX} - r_{VH} = r_{VX} - 1.20 \tag{30}$$

where r_{VX} and r_{VH} are the van der Waals radii of X and H, respectively. Then we may write the steric effect in the form

$$\Delta G_S = Sv + h_S \tag{31}$$

Unfortunately, the steric effect of a nonsymmetric substituent varies with the steric demand for the phenomenon being studied. Steric effects in a data set containing many such groups will often not be well-modeled by a single parameter.

A major disadvantage of monoparametric treatments of the steric effect is that nonsymmetric substituents will exhibit a variable steric effect that depends on the steric demand of the system studied. When the steric requirements of the system studied resemble those of the reference system used to determine the steric parameters for the nonsymmetric groups, then monoparametric steric parameters will provide an effective model. Unfortunately, however, this is often not the case. There are two alternative approaches for dealing with this problem. One of these is the definition of a number of different sets of steric parameters for nonsymmetric groups. The set chosen for a particular problem is that in which the steric demand in the reference system most closely matches that in the data set to be modeled. It is often difficult to determine a priori which set of steric parameters to choose. This approach also results in many different sets of steric parameters which tend to overwhelm the inexperienced user.

2. Branching Equations

The other approach is the use of a multiparameteric model. We have reported on a useful model of this type based on a topological approach called the simple branching (SB) equation (12–14),

$$Q_X = S \sum_{i=1}^{m} a'_i n_i + a_0 = \sum_{i=1}^{m} a_i n_i + a_0 \qquad (32)$$

where n_i is the total number of branches on the ith atoms of the substituent X, a_i is its coefficient, and a_0 is the intercept. Consider, for example, the substituent **1** which is shown bonded to a skeletal group G. As hydrogen atoms are not considered as branches, **1** is shown as a hydrogen-suppressed structure.

$$\begin{array}{c} M^2{-}M^3 \qquad\qquad M^5 \\ \diagup \qquad\qquad\qquad \diagup \\ G{-}M^1{-}M^2{-}M^3{-}M^4{-}M^5{-}M^6 \\ \diagdown \\ M^3{-}M^4{-}M^5 \\ \diagdown \\ M^5 \end{array}$$

1

n_i is equal to the number of M atoms labeled $i+1$. Then for **1** the n_i values are: $n_1 = 2$; $n_2 = 3$; $n_3 = 2$; $n_4 = 4$, $n_5 = 1$.

A major assumption made in this very simple model is that the effect of branching is the same for all branches. This assumption is a crude approximation at best. Generally, the first branch on an atom has little in the way of a steric effect, while the effect of the third banch is large. Thus, for the series CH_3, CH_2Me, $CHMe_2$, and CMe_3, the v values are 0.52, 0.56, 0.76, and 1.24. Thus, the first, second, and third branches increase the size of the methyl group by factors of 1.08, 1.46, and 2.38 respectively. This is the result of the minimum steric interaction (MSI) principle, which states that a nonsymmetric substituent will prefer that conformation which minimizes steric repulsions. The effect of the first branch can be minimized by a suitable conformation of the substituent in which the branch is rotated away from the active site. Such a rotation becomes much more difficult to achieve on introduction of a second branch and impossible on introduction of a third.

When all of the atoms M are not identical, a second approximation often made is that the effect of all second- or third-period elements as either skeletal or branch atoms is about the same. This approximation is also crude. Nevertheless, the simple branching equation can often give a reasonable description of steric effects.

The expanded branching (XB) equation (15, 16) which accounts for the effect of the order of branching gives a much improved description of steric effects. It has the form

$$Q_X = \sum_{i=1}^{m} \sum_{j=1}^{3} a_{ij} n_{ij} + a_{00} \qquad (33)$$

where n_{ij} is the total number of jth branching atoms bonded to the ith atoms in the substituent and a_{ij} is its coefficient. We may consider the same example as before. This time, however, it will be numbered to indicate the order of branching.

$$\begin{array}{c}
\phantom{G-M^1-M^{22}-M^{31}-}M^{21}-M^{31} \phantom{-M^{41}-}M^{51} \\
\phantom{G-M^1-M^{22}-M^{31}-M^{41}-M^{52}-M^{61}}| \\
G-M^1-M^{22}-M^{31}-M^{41}-M^{52}-M^{61} \\
M^{32}-M^{41}-M^{51} \\
\phantom{G-M^1-M^{32}-M^{41}-}M^{52}
\end{array}$$

2

n_{ij} is equal to the number of atoms bearing the label $(i+1)j$. Then for 2; $n_{11} = n_{12} = 1; n_{21} = 2; n_{22} = 1; n_{31} = n_{41} = n_{42} = 2; n_{51} = 1$; all other $n_{ij} = 0$.

Though the XB equation is a far better model of steric effects than is the SB equation, it requires many more independent variables. It is therefore most useful only when the data set to be studied is very large, an extremely rare situation in general and one which does not seem to occur in amino acid, peptide, and protein studies.

When the SB equation is applied to doubly bonded groups, all of which are members of the steric class planar pi bonded groups, it is normally assumed on the basis of the MSI principle that the half-thickness of the group determines is steric effect. An evaluation of the branching parameters for the phenyl group gave $n_1 = n_2 = n_3 = 1$. We have assumed that the values for $C{=}O$, and $N{=}N$ are comparable.

It is often useful to have a parameter that can serve as a measure of group length. By group length we mean the length of the longest chain of atoms in the group from its point of attachment to the rest of the substrate to its terminal atom in its most extended conformation. A crude but useful length parameter is n_b which is defined as the number of bonds in the group skeleton. The group skeleton is the longest chain of atoms in the group. n_b is therefore given by

$$n_b = i_{max} - 1 \tag{34}$$

where i_{max} is the highest value of i for any atom in the group. Thus, the value of n_b for **1** or **2** is 5.

C. The IMF Equation

On substituting Equations 14–16, 19, 20, 24, 28, and 31 into Equation 2, we obtain the IMFU relationship

$$\begin{aligned}\Delta G_{imf} &= L_{dd}\sigma_{lX} + h_{dd} + L_{di}\sigma_{lX} + A_{di}\alpha_X + h_{di} + A_{ii}\alpha_X + h_{ii} + I_{ld}i_X \\ &+ h_{ld} + I_{li}i_X + h_{li} + H_1 n_{HX} + H_2 n_{nX} + h_{hbX} + L_{hbG}\sigma_{lX} \\ &+ L_{hbY}\sigma_{lX} + h_{hbG} + h_{hbX} + C_A n_{AX} + C_D n_{DX} + h_{ctX} \\ &+ L_{ctG}\sigma_{lX} + L_{ctY}\sigma_{lX} + h_{ctG} + h_{ctY} + s\upsilon_X + h_s \end{aligned} \tag{35}$$

$$\begin{aligned} &= L\sigma_{lX} + A\alpha_X + H_1 n_{HX} + H_2 n_{nX} + C_A n_{AX} + C_D n_{DX} \\ &+ I i_X + S\upsilon_X + B^0 \end{aligned} \tag{36}$$

where

$$L = L_{dd} + L_{di} + L_{hbG} + L_{hbY} + L_{ctG} + L_{ctY} \tag{37}$$

$$A = A_{dd} + A_{di} \tag{38}$$

$$I = I_{ld} + I_{li} \tag{39}$$

$$\begin{aligned} B^0 &= h_{dd} + h_{di} + h_{ii} + h_{ld} + h_{li} + h_{hbX} + h_{hbG} + h_{hbY} \\ &+ h_{ctX} + h_{ctG} + h_{ctY} + h_s \end{aligned} \tag{40}$$

As we have already noted that amino acid, peptide, and protein data sets are generally relatively small, we must of necessity avoid the use of the XB equation. Our alternative to the monoparametric steric parameterization therefore is the use of the SB equation. In order to do so, it is necessary, however, to take note of the fact that Gly is the only α-amino acid which is not substituted at the α carbon atom. It is obvious that if we followed the usual procedure and began the numbering of the side-chain from the atom attached to the α-carbon, all amino acids other than Gly would have $n_1 = 1$. We have started the numbering of the side chain from the second atom of the longest chain in order to avoid this problem. Thus, for Ala, Val, Ile, and Leu, the values of n_1 are 0, 2, 2, and 1 respectively. When the data set includes two or more Gly residues, as in the case with some peptide bioactivity series, we have introduced the parameter n_0 which takes the value 0 for Gly and 1 for all other residues. An alternative solution is to use both the v parameter, which for nonsymmetric groups is largely a function of n_1 and n_2, and the branching parameters together. The combination provides a good description of the steric effect while avoiding the problem of insufficient parameter variation. Thus, in place of Equation 31, we write

$$\Delta G_S = \sum_{i=1}^{m} B_i n_i + h_S \tag{41}$$

Then, using the SB equation to represent the steric effect and including the n_b parameter for group length, we obtain the IMFSB equation:

$$\Delta G_{\text{imf}} = L\sigma_{1X} + A\alpha_X + H_1 n_{HX} + H_2 n_{nX} + C_A n_{AX} + C_D n_{AX} \\ + I i_X + B_i n_{iX} + B_b n_b + B^0 \tag{42}$$

D. Problems in the Application of the IMF Model

The problem encountered in the application of the IMF model to amino acid, peptide, and protein properties and bioactivities are largely the result of collinearity among the variables. This is particularly the case in those data sets which consist of the nineteen amino acids other than proline that are commonly found in protein. We will refer to such data sets as *protein basis sets* (PBS). The pairs of inependent variables which are collinear in PBS (confidence levels are in parentheses) are σ_1, n_n (95); α, v (90); n_H, n_n (99); n_H, i(90); and n_n, i(90). Thus, while a good model of a data set can be obtained by correlation with the IMF equation, it is frequently impossible to achieve a clear separation of the contributions from the various intermolecular forces. When the composition of the data set is not limited to the amino acids of the PBS, it is possible to minimize or even exclude collinearity by a careful choice of the side-chains in the set.

Another disadvantage of the PBS is the small number of data points relative to the number of independent variables. Our experience is that when the data to be modeled are of high quality, chance correlations are best avoided if the ratio of the degrees of freedom (DF) to the number of independent variables (NV) is not less than three. DF is given by the relationship

$$DF = n - (NV + 1) = n - NV - 1 \qquad (43)$$

where n is the number of points in the data set. Chemical reactivity and physical property data are generally of high quality. In vivo bioactivity data are generally of lower quality; the quality seems to depend on the complexity of the organism as well as the nature of the activity measured. The quality of in vitro bioactivity data also depends on the nature of the biocomponent. Kinetic and equilibrium data on pure enzymes are comparable in quality to most chemical reactivity data, while measurements on cell extracts, whole cells, and tissues tend to be less reliable. For the PBS, correlation with the IMFU equation gives a value of DF/NV of 2, while correlation with the IMFSB equation gives a value of 1 when all of the steric parameters are used. As studies of the properties generally involve the PBS, there is no simple way to avoid the problem.

E. Method

We have correlated properties and bioactivities of amino acids, peptides and proteins with the IMFU and IMFSB equations by means of multiple linear regression analysis. Values of all the parameters required for correlation with either form of the IMF equation are given in Appendix 1 for more than one hundred amino acid side chains.

The contribution made by each variable is quantitatively described by the quantity C_i, which is defined by the relationship

$$C_i = |b_i x_i| * 100 \Big/ \sum_{i=1}^{n} |b_i x_i| \qquad (44)$$

where the b_i are the coefficients of the independent variables in the IMF equation and the x_i are the values of these variables for some reference side chain X^0. In our applications of the IMF equation to properties of amino acids, peptides, and proteins, we have chosen the side chain of His as the reference side chain.

III. AMINO ACIDS

A. Hydrophobicities

1. Introduction

In 1959 Kauzman (17) proposed that the side chains of amino acids had a major effect upon protein properties due to their "hydrophobicity." This property, also known to those with a more optimistic outlook as lipophilicity, was considered to be a measure of the effect of the side chain on conformation and other properties. In the last thirty years hydrophobicity has become of major interest to biochemists and to those interested in the design of bioactive substances. Its importance is underlined by the publication of two monographs (18, 19) and a Faraday Symposium (20) on the subject. We have classified the various sets of hydrophobicity parameters reported in the literature for amino acids into three categories:

1. Sets based on experimental results, designated E.
2. Sets which are modifications of those based on experimental results. These sets are designated M.
3. Sets which are a composite of two or more M and/or E sets. These composite sets are designated C. The amino acids themselves may be free, in the form of protopeptides with peptide bonds flanking the carbon atom to which the side chain is bound, or in protein.

Experimental methods that have been used for the determination of hydrophobicity parameters for amino acid side chains in water and in organic solvents (Sv) include ΔG_{tr} from (Sv) to (aq); surface tension; partition coefficients; distribution coefficients; and chromatographic properties. A set of "hydration potentials" (HP) has been reported for HX where X is the amino acid side chain. The HP represent the ΔG_{tr} values for the transfer of HX from the gas phase to an aqueous solution.

All of the available data sets are collected in Table 2. We have already reported the correlation of a number of these data sets with the IMF equation. These sets have been rerun, as we have made some minor improvements in the values of some of the parameters. The results of the correlations with the IMFU equation are set forth in Table 3. Zeroth-order correlation coefficients are reported in Table 4, values of C_i are given in Table 5. As we have found that protein data sets are usually best modeled by the IMFSB equation, we have examined the correlation of hydrophobicity scales based on amino acids in protein with this relationship. Results of correlations with the IMFSB equation are reported in Table 6, zeroth-order correlation coefficients in Table 7, and C_i values in Table 8.

TABLE 2
Amino Acid Hydrophobicity Parameters

01. $-\Delta G_{tr}$, (EtOH) to (aq), 25°, kcal mole[a]
Ala, 0.730; Asn, −0.010; Asp, 0.540; Glu, 0.550; Gln, −0.100; Gly, 0; Ile, 2.970; Leu, 2.420; Met, 1.300; Phe, 2.650; Ser, 0.040; Thr, 0.440; Tyr, 2.870.

1. $-\Delta G_{tr}$, (EtOH) to (aq), 25°, kcal mole[b]
Gly, 0; Ala, 0.73; Val, 1.69; Leu, 2.49; Ile, 2.97; Phe, 2.65; Met, 1.30; Cys, 1.00; Trp, 3.00; Tyr, 2.97; His, 1.10; Thr, 0.44; Ser, 0.04; Asn, −0.01; Gln, −0.10; Asp, 0.54; Glu, 0.55; Lys, 1.50; Arg, 0.73.

2. $-\Delta G_{tr}$, (EtOH or dioxane) to (aq), 25°[c]
Trp, 3.400; Nle, 2.600; Phe, 2.500; Tyr, 2.300; Dhp, 1.800; Leu, 1.800; Val, 1.500; Met, 1.300; His(+), 0.500; Ala, 0.500; Thr, 0.400; Ser, −0.300.

3. ΔG, from surface tension, kcal mole[d]
Ala, −0.200; Arg(+), −1.20; Asn, 0.080; Asp(−), −0.200; Cys, −0.450; Gln, 0.160; Glu(−), −0.300; Gly, 0; His(+), −1.20; Ile, −2.260; Leu, −2.460; Lys(+), −0.350; Met, −1.470; Phe, −2.330; Ser, −0.390; Thr, −0.520; Trp, −2.010; Tyr, −2.240; Val, −1.560.

4. log P (octanol/water)[e]
Ada, −0.08; Ala, −2.89; Arg(+), −4.20; Asn, −3.41; Asp, −3.38; Asp(−), −4.25; Bug, −0.176; Car, 0.86; Cit, −3.19; Cys, −2.49; Cys(−), −3.63; Gln, −3.15; Glu, −2.94; Glu(−), −4.19; Gly, −3.25; His, −2.84; His(+), −4.15; Ile, −1.72; Leu, −1.61; Lys, −3.21; Lys(+), −4.44; Met, −1.84; Nle, −1.53; Nva, −1.86; Phe, −1.63; Val, −2.08; Pen, −1.78; Pen(−), −2.41; Lys(Me), −2.77; Lys(Me)(+), −4.21; Try(Me), −1.89; Lys(CHO), −2.84; Pro, −2.50; Hyp, −3.17.

5. Hydrophobicity factors, kcal mole[f]
Trp, 3.00; Ile, 2.95; Tyr, 2.85; Phe, 2.65; Leu, 2.40; Val, 1.70; Lys(+), 1.50; Met, 1.30; Cys, 1.00; Arg(+), 0.75; Ala, 0.75; Thr, 0.45; Gly, 0; Ser, 0; Asp(−), 0; Glu(−), 0; His(+), 0; Asn, 0; Gln, 0.

6. Log P (octanol/water)[g]
Car, 0.99; Ada, 0.43; Trp, −1.06; Phe, −1.35; Neo, −1.42; Ile, −1.69; Bug, −1.77; Met, −1.87; Val, −2.23; Tyr, −2.26; Ala, −2.72; Gly, −3.21; Anp, −0.13; Bnp, 0.06.

7. log P (octanol/water)[h]
Leu, −0.74; Val, −1.14; Ala, −1.60; Gly, −1.81; Nle, −0.51; Nva, −0.98; Abu, −1.34.

8. ΔG_{tr}, ("polar") to ("nonpolar"), kcal mole[i]
Ala, 0.5; Gly, 0; Ile, 2.97; Leu, 1.8; Phe, 2.5; Val, 1.5; Asn, −0.01; Cys, 1.4; Gln, −0.1; Met, 1.3; Ser, −0.3; Thr, 0.4; Trp, 3.4; Tyr, 2.3; Arg(+), 0.73; Asp(−), 0.54; Glu(−), 0.54; His(+), 0.5; Lys(+), 1.5.

9. ΔG_{tr} (ethanol) to (water), estimated values, kcal mole[j]
Ala, 0.5; Gly, 0; Ile, 1.8; Phe, 2.5; Val, 1.5; Asn, −0.2; Cys, 1.0; Gln, −0.2; Met, 1.3; Ser, −0.3; Thr, 0.4; Trp, 3.4; Tyr, 2.3; Arg(+), −3.0; Asp,(−), 2.5; Glu,(−), 2.5; His(+), 0.5; Lys(+) − 3.0.

10. P (BuOH/PhCH$_2$OH-H$_2$O)[k]
Trp, 0.60; Phe, 0.270; Leu, 0.125; Tyr, 0.104; Ile, 0.097; Met, 0.082; Val, 0.044.

11. log D (octanol/water)[l]
Trp, −1.11; His, −1.95; Val, −2.26; Ala, −2.74; Thr, −2.94; Ser, −3.07; Glu, −3.69; Arg, −4.08; Lys, −3.05.

TABLE 2 (Continued)

12. ΔG_{tr}, kcal mole[m]

Trp, 3.4; Phe, 2.5; Tyr, 2.3; Leu, 1.8; Ile, 1.8; Val, 2.5; Met, 1.4; Cys, 1.0; His, 0.5; Ala, 0.5; Thr, 0.4; Ser, 0; Gly, 0; Asn, 0; Asp, 0; Gln, 0; Glu, 0; Lys, 0; Arg, 0.

13. f_{rel}[n]

Trp, 2.31; Phe, 2.24; Leu, 1.99; Ile, 1.99; Tyr, 1.70; Val, 1.46; Met, 1.08; Cys, 0.93; Ala, 0.53; Lys, 0.52; Gly, 0; Asp, −0.02; Glu, −0.07; His, −0.23; Thr, −0.26; Ser, −0.66; Asn, −1.05; Gln, −1.09.

14. Hydration potentials [$\Delta G G_{tr}$ of XH from (g) to (aq)][o]

Ala, 1.94; Arg, −19.92; Asn, −9.68; Asp, −10.95; Cys, −1.24; Gln, −9.38; Glu, −10.20; Gly, 2.39; His, −10.27; Ile, 2.15; Leu, 2.28; Lys, −9.52; Met, −1.48; Phe, −0.76; Ser, −5.06; Thr, −4.88; Trp, −5.88; Tyr, −6.11; Val, 1.99.

15. Hydrophobicity index[p]

Asn, 0.6; Asp, 0.7; Lys, 1.3; Gln, 1.4; His, 1.6; Glu, 1.8; Arg, 2.0; Ser, 3.1; Thr, 3.5; Gly, 4.1; Ala, 5.1; Tyr, 8.0; Val, 8.5; Met, 8.7; Trp, 9.2; Ile, 9.3; Phe, 9.6; Leu, 10.0.

16, 17. $\log K^1$, $\log K_D$, AcNHCHXCONHMe[q]

Phe, 2.304, 0.401; Leu, 1.670, 0.137; Ile, 1.645, 0.135; Tyr, 1.486, −0.321; Vol, 1.154, −0.348; Ala, 0.272, −1.123; Gly, −0.088, −1.560; Lys, −0.073, —.

18. ΔG_{tr}, kcal mole[r]

Ala, 0.83; Arg, 0.83; Asp, 0.64; Asn, 0.09; Cys, 1.48; Glu, 0.65; Gln, 0.00; Gly, 0.10; His, 1.10; Leu, 2.52; Ile, 3.07; Lys, 1.60; Met, 1.40; Phe, 2.75; Ser, 0.14; Thr, 0.54; Trp, 0.31; Tyr, 2.97; Val, 1.74.

19. ΔG kcal mole[s]

Ala, −0.5; Arg, 3.0; Asn, 0.2; Asp, 3.0; Cys, −1.0; Glu, 3.0; Gln, 0.2; Gly, 0; His, −0.5; Ile, −1.8; Leu, −1.8; Lys, 3.0; Met, −1.3; Phe, −2.5; Ser, 0.3; Thr, −0.4; Trp, −3.4; Tyr, −2.3.

20. $\log D$, AcNHCHXCONH$_2$[t]

Ada, 1.90; Ala, −1.52; Arg, −2.84; Asn, −2.41; Asp, −2.60; Car, 2.47; Cys, −0.29; Cys$_2$, −1.70; Glu, −2.47; Gln, −2.05; Gly, −1.83; His, −1.70; Ile, −0.03; Leu, −0.13; Lys, −2.82; Met, −0.60; Phe, −0.04; Ser, −1.87; Thr, −1.57; Trp, 0.42; Tyr, −0.82; Val, −0.61.

21. ΔG_{tr}, (N-cyclohexyl-2-pyrrolidone) to (aq)[u]

Ala, −0.48; Arg, −0.06; Asn, −0.87; Asp, −0.75; Cys, −0.32; Glu, −0.71; Gln, −0.32; Gly, 0; His, −0.51; Leu, 1.02; Ile, 0.81; Lys, −0.09; Met, 0.81; Phe, 1.03; Ser, 0.05; Thr, −0.35; Trp, 0.66; Tyr, 1.24; Val, 0.56.

22, 23. Hydrophobicity parameters, OMH and consensus[v]

Ala, −0.40, 0.62; Arg, −0.59, −2.53; Asn, −0.92, −0.78; Gln, −0.91, −0.85; Gly, −0.67, −0.48; His, −0.64, −0.40; Met, 1.02, 0.64; Phe, 1.92, 1.19; Ser, −0.55, −0.18; Thr, −0.28, −0.05; Trp, 0.50, 0.81; Tyr, 1.67, 0.26; Val, 0.91, 1.08.

24. ΔG_{rel}, chromatographic[w]

Aoc, 6.35; Ala, 1.41; Arg, 3.74; Asn, −0.48; Asp, −6.11; Glu, −8.19; Gln, 0.33; Gly, 0; Ile, 3.85; Leu, 3.30; Lys, 4.74; Met, 3.07; Nva, 3.07; Nle, 3.82; Ser, −0.48; Thr, 0.74; Val, 2.41.

25. ΔG_{tr} kcal mole[x]

Ala, 0.40; Arg, 0.64; Arg(+), −2.64; Asn, −0.28; Asp, 0.28; Asp(−), −1.77; Cys, 0.90; Glu, 0.38; Glu(−), −1.46; Gln, −0.19; Gly, 0; His, 0.60; His(+), −1.57; Ile, 2.13; Leu, 1.91; Lys, 1.04; Lys(+), −2.62; Met, 1.34; Phe, 2.37; Ser, −0.27; Thr, 0.21; Trp, 2.55; Tyr, 2.24; Val, 1.46.

TABLE 2 (Continued)

26. Hydropathy index[y]
Ala, 1.8; Arg, −4.5; Asn, −3.5; Asp, −3.5; Cys, 2.5; Glu, −3.5; Gln, −3.5; Gly, −0.4; His, −3.2; Ile, 4.5; Leu, 3.8; Lys, −3.9; Met, 1.9; Phe, 2.8; Ser, −0.8; Thr, −0.7; Trp, −0.9; Tyr, −1.3; Val, 4.2.

27. ΔG_{tr} (aq) to (oil), kcal mole[z]
Ala, 1.6; Arg, −12.3; Asn, −4.8; Asp, −9.2; Cys, 2.0; Glu, −8.2; Gln, −4.1; Gly, 1.0; His, −3.0; Ile, 3.1; Leu, 2.8; Lys, −8.8; Met, 3.4; Phe, 3.7; Ser, 0.6; Thr, 1.2; Trp, 1.9; Tyr, −0.7; Val, 2.6.

28. Hydrophobicity kcal mole[aa]
Ala, 0.87; Arg, 0.85; Asn, 0.66; Asp, 0.09; Cys, 1.52; Glu, 0.67; Gln, 0.00; Gly, 0.10; His, 0.87; Ile, 3.15; Leu, 2.17; Lys, 1.64; Met, 1.67; Phe, 2.87; Ser, 0.07; Thr, 0.07; Trp, 3.77; Tyr, 2.67; Val, 1.87.

29. Average surrounding hydrophobicity[bb]
Ala, 12.97; Arg, 11.72; Asn, 11.42; Asp, 10.85; Cys, 14.63; Glu, 11.89; Gln, 11.76; Gly, 12.43; His, 12.16; Ile, 15.67; Leu, 14.90; Lys, 11.36; Met, 14.39; Phe, 14.00; Ser, 11.23; Thr, 11.69; Trp, 13.93; Tyr, 13.42; Val, 15.71.

30. ΔG_{tr}, (random coil/aq) to (α-helix/lipid) kcal mole[cc]
Ala, −2.880; Arg, 9.385; Asn, 1.017; Asp, 5.555; Cys, 0.945; Glu, 4.022; Gln, 0.517; Gly, −1.878; His, 1.502; Ile, −4.383; Leu, −4.256; Lys, 2.323; Met, −2.120; Phe, −5.258; Ser, −0.368; Thr, −0.993; Trp, −3.873; Tyr, −0.361; Val, −3.880.

All data sets are from Aax unless otherwise noted.
[a] Ref. 21. [b] Ref. 22.23. [c] Ref. 28. [d] Ref. 44. [e] Ref. 34. [f] Ref. 24. [g] Ref. 35–37. [h] Ref. 38. [i] Ref. 29. [j] Ref. 30. [k] Ref. 39. [l] Ref. 40. [m] Ref. 31. [n] Ref. 41. [o] Ref. 42, 43. [p] Ref. 45. [q] Ref. 50. [r] Ref. 26. [s] Ref. 27. [t] Ref. 51. [u] Ref. 33. [v] Ref. 47, 52. [w] Ref. 46. [x] Ref. 48. [y] Ref. 49. [z] Ref. 53. [aa] Ref. 32. [bb] Ref. 54. [cc] Ref. 55.

All of the 31 sets of hydrophobicity parameters studied gave significant correlations with the IMF equation. Twenty-seven of the sets studied had F test values significant at the 99.9% confidence level, two at the 99.5% level, and two at the 99.0% level. In 19 of the 31 data sets studied, the regression equation accounts for more than 90% of the variance of the data. We conclude that the results show unequivocally the validity of the IMF equation as a quantitative model of amino acid side-chain hydrophobicities.

The first set of amino acid hydrophobicity (AAH) parameters (set AAH 01) was proposed by Tanford (21) who used solubility data in water and ethanol as a basis for their determination. This scale has been modified by a number of other workers (sets AAH 1, 5, 18, 19) (22–27). Examination of the C_i values for these sets shows that they generally exhibit a dependence on σ_1, α, n_H, and v. Nozaki and Tanford (28) published an improved set of parameters based on solubility in water and ethanol or dioxane (set AAH 2). Again, several other groups of workers made modifications (sets AAH 8, 9, 12, 28) (29–32). In this case, all sets showed a dependence on α and n_H. Three of these sets were also dependent on σ_1, the other two on i. Due to the collinearity between these variables, we cannot be sure whether either

TABLE 3
Results of Correlations of Amino Acid Hydrophobicity Data Sets with the IMFU Equation

Set*	01	1	2	3	4	5
L	−9.16	−5.69	−11.5	—	−2.88	−8.63
$S_L{}^a$	5.25^m	2.43^j	2.95^f	—	1.16^h	2.95^h
A	6.87	7.36	7.21	−5.76	4.63	6.81
$S_A{}^a$	2.12^h	1.38	1.05	1.32	0.637	1.67^f
H_1	−3.06	−0.517	—	0.494	−0.535	−0.478
$S_{H1}{}^a$	1.28^k	0.113	—	$0.130^{k,f}$	0.0589	0.137^f
H_2	1.58	—	—	—	—	—
$S_{H2}{}^a$	0.856^m	—	—	—	—	—
I	−3.34	—	−0.717	0.496	−1.08	—
$S_I{}^a$	1.88^m	—	0.409^m	0.288^m	0.131	—
S	1.34	1.15	—	−1.22	0.960	1.13
$S_S{}^a$	0.748^m	0.607^k	—	0.614^k	0.249	0.735^m
B^0	−0.219	−0.133	0.477	0.401	−3.504	−0.0436
$S_{Bo}{}^a$	0.437^r	0.398^s	0.276^m	0.392^o	0.172	0.482^s
$100R^{2b}$	91.41	86.27	92.08	81.25	94.72	81.93
F^c	10.63^q	22.00	30.99	15.16	96.85	15.87
$S_{est}{}^d$	0.493	0.459	0.360	0.468	0.294	0.555
S^{0e}	0.431	0.432	0.345	0.505	0.254	0.495
n^t	13	19	12	19	33	19

Set*	6	7	8	9	10	11
L	−9.86	nt	−4.13	—	nt	7.52
$S_L{}^a$	1.87	nt	2.66^m	—	—	0.665^b
A	8.57	8.62	8.78	10.2	0.836	5.89
$S_A{}^a$	0.341	1.08^f	1.33	1.43	0.213^h	0.296
H_1	−1.22	nt	−0.529	−0.848	0.311	−0.457
$S_{H1}{}^a$	0.120	nt	0.124	0.156	0.0673^q	0.0288
H_2	—	nt	—	—	nt	−0.398
$S_{H2}{}^a$	—	nt	—	—	—	0.0257
I	nt	nt	—	−2.48	nt	—
$S_I{}^a$	—	—	—	0.355	—	—
S	—	−0.581	—	—	—	−0.593
$S_S{}^a$	—	0.252^k	—	—	—	0.333^m
B^0	−3.33	−1.79	0.226	−0.125	−0.0437	−2.65
$S_{Bo}{}^a$	0.0847	0.0764	0.298^p	0.282^r	0.0437^o	0.199
$100R^{2b}$	98.83	97.75	81.56	92.23	95.28	99.82
F^c	281.0	86.77	22.12	59.31	40.41^f	335.3
$S_{est}{}^d$	0.151	0.0849	0.507	0.578	0.0518	0.0620
S^{0e}	0.128	0.199	0.483	0.314	0.287	0.0732
n^t	14	7	19	19	7	9

TABLE 3 (Continued)

Set*	12	13	14	15	16	17
L	—	−4.51	−26.5	−16.2	nt	nt
$S_L{}^a$	—	2.32	4.77	8.98^k	—	—
A	9.09	7.30	−14.8	18.6	8.16	7.84
$S_A{}^a$	0.724	1.14	2.36	3.89	0.431	0.502
H_1	−0.540	−0.974	−3.24	−1.97	−0.932	−0.903
$S_{H1}{}^a$	0.0804	0.149	0.279	0.417	0.0529	0.115^f
H_2	—	—	−0.807	—	nt	nt
$S_{H2}{}^a$	—	—	0.217^f	—	—	—
I	−0.675	—	−3.17	−2.28	—	nt
$S_I{}^a$	0.179f	—	0.474	0.951^j	—	—
S	—	—	2.22	—	0.206	0.303
$S_S{}^a$	—	—	1.00^j	—	0.134^n	0.141^k
B^0	−0.0634	0.283	2.12	4.71	−0.139	−0.149
$S_{Bo}{}^a$	0.142^r	0.259^n	0.639^g	0.869	0.0864^m	0.0723^m
$100R^{2b}$	93.61	87.75	99.05	87.09	99.42	99.54
F^c	733.23	33.43	207.8	21.93	227.9	215.3
$S_{est}{}^d$	0.292	0.434	0.733	1.48	0.0909	0.0809
S^{0e}	0.285	0.397	0.123	0.423	0.108	0.104
n^t	19	18	19	18	8	7

Set*	18	19	20	21	22	23
L	−6.42	—	—	−3.64	—	—
$S_L{}^a$	3.79^m	—	—	2.27^m	—	—
A	—	−10.5	6.45	3.96	6.77	2.41
$S_A{}^a$	—	1.56	0.867	1.10^f	1.36	0.622^f
H_1	−0.342	0.801	−0.653	−0.230	−0.491	−0/744
$S_{H1}{}^a$	0.174^k	0.174	0.0937	0.116^k	0.151^k	0.0615
H_2	—	—	—	—	—	—
$S_{H2}{}^a$	—	—	—	—	—	—
I	—	2.74	−0.903	−0.417	−0.829	−0.753
$S_I{}^a$	—	0.387	0.212	0.266^m	0.336^j	0.136
S	2.63	—	0.620	—	—	1.13
$S_S{}^a$	0.884^g	—	0.382^m	—	—	0.290^f
B^0	0.00223	0.201	−2.07	−0.133	−0.523	−0.392
$S_{Bo}{}^a$	0.660^s	0.308^q	0.229	0.247	$−0.267^k$	0.185^k
$100R^{2b}$	52.63	91.54	95.23	70.19	76.05	96.39
F^c	5.556^g	54.12	84.77	8.241^f	15.88	93.43
$S_{est}{}^d$	0.765	0.630	0.349	0.419	0.547	0.221
S^{0e}	0.775	0.327	0.249	0.636	0.551	0.221
n^t	19	19	22	19	19	19

TABLE 3 (Continued)

Set*	24	25	26	27	28	29	30
L	—	—	—	—	−6.30	—	20.7
$S_L{}^a$	—	—	—	—	2.43^{i-}	—	6.95
A	16.0	3.58	−5.67	5.56	9.28	−6.28	—
$S_A{}^a$	3.58	1.01	3.81^m	3.30^k	1.22	3.37	—
H_1	2.09	−0.582	−0.916	−2.07	−0.555	−0.416	1.98
$S_{H1}{}^a$	0.335	0.0876	0.432^k	0.425	0.113	0.274^m	0.356
H_2	−2.43	—	−1.01	−0.781	—	−0.670	—
$S_{H2}{}^a$	0.246	—	0.297^g	0.292^h	—	0.193^f	—
I	−2.13	−2.58	−1.56	−5.57	—	—	3.63
$S_I{}^a$	0.734^h	0.223	0.763^k	0.782	—	—	3.63
S	—	—	5.45	—	—	2.12	—
S_S	—	—	1.62^h	—	—	0.852^j	—
B^0	0.704	0.0304	−0.579	1.93	0.473	—	−2.800
$S_{B^0}{}^b$	0.608^n	0.193^s	1.00^q	0.679^h	0.272^k	—	0.755
$100R^{2b}$	95.27	93.19	89.08	95.43	86.71	72.12	91.01
F^c	60.47	91.23	21.22	73.04	32.62	12.93	35.42
$S_{est}{}^d$	0.934	0.419	1.19	1.22	1.16	0.912	1.28
S^{0e}	0.259	0.286	0.399	0.249	0.410	0.594	0.349
n^t	17	24	19	19	19	19	19

*$L, A, H_1, H_2, I, S, B^0$, are the regression coefficients, nt = not tested.

[a] Standard errors of the regression coefficients. Superscripts indicate the confidence level (CL) of the t test for the significance of the regression coefficients. In the absence of a superscript, the CL is $>99.9\%$.

[b] Percent of the variance of the data accounted for by the regression coefficients. R is the multiple correlation coefficients.

[c] F test for the significance of the correlation. The superscript indicates the CL. In its absence the CL is 99.9%.

[d] The standard error of the estimate.

[e] S_{est}/(root mean square of the data).

[f] 99.5% CL. [g] 99.0% CL. [h] 98.0% CL. [i] 97.5% CL. [j] 95.0% CL. [k] 90.0% CL. [l] <90.0% CL. [m] 80% CL. [n] 70% CL. [o] 60% CL. [p] 50% CL. [q] 40% CL. [r] 30% CL. [s] ⩽20% CL. [t] Number of data points in the set.

or both are involved in the side-chain structural effect. No dependence on n_n or v was observed.

Set AAH 21 is based on solubility in water and in N-cyclohexylpyrrolidone (33). The latter was considered to be a better model of the interior of protein than ethanol. This may not be the case in view of the fact that N-cyclohexylpyrrolidone is incapable of acting as a hydrogen donor in hydrogen bond formation. The data set is best correlated by the IMFSB equation showing a dependence on σ_1, n_H, i, n_3 and n_b. It should be noted that n_b and α are highly collinear.

Set AAH 4, 6, 7, and 10 (34–39) are based on the determination of partition coefficients, set 11 on distribution coefficients (40). Set AAH 13 is

TABLE 4
Zeroth-Order Partial Correlation Coefficients for the Amino Acid Hydrophobicity Data Sets

Variables i,j	PBS	01	2	4	6	7	10	11	13	15	16	17	20	24	25
σ,α	0.259	0.161	0.242	0.234	0.620[f]	—	—	0.325	0.262	0.232	—	—	0.041	0.248	0.268
σ,n_H	0.164	0.568	0.517	0.026	0.042	—	—	0.077	0.260	0.267	—	—	0.348	0.378	0.067
σ,n_n	0.580[e]	0.775[e]	0.561	0.422[d]	0.120	—	—	0.614	0.610[e]	0.730[b]	—	—	0.672[b]	0.808[b]	0.598[c]
σ,i	0.296	0.569	0.357	0.294[f]	—	—	—	0.087	0.333	0.381	—	—	0.419[f]	0.371	0.220
σ,v	0.021	0.026	0.357	0.294[f]	—	—	—	0.114	0.021	0.008	—	—	0.078	0.002	0.020
α,n_H	0.300	0.012	0.339	0.087	0.202	—	0.548	0.481	0.155	0.283	0.405	0.525	0.143	0.348	0.382
α,n_n	0.085	0.072	0.072	0.231	0.039	—	—	0.095	0.173	0.115	0.184	—	0.059	0.005	0.109
α,i	0.130	0.116	0.068	0.065	—	—	—	0.316	0.025	0.115	—	—	0.072	0.271	0.093
α,v	0.460[f]	0.611	0.326	0.609[a]	0.546	0.891[d]	0.683	0.541	0.479[f]	0.455	0.631	0.640	0.717[b]	0.604[e]	0.418[e]
n_H,n_n	0.661[c]	0.836[c]	0.890[c]	0.676[c]	0.679[f]	—	—	0.519	0.694[c]	0.696[d]	—	—	0.748[a]	0.699[d]	0.559[d]
n_H,i	0.530[f]	0.182	0.196	0.268	—	—	—	0.614	0.386	0.517[f]	0.882[e]	—	0.617[c]	0.595[e]	0.330
n_H,v	0.037	0.067	0.232	0.257	0.171	—	0.316	0.151	0.055	0.023	0.030	0.041	0.182	0.015	0.027
n_n,i	0.528[f]	0.677[f]	0.020	0.164	—	—	—	0.537	0.476[f]	0.515[f]	—	—	0.622[c]	0.621[e]	0.342
n_n,v	0.058	0.080	0.229	0.229	0.116	—	—	0.006	0.061	0.044	—	—	0.173	0.041	0.076
i,v	0.055	0.076	0.011	0.052	—	—	—	0.280	0.061	0.046	0.013	—	0.116	0.044	0.042
n[g]	19	13	12	33	14	7	7	9	18	18	8	7	22	17	24

*PBS refers to sets which consist of the 19 amino acids, other than proline, which are the most common constituents of protein. These sets include 1, 3, 5, 8, 9, 12, 14, 18, 19, 21, 22, 23, 26, 27, 28, 29.
[a] 99.9% CL (confidence level). [b] 99.5% CL. [c] 99.0% CL. [d] 98.0% CL. [e] 95.0% CL. [f] 90% CL. [g] Number of data points in the set. In the absence of a superscript, the confidence level of r_{ij} is <90.0%.

AMINO ACID, PEPTIDE, AND PROTEIN PROPERTIES

TABLE 5
Values of C_i For Amino Acid Hydrophobicity Data Sets

Set	Type*	σ_1	α	n_H	n_n	i	v
01	E	6.52	14.0	27.2	14.1	29.7	8.36
1	M	13.1	48.8	14.9	0	0	23.1
5	M	19.6	44.4	13.6	0	0	22.4
18	M	19.0	0	12.7	0	0	68.3
19	M	0	40.5	13.4	0	46.1	0
2	E	27.9	50.3	0	0	21.8	0
8	M	11.4	70.1	18.4	0	0	0
9	M	0	41.3	15.0	0	43.7	0
12	M	0	63.2	16.3	0	20.4	0
28	M	15.8	66.8	17.4	0	0	0
21	E	15.8	49.2	12.5	0	22.6	0
4	E	6.43	29.7	14.9	0	30.2	18.7
6	E	19.8	49.6	30.6	0	0	0
7	E	nd	83.0	nd	nd	nd	17.0
10	E	nd	38.2	61.8	nd	nd	0
13	M	12.0	55.7	32.3	0	0	0
11	E	18.6	42.0	14.2	12.3	0	12.9
15	E	13.2	43.5	20.1	0	23.2	0
24	E	0	35.7	20.2	23.5	20.6	0
3	E	0	41.8	15.6	0	15.6	27.0
14	E	14.8	23.8	22.7	5.65	22.2	10.9
29	M	0	0	16.2	26.1	0	57.7
22	E	0	54.1	17.1	0	28.8	0
27	E	0	13.2	21.4	8.1	57.4	0
30	E	19.0	16.6	22.7	0	41.7	0
16	E[a]	nd	63.5	31.6	nd	0	4.88
17	E[a]	nd	61.8	30.9	nd	nd	7.27
20	E[b]	0	42.7	18.8	0	26.0	12.5
23	C	0	19.5	26.2	0	26.5	27.8
25	C	0	36.7	11.6	0	51.7	0
26	C	0	15.2	10.7	11.7	44.4	18.1

*Sets designated E are the result of experimental determinations. Sets designated M are modifications of E Sets. Sets designated C are composite sets consisting of a combination of two or more other sets.

[a] The substrate is AcAaxNHMe.

[b] The substrate is AcAaxNH$_2$.

based on a modification of partition coefficient data (41). All of these sets show a dependence on α and n_H, and possibly on σ_1 as well. There is also a dependence on v in some of these sets.

Set AAH 14 represents the transfer of a side-chain from the gas phase to aqueous solution at pH7 (42, 43). It is a function of all of the variables in the IMFU equation giving an excellent fit. Set AAH 3 is based on surface-tension measurements (44). This scale is a function of α, n_H, i, and v.

TABLE 6
Results of Correlations with the IMFSB Equation

Set:	21	21	22	27	29
L	−4.01	—	—	11.7	−9.10
S_L	1.86^k	—	—	7.01^m	4.32^k
A	—	−7.41	9.80	—	−12.8
S_A	—	3.37	2.14	—	3.37^f
H_1	−0.370	−0.260	−0.476	−1.78	−0.940
S_{H1}	0.104^f	0.112	0.142^f	0.372	0.207
H_2	—	−0.287	—	−1.19	—
S_{H2}	—	0.0908	—	0.295^f	—
I	−0.480	−0.589	−0.645	−5.76	—
S_I	0.225^k	0.212	0.332^k	0.709	—
B_1	—	0.267	—	1.03	—
S_{B1}	—	0.177	—	0.484^k	—
B_2	—	—	—	—	—
S_{B2}	—	—	—	—	—
B_3	−0.668	−0.570	−0.710	1.00	—
S_{B3}	0.311^k	0.305	0.404^m	0.584^k	0.615^i
B_b	0.502	0.934	nd	nd	nd
S_{Bb}	0.114	0.223	nd	nd	nd
B^0	−0.178	−0.132	−0.769	1.14	12.7
S_{B^0}	0.189	0.216	0.287^h	0.662^m	0.522
$100R^2$	81.39	87.08	80.38	97.03	78.82
F	11.37	10.59	14.34	65.37	13.02
S_{est}	0.344	0.310	0.512	1.06	0.823
S^0	0.522	0.472	0.516	0.217	0.536
n	19	19	19	19	19

For footnotes see Table 3.

Two sets of data are based on chromatographic quantities. Since one of these (set AAH 15) (45) uses R_F values multiplied by a constant to define the scale, it is inherently in error. The appropriate quantity to use in the definition of the scale would be $\log R_F$, or better still, $\log R_M$. We have nevertheless examined the correlation of this data set with the IMFU equation. It is a function of σ_1, α, n_H, and i. The other scale (set AAH 24) (46) is based on $\log K_i$ values where K_i is the capacity factor in high-performance liquid chromatography (HPLC). It is a function of α, n_H, n_n, and i.

The three composite data sets (AAH 23, 25, and 26) (47–49) are all a function of α, n_H, and i. Sets 23 and 26 also depend on v; the latter set depends on n_n as well.

All three protopeptide sets (AAH 16, 17, and 20) (50, 51) are dependent on α, n_H, and v, with set 20 also dependent on i. The four scales involving amino acid residues in protein (sets AAH 22, 27, 29, and 30) (52–55) are not directly comparable. They are better fit by the IMFSB equation than by the

TABLE 7
Zeroth-Order Partial Correlation Coefficients for Correlations of Amino Acid Hydrophobicities with the IMFSB Equation

Set:	21	22[a]
σ, α	0.259	0.259
σ, n_H	0.164	0.164
σ, n_n	0.580	0.580
σ, i	0.296	0.296
σ, n_1	0.121	0.121
σ, n_2	0.218	0 218
σ, n_3	0.177	0.177
σ, n_b	0.145	—
α, n_H	0.300	0.300
α, n_n	0.085	0.085
α, i	0.130	0.130
α, n_1	0.223	0.223
α, n_2	0.588	0.588
α, n_3	0.808	0.808
α, n_b	0.920	—
n_H, n_n	0.661	0.661
n_H, i	0.530	0.530
n_H, n_1	0.011	0.011
n_H, n_2	0.233	0.233
n_H, n_3	0.383	0.383
n_H, n_b	0.525	—
n_n, i	0.528	0.528
n_n, n_1	0.049	0.049
n_n, n_2	0.171	0.171
n_n, n_3	0.101	0.101
n_n, n_b	0.212	—
i, n_1	0.062	0.062
i, n_2	0.288	0.288
i, n_3	0.333	0.333
i, n_b	0.353	—
n_1, n_2	0.050	0.050
n_1, n_3	0.096	0.096
n_1, n_b	0.100	—
n_2, n_3	0.447	0.447
n_2, n_b	0.873	—

[a]Sets 27, 29 are identical to 22; nd = not determined.

IMFU equation. Set 22 is the optimal matching hydrophobicity (OMH) scale. It is based on residue frequencies in a protein database. Set AAH 27 was derived from calculated values of hydrophobicity and hydrophilicity for amino acid residues in α-helix being transferred from water into oil (53). Set AAH 29 represents the average hydrophobicity of residues in an 800-pm radius of a given residue in protein. Set 30 involves free energies of transfer of a residue from random coil (aq) to α-helix (lipophilic).

TABLE 8
C_i Values for Correlation of Amino Acid Hydrophobicities with the IMFSB Equation

Set:	21	22	27	29
σ_1	13.7	0	8.01	11.7
α	0	52.2	0	47.5
n_H	15.8	11.7	15.2	15.1
n_n	0	0	10.2	0
i	20.5	15.8	49.2	0
n_1	0	0	8.77	0
n_2	0	0	0	0
n_3	28.5	17.4	8.56	25.7
n_b	21.4	nd*	nd	nd

*nd, not determined.

2. Volume and Surface Area Parameters

A number of authors have suggested that volume and surface area can be used as measures of hydrophobicity. In order to see to what extent this is the case, we have examined the correlation of two sets of side-chain volumes (sets AAV 1 and 2) (24, 56) and seven sets of side-chain areas (sets AAA 1–7) (26, 48, 57) with the IMFU equation. Set AAV 1 consists of experimentally determined, limiting partial molar volumes in water (56), set AAV 2 of calculated side-chain volumes (24). The data are reported in Table 9, the results of the correlations in Table 10, the zeroth-order correlation coefficients in Table 11, and the C_i values in Table 12. The limiting partial molar volumes are predominantly a function of α, with some dependence on v, σ_1, and n_n. The calculated volumes are also largely a function of α with some dependence on n_H and i. That the side-chain volumes are strongly dependent on α is to be expected, since we have already noted that α is highly linear in various types of group volume parameters. What is perhaps surprising is the observation of a dependence on n_H, n_n, σ_1, and i as well. Due to collinearity, we can often observe a dependence on v when a set is strongly dependent on α.

Frommel (48) has reported three area scales (sets AAA 1, 2, and 3): the accessible side-chain area A_{acc}, the polar side-chain area A_{pol}, and the contact amino acid area A_{cont}. The data are given in Table 10. All three sets are well-modeled by the IMFU equation (Table 11, 12). Both A_{acc} and A_{cont} are predominantly a function of polarizability. This is not unexpected. We have already noted that group volumes are highly collinear in polarizability. As in general, any cube is a crude linear function of its square, it cannot be surprising to find that volume which is a function of l^3 is linear in area which is a function of l^2. That A_{pol} is independent of polarizability but is a function

TABLE 9
Volume and Surface Area Data Sets

V1. Limiting partial molal volumes of L-Aax side chains (cm³/mol)[a]
Ala[a], 60.42; Abu[b], 75.50; Arg, 123.86; Asp, 74.8; Glu, 89.85; Gln, 93,61; Gly, 43.19; His, 98.3; Ile, 105.80; Leu, 107.83; Met[a], 105.57; Nle[c], 107.93; Nva[c], 91.70; Phe, 122.2; Ser, 60.62; Thr, 76.83; Trp, 143.8; Tyr, 124.4[b]; Val, 90.65.

V2. Aax volumes (Å³)[d]
Ala, 52.6; Arg, 109.1; Asn, 4.0; Asp, 68.4; Cys, 68.3; Glu, 84.7; Gln, 4.0; Gly, 36.3; His, 91.9; Ile, 102.0; Leu, 102.0; Lys, 105.1; Met, 97.7; Phe, 113.9; Ser, 54.9; Thr, 71.2; Trp, 135.4; Tyr, 116.2; Val, 85.1.

A1, 2, 3. Accessible side-chain area, (Aacc); polar side-chain area (Apol); contact amino acid Area, (Acont)[e]
Ala, 93.7, 4.3, 55.6; Arg[f], 250.4, 46.1, 89.8; Asn, 146.3, 78.8, 66.4; Asp[f], 142.6, 32.8, 66.4; Cys, 135.2, 4.5, 63.9; Glu[f], 182.9, 35.7, 74.9; Gln, 177.7, 34.8, 73.1; Gly, 52.6, 2.8, 50.5; His[f], 188.1, 17.2, 78.0; Ile, 182.2, 5.5, 71.8; Leu, 173.7, 5.1, 72.2; Lys[f], 215.2, 31.3, 81.1; Met, 197.6, 9.7, 81.2; Phe, 228.6, 4.0, 83.6; Ser, 109.5, 15.5, 58.3; Thr, 12.1, 13.2, 63.9; Trp, 271.6, 13.6, 99.1; Tyr, 239.9, 18.7, 86.5; Val, 157.2, 5.0, 66.3.

A4, 5, 6. Solvent accessible surface area (standard state), A^0; average solvent-accessible area (folded protein), $\langle A \rangle$; mean area buried on transfer from (standard state) to (folded protein), $A^0 - \langle A \rangle$, all in $Å^{2}$[g]
Ala, 118.1, 31.5, 86.8; Arg, 256.0, 93.8, 162.2; Asn, 165.5, 62.2, 103,3; Asp, 158.7, 60.9, 97.8; Cys, 146.1, 13.9, 132.3; Glu, 186.2, 72.3, 113.9; Gln, 193.2, 74.0, 119.2; Gly, 188.1, 25.2, 62.9; His, 202.5, 46.7, 155.8; Ile, 181, 23.0, 158.0; Leu, 193.1, 29.0, 164.1; Lys, 225.8, 110.3, 115.5; Met, 203.4, 30.5, 172.9; Phe, 222.8, 28.7, 194.1; Ser, 129.8, 44.2, 85.6; Thr, 152.5, 46.0, 106.5; Trp, 266.3, 41.7, 224.6; Tyr, 236.8, 59.1, 177.7; Val, 164.5, 23.5, 141.0.

A7. Bulkiness (V-side chain/l side chain)[h]
Ala, 11.50; Arg, 14.28; Asn, 12.82; Asp, 11.68; Cys, 13.46; Glu, 13.57; Gln, 14.45; Gly, 3.40; His, 13.69; Ile, 21.40; Leu, 21.40; Lys, 15.71; Met, 16.25; Phe, 19.80; Ser, 9.47; Thr, 15.77; Trp, 21.67; Tyr, 18.03; Val, 21.57.

[a] Ref. 56.   [b] DL.   [c] Unknown configuration.   [d] Ref. 24.   [e] Ref. 48.   [f] Uncharged.
[g] Ref. 57.   [h] Ref. 26.

only of n_H and n_n is due to its definition. As there are no ionic side chains in the data set, the dependence on i could not be determined.

The second group of three area scales (AAA 4, 5, and 6) consists of the solvent-accessible surface area of a residue in the standard state, A^0; the average solvent-accessible area of a residue in folded protein, $\langle A \rangle$; and the mean area based on transfer from the standard state to the folded protein $A^0 - \langle A \rangle$. Very good fits were obtained for sets AAA 4 and 6 on correlation with the IMFU equation; set AAA 5 gave a good fit. Both sets 4 and 6 have the same composition for the side-chain structural effect, with polarizability the predominant factor and small contributions from n_H and v. These results are in accord with our expectations. The results for set 5 are surprising, however, set 5 is a function of σ_1, n_H, n_n, and i but is completely independent

TABLE 10
Results of Compilations of AAA and AAV Sets with the IMF Equation*

Set:	V1	V2	A1	A2	A3	A4	A5	A6	A7
L	−101	—	—	—	—	—	−150.9	—	−21.7
$S_L{}^a$	27.2	—	—	—	—	—	70.9	—	9.12
A	217	299	482	—	122.8	400.7	—	386.9	19.9
$S_A{}^a$	9.36	52.4	29.3	—	4.76	22.7	—	29.6	5.17
H_1	—	−18.8	6.96	7.28	—	8.73	12.8	−9.14	−1.13
$S_{H1}{}^a$	—	5.82	2.47	1.18	—	1.92	3.69	3.30	0.425
H_2	1.40	—	—	4.45	0.959	—	5.85	−3.12	—
$S_H{}^{2a}$	0.861	—	—	0.828	0.318	—	3.04	2.22	—
I	—	31.9	nd	nd	nd	nd	17.6	—	—
$S_I{}^a$	—	13.0	nd	nd	nd	nd	7.06	—	—
S	23.8	—	37.6	—	—	25.6	—	27.2	14.2
$S_S{}^a$	4.53	—	13.6	—	—	10.6	—	12.6	2.78
B^0	42.4	34.3	55.3	4.93	49.6	87.1	31.2	60.3	4.01
$S_{B^0}{}^a$	2.83	10.3	8.73	1.26	1.08	6.77	3.95	7.79	1.49
$100R^{2b}$	98.59	71.76	97.05	92.49	97.65	97.50	85.96	96.20	89.75
F^c	244.1	12.70	164.4	98.59	333.1	194.9	21.44	88.69	30.64
$S_{est}{}^d$	3.43	21.1	10.4	3.93	2.01	8.08	11.0	9.26	1.72
S^{0e}	0.139	0.598	0.193	0.299	0.167	0.178	0.436	0.227	0.373
n^f	19	19	19	19	19	19	19	19	19

*For footnotes see Table 3.

TABLE 11
Zeroth-Order Partial Correlation Coefficients for AAV and AAA Sets

	PR	V1	PR[1]
σ, α	0.259	0.122	0.259
σ, n_H	0.164	0.410	0.164
σ, n_n	0.580[e]	0.772[b]	0.580[e]
σ, i	0.296	0.573[e]	—
σ, v	0.021	0.029	0.021
α, n_H	0.300	0.339	0.300
α, n_n	0.085	0.025	0.085
α, i	0.130	0.103	—
α, v	0.460[f]	0.483[f]	0.460[f]
n_H, n_n	0.661[c]	0.694[c]	0.661[c]
n_H, i	0.530[f]	0.564[e]	—
n_H, v	0.037	0.023	0.037
n_n, i	0.528[f]	0.676[c]	—
n_n, v	0.058	0.044	0.058
i, v	0.055	0.080	—
n	19	19	19

For footnotes see Table 4. PR sets include V2, A5, A6, A7. PR[1] sets include A1, A2, A3, A4. For those the zeroth-order correlation coefficients are identical to those of the PR sets except that all involving the i parameter are excluded.

TABLE 12
C_i Values for Volume and Area Data Sets

Set	σ_1	α	n_H	n_n	i	v
V1	10.6	65.6	0	1.84	0	21.9
V2	0	57.6	15.8	0	26.7	0
A1	0	76.9	4.83	0	nd	18.2
A2	0	0	62.0	38.0	nd	0
A3	0	96.7	0	3.28	nd	0
A4	0	77.6	7.34	0	nd	15.1
A5	25.0	0	26.4	12.1	36.5	0
A6	0	76.9	10.4	0	0	12.7
A7	9.99	26.2	6.52	0	0	57.2

of polarizability. It follows that the $\langle A \rangle$ scale is a function of hydrogen bonding, dipole–dipole, charge–dipole, and charge–charge interactions.

Finally, we consider the "bulkiness" parameter proposed by Bigelow (24) (set AAA 7). It is defined as the ratio of the side-chain volume to side-chain length and therefore has the dimension of area. It gives a good fit to the IMFU equation and is largely dependent on v, less so on α, with small contributions from σ_1 and n_H.

The argument that area and volume scales are a measure of hydrophobicity seems to rest on the fact that, like many of the amino acid scales we have examined, they are largely a function of polarizability, with some contribution from n_H.

B. Solution Properties

We have previously reported results (7, 58) for amino acid solubilities (59), enthalpies and free energies of solution (60), and hydration numbers (61, 62). Also studies were partial molar heat capacities at infinite dilution (63). The data used are set forth in Table 13, results of the correlations in Table 14, zeroth-order correlation coefficients in Table 15, and C_i values in Table 16.

The solubilities and thermodynamic properties of amino acid solutions are adequately modeled by the IMFU equation. The aqueous solubilities of Lys and Cys could not be included in the correlation of solubilities, as numerical values are not available. The literature states only that they are very soluble in the case of Lys and freely soluble in that of Cys. From the

TABLE 13
Amino Acid Solution Property Data Sets

S1. Solubility (moles/kg H_2O)·25°[a]
Ala, 1.85; Arg, 0.861; Asn, 0.235; Asp, 0.376; Glu, 0.0571; Gln, 0.248; Gly, 3.33; His, 0.489; Ile, 0.314; Leu, 0.175; Met, 0.235; Phe, 0.180; Ser, 0.476; Thr, 1.72; Trp, 0.0558; Tyr, 0.00276[x]; Val, 0.755.

S2. $H_2^0 - H_1^2$, Aax(s) to Aax(aq), infinite dilution, 25°[b]
Ala, 1.83; Asn, 5.8; Asp, 6.2; Tyr(3.5-I_2), 7.8; Gly, 3.4; His, 3.3; Ile, 0.9; Leu, 1.0; Met, 2.8; Phe, 2.7; Ser, 2.8; Trp, 1.4; Tyr, 6.0; Val, 0.9.

S3. $\Delta G_{\text{solution}}$ (kcal mole)[b]
Ala, −0.368; Arg, (+), −1.03; Asn, 0.983; Asp, 2.06; Cys$_2$, 4.53; Glu, 1.77; Gln, 0.731; Gly, −0.525; Ile, 0.791; Leu, 1.07; Met, 0.656; Phe, 1.06; Ser, −0.524; Trp, 1.60; Tyr, 3.55; Val, 0.461.

S4. C_p°, Partial Molar Heat Capacities at Infinite Dilution, 25°, (kJ/mol)[c]
Gly, 39; Ala, 141.4; Abu, 224; Val, 307; Leu, 382; Phe, 391; Ser, 117.4; Thr, 211; Asn, 158; Gln 246; Nva, 335; Nle, 400.

S5. n_{aq}, Hydration number[d]
Asp(−), 6; Glu(−), 7; Tyr(−), 7; Arg(+), 3; His(+), 4; Lys(+), 4; Asn, 2; Gln, 2; Ser, 2; Thr, 2; Trp, 2; Asp(0), 2; Glu(0), 2; Tyr(0), 3; Arg(0), 3; Lys(0), 4; Ala, 1; Gly, 1; Phe, 0; Val, 1; Ile, 1; Leu, 1; Met, 1.

S6. n_{aq}, Hydration number[e]
Asp(−), 6.5; Glu(−), 6.2; Arg(+), 2.3; His(+), 2.8; Lys(+), 5.3; Asn, 2.2; Gln, 2.1; Ser, 1.7; Thr, 1.5; Asp(0), 3.8; Glu(0), 3.6; Tyr, 2.1; Arg(0), 2.3; Lys(0), 1.2; Ala, 1.0; Gly, 1.1; Phe, 1.4; Val, 0.9; Ile, 0.8; Leu, 0.8; Met, 0.7.

[a] Ref. 59. [b] Ref. 60. [c] Ref 63. [d] Ref. 61. [e] Ref. 62. [x] Excluded.

TABLE 14
Results of Correlations of Amino Acids Solution Property (AAS) Data Sets

Set:	1	2	3	4	5	6
L	—	—	—	—	—	—
$S_L{}^a$	—	—	—	—	—	—
A	−5.13	7.34	10.4	1344.	—	—
S_A	0.768	1.90	1.68	189	—	—
H_1	0.530	—	−1.24	−51.6	—	−0.416
$S_{H1}{}^a$	0.0990	—	0.248	18.2	—	0.178
H_2	−0.420	1.43	0.893	—	0.360	0.696
$S_{H2}{}^a$	0.0660	0.183	0.151	—	0.150	0.131
I	—	—	—	—	3.05	1.12
$S_I{}^a$	—	—	—	—	0.532	0.430
S	—	−3.53	—	—	—	—
$S_S{}^a$	—	1.19	—	—	—	—
B^0	0.518	2.9	−1.17	96.6	1.21	2.44
$S_{B0}{}^a$	0.157	0.824	0.360	29.5	0.347	0.262
$100R^{2b}$	82.84	88.01	83.20	86.50	71.44	81.84
F^c	19.30	26.90	19.81	28.83	25.02	27.04
$S_{est}{}^d$	0.260	0.961	0.671	48.2	1.08	0.787
S^{0e}	0.478	0.404	0.473	0.424	0.573	0.471
n^f	16	15	16	12	23	22

For footnotes see Table 3.

TABLE 15
Zeroth-Order Partial Correlation Coefficients for Solution Properties Data Sets

Set:	1	2	3	4
σ, α	0.206	0.044	0.097	0.053
σ, n_H	0.344	0.344	0.507	0.716
σ, n_n	0.748	0.777	0.836	0.802
σ, i	0.515	0.563	0.659	—
σ, υ	0.008	0.005	0.009	0.059
α, n_H	0.274	0.258	0.422	0.024
α, n_n	0.146	0.064	0.156	0.020
α, i	0.133	0.006	0.211	—
α, υ	0.474	0.365	0.427	0.697
n_H, n_n	0.686	0.722	0.756	0.989
n_H, i	0.480	0.574	0.615	—
n_H, υ	0.024	0.028	0.030	0.164
n_n, i	0.611	0.564	0.778	—
n_n, υ	0.042	0.057	0.047	0.150
i, υ	0.055	0.061	0.055	—
n^g	16	15	16	12

For sets 5 and 6, see the values for the PBS in Table 4.

TABLE 16
Values of C_i for Amino Acid Solution Properties Data Sets

Set	σ_1	α	n_H	n_n	i	v
1	0	55.4	24.9	19.7	0	0
2	0	30.2	0	25.6	0	44.2
3	0	52.8	27.4	19.8	0	0
4	0	85.7	14.3	0	0	0
5	0	0	0	10.6	89.4	0
6	0	0	11.7	19.6	68.7	0

coefficients of the best regression equation (Table 15), we may calculate a solubility for Lys of 1.08 mol/kg water or 15.8 g/100 g water, while for Cys we obtain a solubility of 0.727 mol/kg water or 8.81 g/100 g water. We may similarly estimate the solubility of Cit which is reported to be soluble in water at 0.138 mol/kg water or 2.42 g/100 g water; and that of Orn, which is said to be freely soluble in water, is 1.86 mol/kg water or 24.6 g/100 g water.

Polarizability, hydrogen bonding, and in some cases steric effects are the dominant factors. Although hydration numbers are properties of amino acid residues in peptides or proteins, it is convenient to consider them here. The hydration number of an amino acid residue in a peptide or protein molecule is given by the average number of water molecules associated with the residue. The best scales available are those of Kuntz (61) and Hopfinger (62). Significant correlations were obtained for both of these data sets with the IMFU equation. The hydration number depends on the capacity for hydrogen bonding and the charge of the side-chain. It seems to be independent of van der Waals (dd, di, and ii) interactions.

C. Chromatographic Properties

1. Thin-layer Chromatography

An extensive literature has accumulated over the last 40 years on chromatographic properties of amino acids. In particular, much work has been done on thin-layer chromatography (TLC) of amino acids and their derivatives. We have investigated the applicability of the IMFU equation to 21 data sets consisting of TLC R_F values for amino acids. It is important in carrying out these correlations to take account of the composition of the mobile phase, which frequently contains acids or bases. If there are acids of sufficient strength present in sufficient concentration in the mobile phase, the ionization of side-chain carboxyl groups will be repressed. In that case, the value of i assigned to these side-chains must be zero, whereas otherwise it would have been 1. If bases of sufficient strength are present in sufficient concentration, the ionization of side-chain amino, guanidinyl, and imidazolyl groups will be repressed. Again, in that case, the value of i assigned to these groups will be 0 rather than the normal value of 1. All of the acidic mobile-phase sets we have studied (AAC 1–4, 11, 18, and 21) gave best results when acidic side-chains were assigned an i value of 0. Of the basic mobile-phase sets studied (sets AAC 19, 20, 22–24), only those in which the mobile-phase did not contain ketones gave best results with i equal to 0; the other sets (sets AAC 19, 22) gave best results with i equal to 1. It is conceivable that the dicyclohexylamine in set 19 formed an eneamine with the 2-butanone, and the ammonia in set 22 formed imines with the acetone and 2-butanone. These bases are much weaker than those originally present. The data used in the correlations are presented in Table 17; the results are reported in Table 18. Zeroth-order partial correlation coefficients are set forth in Table 19, and values of the percent contribution C_i are given in Table 20. Significant results were obtained for all of the 21 TLC data sets studied, with 12 sets giving $100R^2$ values greater than 90.0, and all but one set giving values greater than 82.0.

TABLE 17
Amino Acid Chromatographic Properties

AAC1. $100R_F$, TLC, cellulose, 2-BuOH–AcOH–H_2O (3:1:1)^a
Ala, 41.9; Arg, 25.6; Asp, 26.3; Cys, 69^x; Glu, 34.4; Gly, 29.4; His, 20.0; Ile, 73.1; Leu, 75.0; Lys, 18.1; Met, 41.0; Phe, 67.5; Ser, 26.9; Thr, 32.5; Trp, 55.6; Tyr, 50.0; Val, 63.1.

AAC2. $100R_F$, TLC microcrystalline cellulose, 2-BuOH–AcOH–H_2O (3:1:1)^a
Ala, 29.0; Arg, 11.0; Asp, 14.8; Cys, 3.2^x; Glu, 22.6; Gly, 29.4; His, 7.1; Ile, 60.0; Leu, 63.9; Lys, 7.1; Met, 22.5; Phe, 54.8; Ser, 16.1; Thr, 21.3; Trp, 36.1; Tyr, 36.1; Val, 48.4.

AAC3. $100R_F$, TLC silica gel, 1-BuOH–AcOH–H_2O (3:1:1)^a
Ala, 32.4; Arg, 12.9; Asp, 25.3; Cys, 14.1^x; Glu, 30.0; Gly, 25.9; His, 11.7; Ile, 49.4; Leu, 51.8; Lys, 10.0; Met, 47.3; Phe, 52.4; Ser, 26.4; Thr, 30.0; Trp, 54.1; Tyr, 49.4; Val, 43.5.

AAC4. $100R_F$, TLC, silica gel (high-performance) 1-BuOH–AcOH–H_2O(3:1:1)^a
Ala, 28.8; Arg, 10.0; Asp, 21.8; Cys, 7.1^x; Glu, 28.2; Gly, 25.9; His, 7.1; Ile, 47.1; Leu, 48.8; Lys, 7.1; Met, 43.5; Phe, 50.0; Ser, 24.1; Thr, 27.6; Trp, 51.8; Tyr, 45.9; Val, 41.2.

AAC5. $100R_F$, TLC, Fixion strong acid ion exchange sheet (Na form), 84 g citric acid–H_2O + 16.0 g NaOH + 5.8 g NaCl + 54.0 g HOCH$_2$CH$_2$OH + ca. 4 ml conc. HCl^a
Ala, 50.9; Arg, 1.8; Asp, 71.5; Cys, 55.9; Glu, 34.5; Gly, 55.6; His, 10.6; Ile, 27.8; Leu, 21.8; Lys, 7.5; Met, 28.0; Phe, 14.4; Ser, 64.7; Thr, 67.1; Trp, 1.8; Tyr, 11.9; Val, 42.5.

AAC6. $100R_F$, TLC, as in set AAC 5, pre-equilibrated 16 hr.^{a}
Ala, 51.2; Arg, 2.2; Asp, 68.2; Cys, 50.0; Glu, 29.4; Gly, 52.4; His, 8.9; Ile, 22.2; Leu, 17.8; Lys, 5.0; Met, 27.2; Phr, 11.1; Ser, 64.7; Thr, 60.6; Trp, 2.2; Tyr, 13.9; Val, 35.0.

AAC7. $100R_F$, TLC, as in set 5 at 45°^{a}
Ala, 53.6; Arg, 2.2; Asp, 68.6; Cys, 57.9; Glu, 30.6; Gly, 53.6; His, 10.0; Ile, 23.3; Leu, 19.4; Lys, 5.6; Met, 25.0; Phe, 11.7; Ser, 67.1; Thr, 57.2; Trp, 2.2; Tyr, 15.5; Val, 34.4.

AAC8. $100R_F$, RPTLC, MeCN–0.4% aq CF$_3$CO$_2$H (85:15)^b (Aax, $100R_F/C_2$, $100R_F/C_{18}$)
Ala, 22, 12; Asp, 4,—; Glu, 6,—; Gln, 13, 6; Gly, 11, 4; Ile, 59, 56; Leu, 62, 60; Phe, 68, 67; Thr, 20, 10; Tyr, 56, 50; Val, 50, 44.

TABLE 17 (*Continued*)

AAC9. **$100R_F$, RPTLC, MeCN–0.4% aq CF_3CO_2H (80:20)^b (Aax, $100R_F/C_2$, $100R_F/C_{18}$)**
Ala, 41, 35; Arg, 4,—; Asn, 23,—; Asp, 17, 12; Cys, 59,—; Glu, 8,—; Gln, 30, 25; Gly, 11, 4; His, 3,—; Ile, 80, 84; Leu, 88, 89; Phe, 90.97; Ser, 20,—; Thr, 38, 31; Tyr, 78, 82; Val, 74, 71; Cys_2, 6,—.

AAC10. **$100R_F$, TLC, silica gel, EtOH–H_2O(7:3 v/v)^c**
Ala, 60; Arg, 1; Asn, 39; Gln, 53; Gly, 44; His, 9; Ile, 80; Leu, 82; Lys, 1; Met, 77; Phe, 82; Ser, 44; Thr, 58; Trp, 86; Tyr, 81; Val, 74; Abu, 69; Nle, 84; Nva, 77; Orn, 1; Cit, 55; Omt^y, 66.

AAC11. **$100R_F$, TLC, silica gel, 1-BuOH–AcOH–H_2O (4:1:1 v/v)^c**
Ala, 24; Arg, 9; Asn, 14; Asp, 16; Cys, 23; Glu, 24; Gln, 16; Gly, 19; His, 6; Ile, 50; Leu, 52; Lys, 7; Met, 45; Phe, 50; Ser, 19; Thr, 25; Trp, 54; Tyr, 50; Val, 42; Abu, 34; Nle, 57; Nva, 46; Orn, 6; Cit, 17; Omt, 30.

AAC12. **$100R_F$, TLC, silica gel, PhOH–H_2O (3:1 w/w)^c**
Ala, 29; Arg, 6; Asn, 25; Asp, 9; Cys, 27; Glu, 15; Gln, 37; Gly, 22; His, 16; Ile, 49; Leu, 50; Lys, 2; Met, 50; Phe, 52; Ser, 15; Thr, 27; Trp, 54; Tyr, 47; Val, 43; Abu, 37; Nle, 51; Nva, 46; Orn, 2; Cit, 38; Omt, 17.

AAC13. **$100R_F$, TLC, silica gel, EtAc–PyH–AcOH–H_2O (70:15:2:15 v/v)^c**
Ala, 9; Arg, 1; Asn, 5; Arg, 2; Glu, 4; Gln, 7; Gly, 6; His, 1; Ile, 28; Leu, 33; Lys, 1; Met, 29; Phe, 35; Ser, 6; Thr, 11; Trp, 4.0; Tyr, 32; Val, 22; Abu, 15; Nle, 33; Nva, 24; Orn, 1; Cit, 6; Omt, 17.

AAC14. **$100R_F$, TLC cellulose, EtOH–H_2O (7:3 v/v)^c**
Ala, 52; Arg, 19; Asn, 21; Asp, 38; Cys, 46; Glu, 42; Gln, 32; Gly, 32; His, 16; Ile, 72; Leu, 73; Lys, 19; Met, 63; Phe, 66; Ser, 36; Thr, 46; Tyr, 68; Val, 68; Abu, 58; Nle, 74; Nva, 69; Orn, 17; Cit, 32; Omt, 62.

AAC15. **$100R_F$, cellulose, Py–iPeOH–H_2O (7:6:6 v/v)^c**
Ala, 23; Arg, 7; Asn, 12; Asp, 20; Glu, 19; Gln, 20; Gly, 22; His, 20; Ile, 53; Leu, 53; Lys, 7; Met, 49; Phe, 52; Ser, 23; Thr, 29; Trp, 51; Tyr, 51; Val, 43; Abu, 33; Nle, 54; Nva, 48; Orgn, 8; Cit, 22; Omt, 38.

AAC16. **$100R_F$, TLC, kieselbuhr, EtAc–H_2O–PhOH–MeAc–EtOH (1:1:1:1:1 v/v)^c**
Ala, 37; Arg, 1; Asn, 12; Asp, 19; Glu, 28; Gln, 23; Gly, 17; His, 2; Ile, 83; Leu, 83; Lys, 1; Met, 79; Phe, 87; Ser, 13; Thr, 30; Trp, 88; Tyr, 80; Val, 72; Abu, 55; Nle, 81; Nva, 73; Orn, 1; Cit, 29; Omt, 74.

AAC17. **K^1, AAx^+ $nC_{12}H_{25}OSO_3^-$, reversed-phase**
Ala, 6.2; Arg, 28.0; Asp, 4.3; Cys, 0.4^x; Glu, 4.6; Gln, 3.6; Gly, 5.7; His, 17.9; Ile, 18.4; Leu, 21.6; Lys, 16.6; Met, 11.1; Phe, 21.0; Ser, 3.8; Thr, 5.4; Trp, 24.9; Tyr, 10.1; Val, 10.6; Cys_2, 6.1; Nva, 12.0; Orn, 13.3.

AAC18. **$100R_F$, TLC, cellulose, iPrOH–H_2O (20:1:5)^e**
Ala, 34; Arg, 13; Asn, 12; Asp, 21; Glu, 30; Gln, 14; Gly, 22; His, 11; Ile, 65; Leu, 65; Lys, 10; Met, 50; Phe, 54; Ser, 21; Thr, 31; Trp, 41; Tyr, 38; Val, 55; Abu, 42; Cit, 16; Cys_2, 4; Dab, 8; Dap, 5; Dhp, 16; Tyr(3, 5-I_2), 50; Glu(3-OH)^x, 30; Kyn, 39; Lan, 8; Met(O), 20; Met(O_2), 19; Nle, 65; Nva, 55; Orn, 8; Phg 76; Tyr (4, 4'-PnOH), 62; Trx, 73.

AAC19. **$100R_F$, TLC cellulose, 2-BuOH–EtAc–CHx_2NH–H_2O (10:10:2:5)^e**
Ala, 22; Arg, 2; Asn, 7; Asp, 20; Glu, 21; Gln, 18; Gly, 19; His, 34; Ile, 62; Leu, 60; Lys, 6; Met, 54; Phe, 73; Ser, 38; Thr, 67; Trp, 60; Tyr, 56; Val, 44; Abu, 31; Cit, 15; Cys_2, 26; Dab, 20; Dap, 6; Dhp, 10; Tyr(3, 5-I_2), 73; Glu(3-OH), 15; Kyn, 47; Lan, 22; Met(O), 36; Met(O_2)^x, 26; Nle, 60; Nva, 45; Orn, 5; Phg, 81; Tyr (4-HOPn), 82; Trx, 84.

TABLE 17 (Continued)

AAC20. $100R_F$, cellulose, $PhOH-H_2O$ (3:1) + 7.5 mg NaCN, chamber saturated with $NH_3{}^e$
Ala, 61; Arg, 90; Asn, 47; Asp, 20; Glu, 25; Gln, 59; Gly, 46; His, 87; Ile, 88; Leu, 88; Cys, 82; Met, 88; Phe, 88; Ser, 4; Thr, 56; Trp, 88; Tyr, 71; Val, 87; Abu, 78; Cit, 68; Cys_2, 35; Dab, 73; Dap, 36; Dhp^x, 13; Tyr(3,5-I_2), 71; Glu(3-OH), 25; Kyn, 80; Lan, 35; Met(O), 73; Met(O_2), 82; Nle, 88; Nva, 87; Orn, 82; Phg, 77; Tyr (4-HOPn), 88; Trx, 86.

AAC21. $100R_F$, cellulose powder (MN-300) without buffer, washed before use, iPrOH–EtAc–1 m HCl (60:15:25 v/v)^e
Ala, 57; Arg, 19; Asn, 21; Asp, 48; Cys^x, 12; Glu, 56; Gln, 31; Gly, 37; His, 11; Ile, 90; Leu, 90; Lys, 16; Met, 78; Phe, 82; Ser, 39; Thr, 51; Trp, 70; Tyr, 72; Val, 79; Aad, 62; Abu, 65; Aoc, 63; Cit, 34; Cys_2, 6; Dab, 10; Dap, 16; Djk, 9; Dhp, 43; Hcy(Et), 83; Har, 23; Hse, 45; Kyn, 38; Met(O), 50; His(1-Me), 14; His(3-Me), 12; Met(O_2), 39; Nle, 92; Orn, 11; Pen, 34.

AAC22. $100R_F$, cellulose powder (MN-300), without buffer, washed before use, tBuOH–EtAc–MeAc–MeOH–H_2O–conc aq NH_3(40:20:20:1:14:5 v/v)^e
Ala, 23; Arg, 6; Asn, 14; Asp, 1; Cys^x, 5; Glu, 1; Gln, 13; Gly, 16; His, 26; Ile, 63; Leu, 69; Lys, 17; Met, 51; Phe, 67; Ser, 27; Thr^x, 61; Trp, 54; Tyr; 35; Val, 44; Aad, 3; Abu, 29; Aoc, 81; Cit, 12; Cys_2, 3; Dab, 25; Dap, 1; Djk, 5; Dhp, 6; Acy(Et), 66; Har, 12; Hse, 27; Kyn, 67; His(1-Me), 18; His(3-Me), 14; Met(O), 27; Met(O_2), 20; Nle, 73; Orn, 15; Pen^x, 15.

AAC23. $100R_F$, silica gel G, PrOH–34% aq NH_3 (70:30 v/v)^f
Ala, 39; Arg, 10; Asp, 9; Glu, 14; Gly, 29; His, 38; Ile, 52; Leu, 53; Lys, 18; Met, 51; Phe, 54; Ser, 27; Thr, 37; Trp, 55; Tyr, 42; Val, 48; Aoc, 58; Nle, 53; Nva, 49; Cys_x^2, 27.

AAC24. $100R_F$, silica gel G 96% aq EtOH–34% aq NH_3(70:30 v/v)^f
Ala, 40; Arg, 6; Asp, 7; Glu, 15; Gly, 34; His, 42; Ile, 58; Leu, 58; Lys, 11; Met, 60; Phe, 60; Ser, 31; Thr, 40; Trp, 58; Tyr, 51; Val, 56; Aoc, 60; Nle, 59; Nva, 57; Cys_x^2, 22.

AAC25. V_C/V_T, immobilized Ni iminodiacetate, 0.2 M Ethylmorpholine–H_2O, pH = 7.0 + 0.5 M $K_2SO_4{}^g$
Ala, 2.7; Arg, 3.9; Asn, 8.8; Asp, 1.6; Glu, 1.6; Gln, 6; Gly, 3.5; Ile, 3.2; Leu, 3.2; Lys, 3; Met, 4.7; Phe, 7; Ser, 4.6; Thr, 5.9; Tyr, 6.3; Val, 2.7.

AAC26. V_C/V_T, immobilized Ni iminodiacetate, 0.2 M ethylmorpholine–H_2O, pH = 7.0 + 0.5 M $K_2SO_4{}^g$
Ala, 2.7; Arg, 3.9; Asn, 8.8; Asp, 1.6; Glu, 1.6; Gln, 6; Gly, 3.5; Ile, 3.2; Leu, 3.2; Lys, 3; Met, 4.7; Phe, 7; Ser, 4.6; Thr, 5.9; Tyr, 6.3; Val, 2.7.

AAC27. V_C/V_T, immobilized Ni iminodiacetate, 0.2 M ethylmorpholine–H_2O, pH = 7.0 + 4 M NaCl^g
Ala, 2; Arg, 11; Asn, 9; Asp, 2; Glu, 2.2; Gln, 7.5; Gly, 2; Ile, 3.3; Leu, 3.3; Lys, 3; Met, 5; Ser, 5; Thr, 6; Val, 2.8.

AAC28. K', reversed-phase (ODS Partisil 10) column, 3×10^{-4} M aq Cu(II), $34°{}^h$
Ala, 1.12; Arg, 7.0; Asn, 12.09; Asp, 0.24; Glu, 0.29; Gly, 0.88; His, 1.86; Leu, 7.5; Lys, 3.9; Met, 4.2; Phe, 28.0; Ser, 0.97; Thr, 1.24; Trp, 67.0; Tyr, 9.6; Val, 2.47; Nva, 3.5; Ohp, 4.7.

AAC29. $100R_F$, Whatman 3 MM, 1.0 M NH_4O Ac-satd. aq $(NH_4)_2SO_4$ (10:90) pH 7.0, ca. $25°{}^i$
Ala, 89; Arg, 88; Asn, 89; Asp, 87; Cys, 85; Glu, 84; Gln, 82; Gly, 92; His, 83; Ile, 76; Leu, 73; Lys, 97; Met, 74; Phe, 52; Ser, 96; Thr, 92; Trp, 20; Tyr; 49; Val, 85.

---

^{a}Ref. 64. ^{b}Ref. 68. ^{c}Ref. 65. ^{d}Ref. 69.^{e} ^{e}Ref. 66. ^{f}Ref. 67. ^{g}Ref. 70.
^{h}Ref. 71. ^{i}Ref. 72. ^{x}Excluded from the correlation. ^{y}Omt = Thr(Me).

TABLE 18
Results of Correlations of Amino Acid Chromatographic Properties with the IMF Equation

Set:	1A	2A	3A	4A	5	6
L	−2.07	−2.69	—	−1.21	−1.84	1.91
$S_L{}^a$	0.420	0.593	—	0.265	0.648^h	0.522^f
A	—	—	0.721	0.739	−3.38	−3.08
$S_A{}^a$	—	—	0.127	0.143	0.327	0.263
H_1	—	—	—	—	−0.218	−0.191
$S_{H1}{}^a$	—	—	—	—	0.0405	0.0326
H_2	—	—	−0.0357	—	0.0810	0.107
$S_{H2}{}^a$	—	—	0.00788	—	0.0313^j	0.0252^f
I	−0.325	−0.560	−0.559	−0.704	−0.159	−0.280
$S_I{}^a$	0.0506	0.0715	0.0301	0.0327	0.0799^k	0.0643^f
S	0.397	0.588	0.157	0.166	0.346	0.218
$S_S{}^a$	0.0918	0.130	0.0583^i	0.0646^j	0.138^i	0.111^k
B^0	1.46	1.18	1.39	1.36	1.76	1.74
$S_{B0}{}^a$	0.0680	0.0961	0.0374	0.0417	0.0861	0.0693
$100R^{2b}$	87.53	89.51	97.63	98.01	97.64	98.34
F^c	28.08	34.14	113.3	135.8	69.01	98.94
$S_{est}{}^d$	0.790	0.112	0.0436	0.0481	0.0986	0.0793
S^{0e}	0.408	0.374	0.186	0.170	0.200	0.168
n^f	16	16	16	16	17	17

Set:	7	8^{aa}	9	10	11A	12
L	1.99	—	—	—	−1.26	—
$S_L{}^a$	0.504^f	—	—	—	0.450^h	—
A	−3.05	2.61	2.16	1.28	1.40	1.77
$S_A{}^a$	0.254	0.384	0.280	0.401^g	0.252	0.419
H_1	−0.196	−0.312	—	−0.211	−0.120	−0.230
$S_{H1}{}^a$	0.0315	0.0418	—	0.0779^f	0.0239	0.0499
H_2	0.0991	—	−0.113	0.109	—	0.112
$S_{H2}{}^a$	0.0243^f	—	0.0119	0.0555^k	—	0.0327^f
I	−0.250	−0.526	−1.15	−1.26	−0.537	−0.698
$S_I{}^a$	0.0621^f	0.0964	0.0756	0.129	0.0663	0.088
S	0.212	0.506	0.238	—	—	—
$S_S{}^a$	0.107^k	0.127^f	0.100^j	—	—	—
B^0	1.75	−3.17	1.29	1.65	1.37	1.30
$S_{B0}{}^a$	0.0670	1.27	0.0602	0.0803	0.0511	0.0809
$100R^{2b}$	98.44	94.72	95.27	95.59	91.07	86.90
F^c	104.8	50.23	115.9	92.08	51.01	33.17
$S_{est}{}^d$	0.0766	0.117	0.988	0.153	0.100	0.163
S^{0e}	0.163	0.275	0.240	0.239	0.334	0.405
n^f	17	20	28	22	25	25

TABLE 18 (Continued)

Set:	13	14	15	16	17	18A
L	−1.93	−1.07	—	−3.78	—	—
$S_L{}^a$	1.02[k]	0.603[k]	—	1.23[g]	—	—
A	2.57	0.933	1.54	2.59	1.68	0.753
$S_A{}^a$	0.370	0.321[h]	0.232	0.446	0.298	0.144
H_1	−0.349	−0.203	−0.238	−0.614	—	—
$S_{H1}{}^a$	0.0497	0.350	0.0275	0.0599	—	—
H_2	0.114	0.0735	0.0549	0.334	−0.113	−0.102
$S_{H2}{}^a$	0.0404[h]	0.0262[h]	0.0181[g]	0.0486	0.0193	0.0156
I	−0.767	−0.185	−0.277	−0.788	−0.269	−0.551
$S_I{}^a$	0.0827	0.0597	0.0410	0.0996	0.0764[g]	0.0609
S	—	—	—	—	—	—
$S_S{}^a$	—	—	—	—	—	—
B^0	0.908	1.63	1.36	1.36	0.778	1.53
$S_{B0}{}^a$	0.0744	0.0581	0.0458	0.0896	0.0646	0.0452
$100R^{2b}$	94.72	82.57	91.84	94.60	82.15	84.74
F^c	64.52	17.06	53.48	63.07	24.54	57.39
$S_{est}{}^d$	0.144	0.105	0.0900	0.173	0.135	0.146
S^{0e}	0.265	0.482	0.321	0.268	0.472	0.415
n^f	24.	24.	24.	24.	20.	35.

Set:	19N	20B	21A	22N	23B	24B
L	2.40	—	−2.08	2.28	—	—
$S_L{}^a$	0.727[f]	—	0.435	0.904[h]	—	—
A	0.640	0.303	0.594	0.900	0.974	1.16
$S_A{}^a$	0.194[f]	0.0695	0.233[h]	0.383[j]	0.144	0.233
H_1	−0.228	0.0317	−0.0535	0.125	−0.241	−0.373
$S_{H1}{}^a$	0.0398	0.0138[f]	0.0212[h]	0.0446[g]	0.253	0.0407
H_2	—	−0.0477	—	−0.301	0.0522	0.125
$S_{H2}{}^a$	—	0.0115	—	0.0333	0.0254[k]	0.0409[g]
I	−0.267	−0.290	−0.574	−0.530	−0.564	−0.804
$S_I{}^a$	0.0961[g]	0.0446	0.0556	0.0772	0.0852	0.137
S	—	0.272	—	0.537	—	—
$S_S{}^a$	—	0.0716	—	0.218[i]	—	—
B^0	1.53	1.64	1.75	1.17	1.52	1.54
$S_{B0}{}^a$	0.0661	0.0518	0.0504	0.148	0.0306	0.0493
$100R^{2b}$	76.94	88.67	87.27	89.62	96.63	94.65
F^c	23.03	45.39	56.54	41.71	100.4	61.90
$S_{est}{}^d$	0.200	0.0674	0.128	0.190	0.0536	0.0863
S^{0e}	0.519	0.370	0.383	0.359	0.214	0.269
n^f	35	35	38	36	19	19

TABLE 18 (Continued)

Set:	25	26	27	28	29
L	3.91	—	1.41	—	—
$S_L{}^a$	1.18^g	—	0.321^f	—	—
A	1.64	0.684	1.47	4.45	−1.80
$S_A{}^a$	0.341	0.322^k	0.233	0.379	0.246
H_1	0.290	0.125	0.183	0.154	0.0746
$S_{H1}{}^a$	0.0466	0.0301^f	0.0163	0.0526^h	0.0280
H_2	−0.160	—	—	−0.201	−0.0334
$S_{H2}{}^a$	0.0473^g	—	—	0.0318	0.0192
I	−0.380	−0.491	−0.465	−0.396	0.103
$S_I{}^a$	0.0922^f	0.0724	0.0372	0.0930^f	0.0494
S	—	—	—	—	0.266
$S_S{}^a$	—	—	—	—	0.105
B^0	0.0879	0.479	0.268	−0.134	1.96
$S_{B0}{}^a$	0.0744	0.0563	0.0352	0.0813^m	0.0645
$100R^{2b}$	90.51	81.37	97.31	95.76	83.36
F^c	20.58	17.47	81.38	73.44	13.02
$S_{est}{}^d$	0.124	0.104	0.0499	0.148	0.0769
S^{0e}	0.383	0.498	0.205	0.242	0.493
n^f	17	16	14	18	19

For footnotes see Table 3. $^{aa}Z = 2.49$, $S_Z = 0.789^g$.

TABLE 19
Zeroth-Order Partial Correlation Coefficients for Amino Acid Chromatographic Properties Data Sets

Set:	1, 2, 3, 4A	5, 6, 7	10	11A	12	13, 15, 16
σ, α	0.222	0.253	0.088	0.203	0.203	0.177
σ, n_H	0.224	0.147	0.208	0.106	0.106	0.181
σ, n_n	0.750	0.602	0.518	0.519	0.519	0.638
σ, i	0.013	0.319	0.103	0.021	0.275	0.343
σ, υ	0.002	0.028	0.010	0.029	0.029	0.058
α, n_H	0.340	0.355	0.354	0.356	0.356	0.344
α, n_n	0.089	0.061	0.127	0.032	0.032	0.013
α, i	0.289	0.118	0.240	0.263	0.125	0.114
α, υ	0.470	0.475	0.480	0.458	0.458	0.455
n_H, n_n	0.587	0.603	0.798	0.689	0.689	0.679
n_H, i	0.721	0.667	0.562	0.559	0.501	0.491
n_H, υ	0.001	0.014	0.027	0.040	0.040	0.029
n_n, i	0.162	0.679	0.151	0.066	0.406	0.395
n_n, υ	0.022	0.035	0.043	0.082	0.082	0.071
i, υ	0.013	0.071	0.017	0.014	0.056	0.048
n^g	16	17	22	25	25	24

TABLE 19 (Continued)

Set:	14	17	18a	19n	20B	21A	22N
σ, α	0.138	0.092	0.193	0.180	0.180	0.242	0.276
σ, n_H	0.111	0.418	0.253	0.316	0.275	0.167	0.197
σ, n_n	0.505	0.815	0.586	0.603	0.614	0.560	0.603
σ, i	0.262	0.068	0.117	0.306	0.482	0.145	0.304
σ, υ	0.034	0.010	0.156	0.155	0.156	0.225	0.188
α, n_H	0.409	0.447	0.203	0.193	0.186	0.419	0.428
α, n_n	0.157	0.182	0.388	0.334	0.332	0.318	0.331
α, i	0.225	0.176	0.015	0.099	0.021	0.372	0.225
α, υ	0.530	0.429	0.186	0.186	0.185	0.290	0.331
n_H, n_n	0.704	0.687	0.672	0.703	0.662	0.687	0.678
n_H, i	0.507	0.571	0.657	0.564	0.414	0.551	0.492
n_H, υ	0.039	0.009	0.036	0.029	0.035	0.145	0.083
n_n, i	0.395	0.031	0.238	0.414	0.681	0.227	0.410
n_n, υ	0.088	0.029	0.158	0.154	0.154	0.009	0.106
i, υ	0.059	0.009	0.040	0.014	0.014	0.115	0.073
n^g	24	20	35	35	35	38	36

Set:	23, 24	25	26	27	28	8	9
σ, α	0.223	0.265	0.179	0.157	0.223	0.038	0.123
σ, n_H	0.331	0.271	0.280	0.275	0.254	0.648	0.479
σ, n_n	0.778	0.764	0.751	0.771	0.681	0.831	0.734
σ, i	0.580	0.331	0.311	0.300	0.389	0.635	0.593
σ, υ	0.013	0.004	0.008	0.011	0.122	0.049	0.031
α, n_H	0.295	0.284	0.336	0.529	0.342	0.187	0.434
α, n_n	0.102	0.108	0.033	0.130	0.007	0.112	0.248
α, i	0.204	0.074	0.189	0.395	0.068	0.105	0.236
α, υ	0.468	0.457	0.543	0.645	0.484	0.604	0.504
n_H, n_n	0.627	0.648	0.670	0.652	0.650	0.876	0.771
n_H, i	0.111	0.556	0.564	0.558	0.517	0.248	0.515
n_H, υ	0.008	0.022	0.022	0.027	0.124	0.051	0.058
n_n, i	0.748	0.588	0.577	0.585	0.467	0.669	0.697
n_n, υ	0.012	0.045	0.051	0.057	0.149	0.068	0.089
i, υ	0.064	0.042	0.045	0.053	0.109	0.053	0.077
σ, ζ	—	—	—	—	—	0.151	0.187
α, ζ	—	—	—	—	—	0.105	0.093
n_H, ζ	—	—	—	—	—	0.039	0.215
n_n, ζ	—	—	—	—	—	0.157	0.098
i, ζ	—	—	—	—	—	0.095	0.012
v, ζ	—	—	—	—	—	0.270	0.123
n_g	19	17	16	14	18	20	28

TABLE 20
Values of C_i for Amino Acid Chromatographic Properties

Set:	σ_I	α	n_H	n_n	i	v
1A	21.5	0	0	0	42.3	36.2
2A	18.1	0	0	0	47.2	34.7
3A	0	19.1	0	4.10	64.2	12.6
4A	8.92	15.6	0	0	64.8	10.7
5	9.04	47.8	13.4	4.98	9.79	14.8
6	9.59	44.5	12.0	6.74	17.6	9.59
7	10.2	45.2	12.6	6.38	16.1	9.53
8[a]	0	14.0	7.30	0.	12.3	8.27
9	0	25.8	0	5.86	59.7	8.63
10	0	15.2	13.9	5.62	65.2	0
11A	9.32	29.8	11.1	0.	49.8	0
12	0	28.2	15.9	7.74	48.2	0
13	7.83	29.9	17.7	5.75	38.9	0
14	11.3	28.2	26.6	9.65	24.3	0
15	0	38.3	25.8	5.95	30.0	0
16	11.5	22.6	23.3	12.7	29.9	0
17	0	50.3	0	14.7	35.0	0
18A	0	21.0	0	12.3	66.7	0
19N	23.0	17.6	27.3	0	32.0	0
20B	0	11.1	5.03	7.56	46.1	30.3
21A	17.9	14.7	5.76	0	61.6	0
22N	10.6	12.0	7.28	17.5	30.8	21.8
23B	0	20.7	22.3	4.83	52.2	0
24B	0	17.0	23.8	7.94	51.3	0
25	20.6	24.8	19.1	10.5	25.0	0
26	0	20.3	16.2	0	63.5	0
27	10.3	30.7	16.6	0	42.4	0
28	0	57.6	8.70	11.3	22.3	0
29	0	51.0	9.19	4.11	12.7	23.0

[a] $C_z = 58\%$

We have examined the dependence of the C_i values on the nature of the layer and of the mobile-phase by correlating them with the equation

$$C_i = a_c p_c + a_s p_s + a_i p_i + a_A p + a_B p_B + a_0 \tag{45}$$

where p_c, p_s, and p_i are layer parameters which take the value i if the layer is cellulose, silica gel, or the ion-exchange resin Fixion, respectively, and 0 otherwise; p_A and p_B are mobile-phase parameters which take the value 1 if the mobile-phase is acidic or basic, respectively, and 0 otherwise. a_c, a_s, a_i, a_A, and a_B are the coefficients of the corresponding parameters and a_0 is the

TABLE 21
Results of Correlations with Equation 45

Set:	$C(\sigma_1)$	$C(\alpha)$	$C(n_H)$	$C(n_n)$	$C(i)$	$C(v)$
a_c	—	—	—	—	—	10.2
s_{ac}	—	—	—	—	—	5.42^k
a_s	−6.47	6.58	—	—	12.4	—
s_{as}	2.97^j	4.07^m	—	—	4.27^h	—
a_l	—	24.3	−6.33	—	−18.3	11.2
s_{al}	—	5.88	4.07^m	—	6.18^g	7.84^m
a_A	—	−10.0	−16.3	−5.04	18.6	7.52
s_{aA}	—	4.35^j	3.02	1.96^h	4.57	5.56^m
a_B	−7.22	−9.65	—	—	8.85	—
s_{aB}	4.12^k	5.80^m	—	—	6.10^m	—
a^0	11.5	21.5	19.0	7.38	32.8	0.0810
s_a^0	1.80	3.34	1.89	1.13	3.50	4.24
$100R^2$	36.33	68.24	61.75	25.73	79.35	26.91
F	5.136^i	8.593	14.53	6.583^i	15.37	2.078^l
S_{est}	6.41	8.39	6.25	4.24	8.81	11.4
s^0	0.862	0.646	0.668	0.906	0.521	0.950
n	21	21	21	21	21	21

Zeroth-order correlation coefficients are r_{cs}, 0.679; r_{cI}, 0.354; r_{cA}, 0.204; r_{cB}, 0.079; r_{sI}, 0.320; r_{sA}, 0.069; r_{sB}, 0.240; r_{IA}, 0.289; r_{IB}, 0.167; r_{AB}, 0.289.

Footnotes are from Table 3.

intercept. The results of the correlations are reported in Table 21. From these results we may draw the following conclusions:

1. The contributions of the σ_1 and α parameters which represent the van der Waals interactions are a function of the presence of a silica gel layer and possibly of that of an ion-exchange layer as well. They also depend on the presence of an acidic and that of a basic mobile-phase.
2. The contributions of the n_H and n_n parameters which represent hydrogen bonding are certainly dependent on the presence of an acidic mobile-phase; they may possibly be dependent on that of a basic one as well.
3. The contribution of the i parameter which represents the ion–dipole and ion–induced dipole interactions depends on the presence of a silica gel or of an ion-exchange layer. It is also dependent on the presence of an acidic mobile-phase and may possibly be dependent on the presence of a basic mobile-phase as well.
4. The nature of the dependence of the steric effect contribution on the type of layer or mobile phase is uncertain, as the correlation of the C_i values with Equation 45 is not significant. There may possibly be a

dependence on the presence of cellulose and ion-exchange layers and on a basic mobile-phase.

Overall, an acidic mobile-phase seems to have the greatest effect on the intermolecular forces involved in TLC.

2. Other Chromatographic Methods

In addition to the TLC data, we have studied other chromatographic properties of amino acids as well. These properties include 100 R_F values for reversed-phase TLC (68), k' values for reversed-phase chromatography of amino acid–dodecyl sulfate ion pairs (69), v_c/v_t ratios for chromatography on immobilized nickel iminodiacetate (70), k' values for reversed-phase chromatography with a copper II mobile-phase (71), and 100 R_F values for paper chromatography with a saturated ammonium sulfate mobile phase (72). In order to obtain usable data sets for the reversed-phase TLC data, we have applied the Zeta method (73, 74). This method permits the combination of subsets that differ in a single experimental condition into a single data set. The method is described in Appendix 2. Again, the data used, results of correlations, zeroth-order partial correlation coefficients, and C_i values are given in Tables 17, 18, 19, and 20, respectively. Of the eight sets of chromatographic properties which are not derived from TLC, five had $100R^2$ values greater than 90.00.

It is interesting to note that in seven of the sets studied (sets AAc 1–4, 17, 21, 22) and in set 22, the data point for Cys had to be excluded from the correlation, and that in set 22 the data point for Pen had to be excluded as well. It is possible that, under the experimental conditions used for these sets, oxidation of the sulfhydryl group with the formation of the disulfide bond occurs, resulting in the formation of cystine and of the corresponding Pen disulfide. The disulfide can be expected to have very different properties from those of the sulfhydryl amino acids, as is shown by a comparison of the parameters given in Appendix 1 for cystine and cysteine.

D. Other Amino Acid Properties

Saringa (73) has reported selectivity coefficients K_P^N for binding of ribonucleoside 5'-phosphates versus phosphate to immobilized L-amino acids. We have used the Zeta method (Appendix 2) to combine the data for the five bases studied into a single data set (set AAO1, Table 22). The results of the correlation are given in Table 23, the zeroth-order partial correlation coefficients in Table 24, and the C_i values in Table 25. The results show that the selectivity is a function of all the variables in the IMFU equation with

TABLE 22
Other Amino Acid Properties

AAO1. Selectivity coefficients K_p^N for binding of ribonuclease-5'-monophosphate and phosphate to immobilized Aax; Aax, 5'-Up, 5'-Cp, 5'-Ip, 5'-Ap, 5'-Gp[a]
Gly, 0.48, 0.58, 0.80, 0.52, 1.00; Lys, 0.56, 0.66, 0.90, 0.62, 0.92; Met, 1.20, 1.10, 1.37, 1.62, 2.26; Arg, 1.38, 1.38, 2.06, 2.19, 2.98; His, 1.45, 1.51, 2.19, 2.30, 3.20; Phe^z, 1.50, 1.39, 2.74, 3.00, 4.30; Trp, 1.99, 1.48, 5.48, 4.17, 6.61; Tyr, 2.38, 2.24, 8.96, 4.61, 6.00.

AAO2. Capacity factors K' for aax in water or aqueous MeCN; % MeCN = 0, 10, 15[b]
Asp, 1.44, 0.10, 0.05; Glu, 2.20, 0.28, 0.10; Asn, 2.70, 0.60, 0.18; Gln^z, 3.52, 0.68, 0.20; Ala, 3.24, 1.10, 0.12; Val, 16.90, 6.40, 1.54; Tyr, 21.86, 8.00, 1.90; Nva, 25.00, 9.24, 1.96; PhGly, 52.00, 21.80, 5.00; Ile, —, 19.74, 4.40; Leu, —, 26.60, 5.64; Nle, —, 31.50, 6.60; Phe, —, 71.20, 14.70; Trp, —, —, 20.80.

AAO3. Capacity factors, K', for Aax in water or aqueous MeCN; % MeCN = 0, 10, 15[b]
Asp, 2.00, 0.29, 0.05; Glu, 2.63, 0.58, 0.10; Asn, 2.93, 0.60, 0.18; Gln^z, 3.52, 0.68, 0.20; Ala, 2.40, 0.78, 0.12; Val, 8.25, 3.70, 1.00; Tyr, 10.51, 4.50, 1.20; Nva, 12.00, 5.20, 1.22; PhGly, 22.81, 9.20, 2.60; Ile, —, 9.40, 2.60; Leu, —, 16.00, 3.82; Nle, —, 16.80, 4.10; Phe, —, 37.40, 9.50; Trp, —, —, 13.20.

AAO4. Rate constants for hydrolysis of $PhCH_2O_2AaxCOC_6H_4NO_2$-4 and $PhCH_2O_2CaaxC_6H_4NO_2$-4 by NoCO–His, 1, and PhO_2C–Leu–His, 2
Gly, 162, 72.0; gly, 161, 64.6; Ala^z, 280, 202; ala^z, 141, 52.8; Val, 58.1, 51.9; val, 20.5, 9.03; Leu, 41.5, 36.0; leu, 134, 52.3; Ile, 36.1, —; ile, 13.1, —; Phe, 572, 473; phe, 228, 78.4; Trp, 42.3, 19.0; trp, 36.5, 11.6.

AAO5, 6. Electrostatic (E) and van der Waals (V) free energies for the transformation of Ala into Aax. Aax, ΔG_E, ΔG_V[d]
Arg, −12.51, −1.89; Cys, −2.02, −1.02; His, −11.02, −2.30; Leu, 0.30, 0.35; Lys, −7.03, 0.28; Phe, −2.35, −0.51; Ser, −6.16, −0.88; Trp, −4.10, −3.19.

AAO7, 8. Polarity and R_F rank index parameters for amino acids[e]**. Aax, P, RI**
Ala, 0.00, 9.9; Arg, 52.00, 4.6; Asp, 49.70, 2.8; Asn, 3.38, 5.4; Cys, 1.48, 2.8; Glu, 49.90, 3.2; Gln, 3.53, 9.0; Gly, 0.00, 5.6; His, 51.60, 8.2; Ile, 0.13, 17.1; Leu, 0.13, 17.6; Lys, 49.50, 3.5; Met, 1.43, 14.9; Phe, 0.35, 18.8; Ser, 1.67, 6.9; Thr, 1.66, 9.5; Trp, 2.10, 17.1; Tyr, 1.61, 15.0; Val, 0.13, 14.3.

AAO9. Bonding of $AaxNH_2$ to polyribonucleotides. $AaxNH_2$, K_h/K_c (poly(A)·poly(U)), K_h/K_c (poly(I)·poly(C))
Ala, 1.30, 1.20; Asn, 1.00, 1.00; Gly, 1.15, 1.15; Ile, 0.76, 0.78; Leu, 0.75, 0.75; Met, 0.80, 0.76; Phe, 0.70, 0.59; Ser, 1.07, 1.16; Trp, 0.75, 0.71; Tyr, 0.71, 0.71; Val, 0.80, 0.85.

AAO10. Binding of $AaxNH_2$ to polyribonucleotides. $AaxNH_2$, k_H (poly(A)·poly(U)), K_h(poly(I)·poly(L)))[f]
Ala, 150^x, 210; Asn 260, 360; Gly^z, 350, 480; Ile, 440, 470; Leu, 430, 440; Met, 460, 380; Phe, 480, 290^x, Ser, 380, 410; Trp, 1480, 1060; Tyr, 750, 750; Val, 420, 490.

[a] Ref. 73. [b] Ref. 74. [c] Ref. 75. [d] Ref. 76. [e] Ref. 26, P is defined as M_x for nonionic side chains, +48 for ionic side chains. RI is defined as ranked values of R_F averaged over seven solvent systems. [f] Ref. 76a. [x] Excluded. [z] Defined as zeta.

TABLE 23
Results of Correlations of Other Amino Acid Properties with the IMFU and IMFSH Equations*

Set:	1	2	3	4	5	6
Z	0.754	0.965	0.915	0.943	—	—
$S_Z{}^a$	0.0840	0.0762	0.0750	0.138	—	—
L	2.85	—	—	13.5	−33.2	−21.9
$S_L{}^a$	0.832	—	—	2.81	11.2	6.76
A	3.46	5.45	4.88	—	−15.9	−13.2
$S_A{}^a$	0.286	0.473	0.478	—	5.33	3.24
H_1	−0.195	—	—	−0.634	—	—
$S_{H1}{}^a$	0.0264	—	—	0.104	—	—
H_2	0.202	−0.342	−0.294	—	−1.83	—
$S_{H2}{}^a$	0.0326	0.0223	0.0312	—	0.506	—
I	—	—	0.240	—	−4.84	—
$S_I{}^a$	—	—	0.141	—	1.08	—
S	−0.724	1.11	1.03	—	—	—
$S_S{}^a$	0.144	0.299	0.293	—	—	—
B_1	—	—	—	−0.473	—	—
$S_{B1}{}^a$	—	—	—	0.0542	—	—
B_2	—	—	—	0.321	—	—
$S_{B2}{}^a$	—	—	—	0.0539	—	—
B^0	−0.474	−0.637	−0.725	0.135	2.94	2.87
$S_{B0}{}^a$	0.0542	0.228	0.224	0.304	1.64	0.999
$100R^2$ [b]	92.00	94.40	92.61	89.43	97.26	77.53
F^c	63.26	130.6	75.14	33.85	26.59	8.626
$S_{est}{}^d$	0.0992	0.216	0.212	0.187	1.13	0.697
S^0 [e]	0.144	0.255	0.298	0.371	0.270	0.600
n^f	40	36	36	26	8	8

Set:	7	8	9	10
L	10.6	−32.8	—	—
$S_L{}^a$	3.55[g]	13.3[j]	—	—
A	2.95	25.8	−0.782	1.63
$S_A{}^a$	1.72[m]	7.31[f]	0.0978	0.228
H_1	0.899	−1.98	0.116	—
$S_{H1}{}^a$	0.182	0.681[h]	0.0317	—
H_2	—	—	−0.0519	—
$S_{H2}{}^a$	—	—	0.0193	—
I	47.8	−4.69	nd	nd
$S_I{}^a$	0.416	1.54	nd	nd
S	—	6.82	−0.0567	−0.281
$S_S{}^a$	—	3.22[i]	0.0365	0.0974
B	−0.147	5.030	0.0949	2.57
$S_0{}^a$	0.385	2.112[j]	0.0222	0.0618
$100R^2$ [b]	99.93	86.75	88.61	75.36
F^c	522.8	17.03	33.05	26.00
$S_{est}{}^d$	0.655	2.43	0.0362	0.101
S^0 [e]	0.0301	0.440	0.384	0.538
n^f	19	19	22	20

*For footnotes see Table 3.

TABLE 24
Zeroth-Order Partial Correlation Coefficients for Other Amino Acid Property Data Sets

Set:	1	2, 3	4	5, 6	9	10
ζ, σ	0.000	0.140	0.037	—	—	—
ζ, α	0.000	0.164	0.002	—	—	—
ζ, n_H	0.000	0.103	0.017	—	—	—
ζ, n_n	0.000	0.148	—	—	—	—
ζ, i	0.000	0.082	—	—	—	—
ζ, υ	0.000	0.153	—	—	—	—
ζ, n_1	—	—	0.060	—	—	—
ζ, n_2	—	—	0.017	—	—	—
σ, α	0.118	0.012	0.350	0.649	0.138	0.211
σ, n_H	0.086	0.393	0.072	0.136	0.573	0.574
σ, n_n	0.310	0.634	—	0.238	0.723	0.730
σ, i	0.374	0.618	—	0.101	—	—
σ, v	0.422	0.191	—	0.685	0.097	0.128
σ, n_1	—	—	0.249	—	—	—
σ, n_2	—	—	0.084	—	—	—
α, n_H	0.303	0.016	0.732	0.257	0.168	0.174
α, n_n	0.223	0.187	—	0.169	0.047	0.111
α, i	0.014	0.287	—	0.153	—	—
α, υ	0.823	0.089	—	0.244	0.519	0.514
α, n_1	—	—	0.435	—	—	—
α, n_2	—	—	0.570	—	—	—
n_H, n_n	0.854	0.817	—	0.854	0.885	0.881
n_H, i	0.738	0.230	—	0.738	—	—
n_H, υ	0.251	0.000	—	0.229	0.034	0.017
n_H, n_1	—	—	0.045	—	—	—
n_H, n_2	—	—	0.182	—	—	—
n_n, i	0.582	0.723	—	0.582	—	—
n_n, υ	0.251	0.007	—	0.408	0.000	0.016
n_n, n_1	—	—	—	—	—	—
n_n, n_2	—	—	—	—	—	—
i, υ	0.228	0.013	—	0.078	—	—
i, n_1	—	—	—	—	—	—
i, n_2	—	—	—	—	—	—
n_1, n_2	—	—	0.247	—	—	—

TABLE 25
C_i Values for Other Amino Acid Properties

Set	ζ	σ_1	α	n_H	n_n	i	υ	n_1	n_2
1	14.6	10.1	35.3	8.65	8.94	—	22.4	—	—
2	18.2	—	43.2	—	11.8	—	26.7	—	—
3	17.4	—	39.0	—	10.2	8.33	25.0	—	—
4	44.7	23.9	—	14.0	—	—	—	10.4	7.07
5	—	20.4	28.2	—	14.1	37.3	—	—	—
6	—	36.6	63.4	—	—	—	—	—	—
7	—	1.69	1.35	1.79	—	95.2	—	—	—
8	—	13.1	29.7	9.88	—	23.5	23.9	—	—
9	—	0	44.5	28.7	12.8	—	14.0	—	—
10	—	0	57.1	0	0	—	42.9	—	—

the exception of the side-chain charge parameter i. The largest contributions to the side-chain effect are from polarizability and steric effects, with hydrogen bonding also of considerable importance.

In the course of a study on the chromatographic resolution of amino acid enantiomers, Nimura and coworkers (74) have reported capacity factors for the reversed-phase HPLC of D- and L-amino acids using the copper II complex of chiral N-(p-toluenesulfonyl)-L-phenylalanine as the mobile-phase. The measurements were made in pure water and in 10% and 15% aqueous acetonitrile. We have used the Zeta method to combine all of the data into a single data set. Very good results were obtained for both the D-amino acids (set AAO2, Table 22) and the L-acids (set AAO3). A comparison of the coefficients shows that the values of Z, A, S, and B^0 are not significantly different. There may be a difference in the values of H_2 with that for the D-acids being slightly larger. The L-acids show a borderline dependence on i, significant at the 90.0% confidence level, while no such dependence is observed for the D-acids. For both enantiomers, the predominant contributions are from polarizability and steric effects.

Ihara and coworkers (75) have studied the hydrolysis of D- and L-4-nitrophenyl amino acid benzylcarbamates catalyzed by decanoylhistidine or by leucylhistidine benzylcarbamate ($PhCH_2O_2C$–Leu–His). Again, using the Zeta method we have combined all of the data into a single data set (set AAO4, Table 22). Best results were obtained by correlation with the IMFSB equation (Table 23). Although no dependence on polarizability is observed for this set, this may well be due to the lack of variaton of side-chain structure within the set. The largest contribution to the effect of side-chain structure in this set is made by steric effects, with hydrogen bonding likely to also be of major importance.

Bash and coworkers (76) have reported calculations of electrostatic (E) and van der Waals (V) free energies of solvation for amino acids. Although values are reported for only eight amino acids, using the same reference substance, we thought it would be interesting to determine the composition of these free energies in terms of the IMF parameters. They were correlated with the IMFU equation (sets AAO 5 and 6). The ΔG_E values gave a good correlation; they seem to be a function of i, α, σ_l, and n_n in order of decreasing importance. The ΔG_V values gave a poorer but still significant fit. They are dependent only on the localized electrical effect and polarizability parameters, with the latter predominating, in accord with what is expected for a composite parameter which is a measure of van der Waals interactions. Zimmerman and coworkers (26) proposed a polarity parameter and a chromatographic parameter for the characterization of the properties of amino acid residues. We have successfully correlated these sets with the IMFU equation (sets AAO 7 and 8). The former set is largely a function of i, which, in view of the definition of the parameter, is hardly surprising. For nonpolar groups, they were obtained from dipole moments, for ionizable groups from dipole moments with the addition of a very large constant (48) which therefore dominates the parameter. There are also significant contributions from σ_1 and n_H, and possibly from α as well. In the latter set, there is a dependence in order of importance on α, v, i, σ_1, and n_H.

Porschke (76a) has determined the binding constant ratio K_h/K_c and the binding constant K_h for binding of amino acid amides to α-helix and coil in the polyribonucleotides poly (A)–poly(U) and poly(I)–poly(C). He states that the K_h/K_c values are much more reliable than are the K_h values. We have made use of the Zeta method (Appendix 2) to combine the data for the two polynucleotides into single sets (sets AAO 9 and 10). In agreement with Porschke's statement regarding the relative quality of the data, the $\log(K_h/K_c)$ values gave much better results on correlation with the IMFU equation, showing a dependence on α, n_H, and n_n, and possibly on v as well. The results obtained for the correlation of $\log K_h$ were poorer even after the exclusion of the Ala value for poly(A)–poly(C) and the Phe value for poly(I)–poly(U). Log K_h is a function of polarizability and of steric effects. As the ratio K_h/K_c represents the difference in intermolecular forces between binding to alpha helix and binding to coil, and the K_h values describe the binding to α-helix, we can use the difference to determine the characteristics of the intermolecular forces involved in binding to coil. It appears that $\log K_c$ is predominantly a function of n_H and n_n and is therefore strongly dependent on hydrogen bonding. This is reasonable in view of the availability of hydrogen bonding sites on the polyribonucleotide in the α-helix conformation as compared to that available in the coil conformation.

E. Amino Acid Bioactivities

Having shown that amino acid hydrophobicities and other properties are successfully modeled by the IMF equation (7, 58), we then demonstrated the applicability of the IMF equation to amino acid bioactivities (58, 77).

1. Bioactivity Model

In order to justify the application of the IMF model to bioactivities, we must consider the manner in which a bioactive substrate (bas) expresses its bioactivity. For this purpose, we have chosen a modification of the bioactivity model proposed by McFarland (78, 16). In this model, bioactivity is considered to be dependent on any one or some combination of the following sequences:

1. Transport. The bas moves from its point of entry into the organism to a receptor (rcp) site with which it is to interact. This movement is through an aqueous phase. It may involve diffusion through the medium or random binding to a biopolymer molecule such as plasma protein which carries it. During its passage, it is likely to cross one or more biomembranes. The crossing of a biomembrane begins with the transfer of the bas from the initial aqueous phase, ϕ_1 to the anterior membrane surface (ams). It then proceeds to the posterior membrane surface (pms) either by diffusion or by complexing with a lipid-soluble membrane carrier molecule (mcm) which transports it. The bas is then transferred from the surface of the pms to a second aqueous phase ϕ_2. Each step in this process is equivalent to a transfer from one phase to another and is therefore a function of the difference between intermolecular forces involving medium and bas in initial and final phases.

2. *Receptor–substrate binding.* The interaction between receptor and substrate can conveniently be divided into two parts, recognition and tight complex formation:

> a. *Recognition.* It is necessary for the receptor to distinguish the substrate from all of the other chemical species present in the medium which surrounds it. The receptor consists of some number of functional groups which have a particular orientation in space. To be recognized, a substrate must have functional group that are capable of interacting with those of the receptor and have the proper spatial arrangement to do so. The result of recognition is the formation of a loose complex bas...rcp held together by intermolecular forces. Recognition is therefore a function of the difference between the intermolecular forces involving medium and the substrate in the aqueous phase surrounding the receptor site, and those between receptor site and substrate in the loose complex.
>
> b. *Tight complex formation.* Conformational changes occur in the substrate and/or in the receptor which maximize the intermolecular forces between the two. The resulting increase in binding energy causes the formation of a tight complex. Again, the process is a function of a difference in intermolecular forces between an initial and a final state, in this case between the loose and the tight complexes.

3. *Chemical reaction.* The tight complex forms product and receptor by the formation and/or cleavage of covalent bonds. The model is summed up in Scheme 1.

We have carefully noted that each step in the sequences described above involves a difference in intermolecular forces between an initial and a final state. The IMF equation was expressly designed to model such differences. It is clear, then, that it should be capable of modeling bioactivities.

Correlation of amino acid bioactivities has been carried out with either the IMFU or the IMFSB equations, the choice between them generally being made on the basis of the number of data points in the set. The data used in

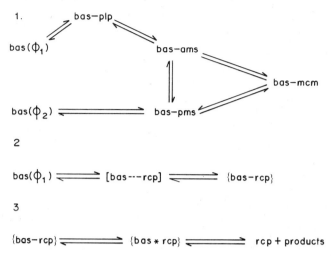

bas, bioactive substrate; ams, anterior membrane surface; pms, posterior membrane surface; plp, plasma protein; mcm, membrane carrier molecule; rcp, receptor; [bas–rcp], weak complex; {bas–rcp}, strong complex; {bas∗rcp}, transition state.

Scheme 1.

the correlations are given in Table 26, results of the correlatons in Table 27, zeroth-order partial correlation coefficients in Table 28, and C_i values in Table 29.

2. Hydrolytic Enzymes

Knowles (79) has reported values of V_{max} for the reaction of N-acetylamino acid methyl esters with chymotrypsin. Correlation with the IMFU equation gave significant results. V_{max} is primarily a function of polarizability and to a lesser extent of the number of long pairs in the side-chain, with steric effects accounting for about a third of the overall substituent effect. Unfortunately, only two set members are capable of hydrogen bonding while none has an ionizable side-chain. Although the results obtained are very good, the set does not provide a good test of the model. Jones and Beck (80) in their review have reported values of K_M and k_{cat} for the hydrolysis of N-acetylamino acid methyl esters by chymotrypsin in 0.1 M aqueous NaCl at 25° (sets AAB 7 and 8). Regrettably, again most of the amino acids studied are nonpolar. The results are in accord with those obtained for V_{max}: polarizability makes the major contribution, with steric effects next in importance and a small but significant dependence on the number of line

TABLE 26
Amino Acid Bioactivity Data Sets

AAB1, 2. Percent inhibition of Gly–Leu hydrolase (monkey intestine) by Aax in:
1. cytosol. 2. brush border. Aax, PI$_1$, PI$_2$[a]
Arg, 10, 23; Asp, 45, 39; Cys, 100, 100; His, 95, 74; Ile, 59, 22; Leu, 64, 36; Lys, 34, 0^x;
Met, 73, 31; Phe, 62, 47; Ser, 8, 19; Thr, 21, 43; Trp, 63, 25; Val, 31, 16; Dhp, 26, 38;
Nle, 63, 25.

AAB3. Percent inhibition of Gly–(^{14}C-Leu) uptake by monkey intestinal strip. P$_1$, Aax[a]
Ala, 59; Arg, 30; Asp, 5; Cys, 53; Glu, 7; Gly, 27; His, 42; Leu, 56; Lys, 49; Met, 58; Phe,
62; Ser, 28; Thr, 47; Trp, 31; Val, 43.

AAB4. V_{max} for hydrolysis of AcAaxOMe by α-chymotrypsin. V_{max}, Aax[b]
Ala, 1.72; Gly, 0.0098; Leu, 1590; Met, 2300; Phe, 42,000; Trp, 420,000; Tyr, 365,000; Val,
1.35; Abu, 19.6; Ahp, 8170; Aoc, 2120; Cha, 80,000; Nle, 1250; Nva, 265.

AAB5. Effect of Aax on relative rates of transport of Na$^+$-dependent-5-(O)-Pro. T$_{rel}$, Aax[c]
Ala, 40; Arg, 116; Asp, 104; Glu, 102; Gln, 57; Gly, 104; Ile, 37; Leu, 80^x; Lys, 113; Met, 67;
Phe, 58; Ser, 51; Thr, 61; Trp, 49; Val, 46.

AAB6. Percent relative activity for the oxidation of Aax by L-lysine-α-oxidase. Aax, P$_{rel}$[d]
Arg, 6.1; His, 3.8; Phe, 8.3; Lys, 100; Tyr, 5.5; Dhp, 2.5; Cys(CH$_2$CH$_2$NH$_2$), 9.8;
Cys(CH$_2$CH$_2$Py), 2.7; Met,(CH$_2$NH$_2$), 34.8; Orn, 18.2; Phe(4-NO$_2$), 2.8; Ahp(5-NH$_2$),
31.1; Lys(4-OH), 37.1.

AAB7, 8. K_M(mM), K_{cat}(s^{-1}) for hydrolysis of L-AcAaxOMe by chymotrypsin in
0.1 M NaCl, pH = 7.9, at 25°. Aax, K_M, K_{cat}[e]
Ala, 739, 1.27; Ile, 3.76, 4.98; Phe, 1.25, 52.5^z; Trp, 7.4, 49.5^y; Val, 112, 0.15; Abu, 53.5, 1.05;
Ahp, 1.64, 13.4; Ana, 25.1 49.2^y; Aoc, 2.94, 6.23; Bna, 13, 0.32^y; Bza, 7.9, 11.1^y; Cha, 0.19,
15.2; Nle, 6.7, 2.7; Nva, 10.2, 2.7; Pea, 26.4, 0.36^y; Ppa, 24.4, 0.053^y; Tyr(Me), 2.6, 0.49;
Tyr(Et), 12.4, 10.2; Tyr(iPr), 1.6, 0.012^x.

AAB9, 10. Reactions of ZCH$_2$-CO-Aax with hog kidney renal acylase. 9, Z = Cl; 10,
Z = H. Aax, K(Cl), K(H)[e]
Ala, 11,700, 3200; Glu, 12,700, 3080; Gly, 2640, —; His, —, 150^y; S-Ile, —; 1010; Leu,
16,500, 5400; Met, 100,000^z, 24,200; Phe, 460^y, 138^y; Ser, 11,600, 4960; R-Thr, 25, 8; Trp,
12^y, 5^y; Tyr, 330^y, —; Val, 4970, 1660; Abu, 33,600, 9500; Ade, 120, —; Ahp, 28,200^y, —;
Ano, 1600^y, —; 4, Δ^4, —, Ape, —, 8300; Ahx(6-OH), 10,000, —; Ape(5-OH), 23,800, —; Aoc,
7700^y, —; Aud, 9^y, —; Ceg, 132^y, —; Cha, 350^y, —; Chg, 4600, —; Cys(Me), —, 9500;
Cys(Bzl), —, 100^y; Hcy(Et), —, 15,400; Hcy(Bzl), —, 170^y; Hse, 12,600, —; Nle, 30,400,
14,400; Nva, 40,500, 9800; PhGly, 4500^y, —.

AAB11, 12, 13, 14. K_{cat}(s^{-1}) and M_M(mM) for aminoacylation of tRNA^{Ile}-C-C-A(3'-NH$_2$)
by isoleucyl-tRNA-synthetase from yeast (Y) or E. coli (B), Aax, $K_{cat,Y}$, $K_{M,Y}$, $K_{cat,B}$, $K_{M,B}$[f]
Ala, 0.0053, 0.2, 0.00016, 0.06; Arg, 0.0025, 0.02, 0.00099, 0.1; Asn, 0.0072, 0.06, 0.00055,
0.04; Asp, 0.020, 0.2, 0.00036, 0.07; Cys, 0.15, 0.20, 0.19^x, 0.05; Glu, 0.0041, 0.1, 0.00025,
0.04; Gln, 0.0077, 0.2, 0.00037, 0.09; Gly, 0.0068, 0.5, 0.00046, 0.2; His, 0.024, 0.1, 0.00089,
0.08; Ile, 0.10^x, 0.006, 0.02, 0.002; Leu, 0.0015, 0.05, 0.0013, 0.09; Lys, 0.047, 0.05, 0.00031,
0.05; Met, 0.0027, 0.1, 0.00029, 0.06; Phe, 0.0059, 0.2, 0.00051, 0.1; Ser, 0.0077, 0.3, 0.00049,
0.08; Thr, 0.0041, 0.09, 0.0025, 0.2; Trp, 0.012, 0.05, 0.00029, 0.006; Tyr, 0.0039, 0.09,
0.00081, 0.2; Val, 0.0020, 0.02, 0.0080, 0.08.

AAB15, 16, 17, 18. K_{cat}(s^{-1}) and K_M(mM) for aminoacylation of tRNA Ile-C-C-A by
isoleucyl-tRNA-synthetase from yeast (Y) or E. coli (B). Aax, $K_{cat,Y}$, $K_{M,Y}$, $K_{cat,B}$, $K_{M,B}$[g]
Ala, 0.0018, 0.08, 0.0012, 0.6; Arg, 0.0046, 0.2, 0.0048, 0.1; Asn, 0.0065, 0.1, 0.00034, 0.06;
Asp, 0.0019, 0.03, 0.00035, 0.1; Cys, 0.10, 0.7, 0.011, 0.05; Glu, 0.00034, 0.05, 0.00025,

TABLE 26 (Continued)

0.07; Gln, 0.00045, 0.1, 0.00028, 0.1; Gly, 0.00075, 0.3, 0.0013, 0.6; His, 0.00069, 0.06, 0.00048, 0.09; Ile, 0.83^x, 0.02, 0.82^x, 0.006; Leu, 0.00017, 0.05, 0.00057, 0.09; Lys, 0.00016, 0.03, 0.00039, 0.06; Met, 0.00025, 0.1, 0.00036, 0.07; Phe, 0.00066, 0.1, 0.00026, 0.08; Ser, 0.00065, 0.02, 0.0016, 0.5; Thr, 0.00040, 0.2, 0.00032, 0.06; Trp, 0.0019, 0.03, 0.00028, 0.007; Tyr, 0.00066, 0.05, 0.00024, 0.05; Val, 0.0001, 0.09, 0.00019, 0.1.

AAB19, 20, 21, 22. α-chymotrypsin hydrolysis of BzlO$_2$CAaxOPnNO$_2$-4 in 4.5% aq acetonitrile 0.50 M in NaCl at 25°. Aax, p$K_{1''}$, $10^6 K_M$, $10^2 K_3$, $10^4 K_{cat}/K_M^h$
Ala, 7.21, 20, 304, 15.2; Asn, 7.25, 13.6, 3370, 248; Gly, 7.01, 4.9, 41.1, 8.38; Leu, 7.59, 4.8, 834, 174; Lys, 6.91, 43, 294, 6.79; Phe, 7.53, 0.48, 3915, 8156; Trp, 7.51, —, 1782, —; Val, 7.27, 17.8^y, 11.85, 0.66; Abu, 7.48, 1.8, 271, 150; Nle, 7.45, 5.8, 1230, 212; Nva, 7.57, 2.67, 8.18, 306.

AAB23. Mean % decrease from preinjection level after human neurophysis II in:
1. rabbits (R) after 2 hours. 2. Boys (B) after 3 hours. Aax, P_R, P_B^i
Ala, 31.8, 67; Arg, 44.2, 84; Asp, —, 3; Glu, 44.9, 33; Gly, 40.5, 74; His, 19.6, —; Ile, 45.1, 63; Leu, 42.9, 66; Met, 40.7, 70; Val, 43.0, 78.

^{a}Ref. 85. ^{b}Ref. 79. ^{c}Ref. 86. ^{d}Ref. 87. ^{e}Ref. 80, ^{f}Ref. 83. ^{g}Ref. 84.
^{h}Ref. 81. ^{i}Ref. 88. ^{x}Excluded. ^{y}Member of B. set. Other Aax belong to A set.
^{z}Defined as zeta.

TABLE 27
Results of Bioactivity Correlations with the IMFU and IMFSB Equations

AAB\Set:	1	2	3	4	5	6
L	—	4.19	—	—	—	−6.23
S_L	—	1.41^h	—	—	—	2.30^i
A	2.03	1.09	−1.09	13.5	—	—
S_A	0.507^f	0.662^m	0.570^k	1.39	—	—
H_1	−0.355	—	0.181	—	—	—
S_{H1}	0.0532	—	0.0586^h	—	—	—
H_2	—	−0.700	−0.273	0.652	—	—
S_{H2}	—	0.0407^m	0.0400	0.228^h	—	—
I	0.437	—	—	—	0.319	0.558
S_I	0.124^f	—	—	—	0.0479	0.187^g
S	—	—	0.545	2.78	−0.347	—
S_S	—	—	0.247^k	0.650^f	0.101^g	—
B_1	—	—	—	—	—	—
S_{B1}	—	—	—	—	—	—
B_2	—	—	—	—	—	—
S_{B2}	—	—	—	—	—	—
B_3	—	—	—	—	—	—
S_{B3}	—	—	—	—	—	—
B^0	1.38	1.18	1.52	−1.711	1.96	0.96
B_{B^0}	0.107	0.176	0.133	0.358	0.0694	0.198
$100R^2$	81.39	46.88	84.85	96.93	82.29	68.66
F	16.04	2.941^k	14.00	105.4	25.56	10.95^f
S_{est}	0.161	0.185	0.151	0.420	0.0803	0.317
S^0	0.504	0.862	0.477	0.207	0.475	0.638
n	15	14	15	14	14	13

TABLE 27 (Continued)

AAB\Set:	7	8	9A	9B	10A	10B
L	—	—	6.53	−6.53	—	—
S_L	—	—	4.51	3.14	—	—
A	−5.42	12.5	6.93	−20.3	—	−8.67
S_A	2.08	3.77^h	2.08	1.84	—	3.10
H_1	—	—	2.18	—	2.79	−1.66
S_{H1}	—	—	0.475	—	0.574	0.633
H_2	0.467	−1.11	−1.34	−0.295	−1.38	—
S_{H2}	0.268	0.312^g	0.256	0.180	0.193	—
I	—	—	2.48	—	1.95	—
S_I	—	—	0.687	—	0.727	—
S	—	—	—	—	—	—
S_S	—	—	—	—	—	—
B_1	—	−0.813	−0.555	1.31	—	—
S_{B1}	—	0.327^j	0.241	0.808	—	—
B_2	−0.934	—	—	−0.724	—	—
S_{B2}	0.293	—	—	0.367	—	—
B_3	—	−0.714	—	—	0.605	—
S_{B3}	—	0.467^m	—	—	0.287	—
B^0	2.685	−0.466	3.863	11.396	3.666	5.869
S_{B^0}	0.274	0.426^o	0.282	1.37	0.128	1.08
$100R^2$	90.23	76.61	87.74	95.42	88.56	77.61
F	27.70	5.732^i	9.542	29.14	15.48	6.933
S_{est}	0.340	0.430	0.402	0.341	0.363	0.755
S^0	0.376	0.633	—	—	—	—
n	13	12	15	13	13	7

AAB\Set:	11	12	13	14^v	15	16	17
L	10.4	5.47	−3.79	3.42	10.0	5.64	7.42
S_L	1.97	1.10	1.40^h	2.35^m	2.62^f	1.82^g	2.15^f
A	1.10	−2.67	—	—	9.28	−1.93	—
S_A	0.770^m	0.990^i	—	—	2.48^f	0.764^i	—
H_1	—	—	—	—	—	0.308	0.414
S_{H1}	—	—	—	—	—	0.102^h	0.106^f
H_2	−0.157	—	—	—	—	−0.221	−0.332
S_{H2}	0.0575^i	—	—	—	—	0.0804^h	0.0842^f
I	—	−0.439	—	—	—	−0.365	—
S_I	—	0.121^f	—	—	—	0.181^k	—
S	—	—	—	—	—	—	—
S_S	—	—	—	—	—	—	—
B_1	−0.423	−0.509	0.799	—	−1.13	—	−0.494
S_{B1}	0.163^j	0.108	0.120	—	0.280^f	—	0.161^g
B_2	—	—	—	—	−0.550	—	—
S_{B2}	—	—	—	—	0.246^j	—	—

TABLE 27 (Continued)

AAB\Set:	11	12	13	14[v]	15	16	17
B_3	—	0.490	−0.308	—	−1.15	—	−0.331
S_{B3}	—	0.179[h]	0.115[h]	—	0.394[h]	—	0.151[j]
B^0	−2.23	−0.42	−3.68	0.70	−3.05	−0.86	−2.86
$S_B{}^0$	0.200	0.135[h]	0.160	0.469[m]	0.320	0.167	0.192
100R^2	69.02	86.80	80.46	17.83	69.99	63.18	71.69
F	7.241[f]	17.10	19.22	1.736[l]	5.598[g]	4.461[i]	6.079[f]
S_{est}	0.295	0.199	0.262	0.469	0.461	0.280	0.306
S^0	0.655	0.439	0.501	0.988	0.671	0.734	0.652
n	18	19	18	19	18	18	18

AAB\Set:	18	19	20A	20B[u]	21	22	23[w]
L	−2.92	−8.60	−14.5	−13.6	−22.7	—	−9.79
S_L	0.762[f]	2.77[i]	5.05[j]	6.22[k]	11.6[m]	—	1.25
A	—	2.56	—	−2.27	9.76	14.4	4.93
S_A	—	0.386	—	1.41[m]	1.77[f]	3.57[h]	0.928
H_1	—	−5.95	0.742	0.691	−1.41	−1.74	−0.201
S_{H1}	—	0.0929	0.143[f]	0.171[g]	0.397[h]	0.607[j]	0.0609[g]
H_2	—	0.525	—	—	1.81	1.39	0.144
S_{H2}	—	0.113[f]	—	—	0.480[h]	0.520[j]	0.0450[g]
I	—	—	—	—	—	—	—
S_I	—	—	—	—	—	—	—
S	—	—	—	—	—	—	−1.02
S_S	—	—	—	—	—	—	0.215
B_1	−0.462	—	−0.591	—	−0.815	−1.28	—
S_{B1}	0.145[g]	—	0.237[k]	—	0.254[i]	0.503[k]	—
B_2	—	—	—	—	—	—	—
S_{B2}	—	—	—	—	—	—	—
B_3	—	—	—	—	—	—	—
S_{B3}	—	—	—	—	—	—	—
B^0	−0.092	7.03	0.92	0.81	1.858	1.068	0.576
$S_B{}^0$	0.194	0.0646	0.201[g]	0.233[h]	0.296	0.498[k]	0.393[m]
100R^2	66.64	90.87	85.37	80.12	88.11	79.19	91.20
F	15.98	14.93[f]	9.727[i]	5.038[k]	7.410[i]	4.757[k]	19.00
S_{est}	0.314	0.0903	0.282	0.344	0.378	0.704	0.121
S^0	0.629	0.409	0.573	0.631	0.511	0.645	0.379
n	19	11	9	10	11	10	18

For footnotes not given below, see Table 3.

[u]Values of B_{12}, the coefficients of n_{12} and S_{B12}, are 0.895 and 0.376[k], respectively.
[v]Values of B_0, the coefficients of n_0, and S_{B0}, are −0.724 and 0.495[m], respectively.
[w]Values of O, the coefficients of ω and S_0, are 0.704 and 0.223[g] respectively.

TABLE 28
Zeroth-Order Partial Correlation Coefficients for Amino Acid Bioactivity

Set:	1	2	3	4	5	6
σ, α	0.528	0.533	0.245	0.275	0.256	0.550
σ, n_H	0.097	0.175	0.106	0.302	0.207	0.398
σ, n_n	0.531	0.541	0.603	0.444	0.762	0.017
σ, i	0.263	0.423	0.280	—	0.325	0.260
σ, υ	0.485	0.532	0.100	0.047	0.101	0.285
σ, n_1	—	—	—	—	—	—
σ, n_2	—	—	—	—	—	—
σ, n_3	—	—	—	—	—	—
α, n_H	0.324	0.324	0.367	0.576	0.344	0.486
α, n_n	0.089	0.088	0.108	0.249	0.119	0.204
α, i	0.046	0.030	0.182	—	0.117	0.675
α, υ	0.057	0.063	0.521	0.620	0.508	0.361
α, n_1	—	—	—	—	—	—
α, n_2	—	—	—	—	—	—
α, n_3	—	—	—	—	—	—
n_H, n_n	0.686	0.721	0.587	0.679	0.605	0.260
n_H, i	0.629	0.587	0.676	—	0.616	0.700
n_H, υ	0.335	0.317	0.100	0.047	0.084	0.182
n_H, n_1	—	—	—	—	—	—
n_H, n_2	—	—	—	—	—	—
n_H, n_3	—	—	—	—	—	—
n_n, i	0.463	0.528	0.713	—	0.675	0.272
n_n, υ	0.282	0.287	0.123	0.032	0.110	0.192
n_n, n_1	—	—	—	—	—	—
n_n, n_2	—	—	—	—	—	—
n_n, n_3	—	—	—	—	—	—
i, υ	0.139	0.098	0.165	—	0.129	0.035
i, n_1	—	—	—	—	—	—
i, n_2	—	—	—	—	—	—
i, n_3	—	—	—	—	—	—
n_1, n_2	—	—	—	—	—	—
n_1, n_3	—	—	—	—	—	—
n_2, n_3	—	—	—	—	—	—

Set:	7	8	9A	9B	10A	10B
σ, α	—	—	0.074	0.454	0.048	0.146
σ, n_H	—	—	0.662	0.323	0.598	0.174
σ, n_n	—	—	0.720	0.042	0.515	—
σ, i	—	—	0.357	—	0.277	—
σ, υ	—	—	—	—	—	—
σ, n_1	—	—	0.255	0.093	0.012	—
σ, n_2	—	—	0.107	0.143	0.087	—
σ, n_3	—	—	0.054	0.358	0.078	0.739
α, n_H	—	—	0.009	0.041	0.321	0.062
α, n_n	0.751	0.659	0.090	0.148	0.145	—
α, i	—	—	0.096	—	0.044	—

TABLE 28 (Continued)

Set:	7	8	9A	9B	10A	10B
α, υ	—	—	—	—	—	—
α, n_1	0.064	0.103	0.466	0.235	0.431	—
α, n_2	0.635	0.693	0.733	0.038	0.748	—
α, n_3	0.724	0.464	0.630	0.164	0.483	0.356
n_H, n_n	—	—	0.836	0.677	0.839	—
n_H, i	—	—	0.327	—	0.677	—
n_H, υ	—	—	—	—	—	—
n_H, n_1	—	—	0.142	0.123	0.123	—
n_H, n_2	—	—	0.000	0.088	0.199	—
n_H, n_3	—	—	0.289	0.118	0.272	0.645
n_n, i	—	—	0.634	—	0.807	—
n_n, υ	—	—	—	—	—	—
n_n, n_1	0.089	0.076	0.276	0.083	0.092	—
n_n, n_2	0.158	0.135	0.121	0.059	0.017	—
n_n, n_3	0.278	0.239	0.305	0.068	0.324	—
i, υ	—	—	—	—	—	—
i, n_1	—	—	0.031	—	0.083	—
i, n_2	—	—	0.149	—	0.116	—
i, n_3	—	—	0.378	—	0.527	—
n_1, n_2	0.047	0.051	0.065	0.317	0.116	—
n_1, n_3	0.185	0.181	0.082	0.662	0.158	—
n_2, n_3	0.807	0.806	0.395	0.657	0.220	—

Set:	11, 15, 17	12, 14, 16, 18	13	19, 21	20	22	23*
σ, α	0.264	0.259	0.231	0.131	0.257	0.254	0.232
σ, n_H	0.114	0.164	0.267	0.537	0.590	0.568	0.381
σ, n_n	0.555	0.580	0.730	0.793	0.815	0.793	0.779
σ, i	0.270	0.296	0.381	0.068	0.147	0.076	0.794
σ, υ	—	—	—	0.151	0.068	0.158	0.056
σ, n_1	0.280	0.121	0.141	—	0.218	—	—
σ, n_2	0.195	0.218	0.112	—	—	—	—
σ, n_3	0.253	0.177	0.104	—	—	—	—
α, n_H	0.311	0.300	0.283	0.300	0.209	0.208	0.584
α, n_n	0.083	0.085	0.115	0.047	0.064	0.065	0.227
α, i	0.134	0.130	0.115	0.151	0.709	0.317	0.406
α, υ	—	—	—	0.549	0.317	0.695	0.682
α, n_1	0.239	0.223	0.222	—	0.768	0.553	—
α, n_2	0.589	0.588	0.583	—	—	—	—
α, n_3	0.833	0.808	0.807	—	—	—	—
n_H, n_n	0.646	0.661	0.646	0.815	0.870	0.873	0.671
n_H, i	0.517	0.530	0.517	0.625	0.218	0.667	0.756
n_H, υ	—	—	—	0.196	0.661	0.182	0.009
n_H, n_1	0.115	0.011	0.005	—	0.286	0.093	—
n_H, n_2	0.264	0.233	0.181	—	—	—	—
n_H, n_3	0.354	0.383	0.354	—	—	—	—
n_n, i	0.515	0.528	0.515	0.228	0.227	0.218	0.924

TABLE 28 (Continued)

Set:	11, 15, 17	12, 14, 16, 18	13	19, 21	20	22	23*
n_n, v	—	—	—	0.182	0.205	0.195	0.034
n_n, n_1	0.162	0.049	0.044	—	—	0.081	—
n_n, n_2	0.201	0.171	0.111	—	—	—	—
n_n, n_3	0.057	0.101	0.058	—	—	—	—
i, v	—	—	—	0.057	0.088	0.066	0.033
i, n_1	0.000	0.062	0.066	—	—	—	—
i, n_2	0.309	0.288	0.259	—	—	—	—
i, n_3	0.313	0.333	0.313	—	—	—	—
n_1, n_2	0.000	0.050	0.044	—	—	—	—
n_1, n_3	0.000	0.096	0.114	—	—	—	—
n_2, n_3	0.487	0.447	0.408	—	—	—	—

*For Set 23: $\omega, \sigma = 0.091$; $\omega, \alpha = 0.082$; $\omega, n_H = 0.000$; $\omega, n_n = 0.103$; $\omega, i = 0.000$; $\omega, v = 0.012$.

TABLE 29
C_i Values for Amino Acid Bioactivities

Set:	σ_1	α	n_H	n_n	i	v	n_1	n_2	n_3
1	0	37.1	28.2	0	34.7	0	—	—	—
2	51.1	38.2	0	10.7	0	0	—	—	—
3	0	23.1	16.6	25.2	0	35.1	—	—	—
4	0	54.4	0	11.4	nd	34.1	—	—	—
5	0	0	0	—	56.8	43.2	—	—	—
6	47.2	0	0	—	52.8	0	—	—	—
7	—	47.1	—	17.6	—	—	0	35.3	0
8	—	52.0	—	20.2	—	—	14.8	0	13.0
9A	—	—	—	—	—	—	—	—	—
9B	—	—	—	—	—	—	—	—	—
10A	—	—	—	—	—	—	—	—	—
10B	—	—	—	—	—	—	—	—	—
11	49.9	15.2	0	9.43	0	—	25.5	0	0
12	17.6	24.7	0	0	17.6	—	20.4	0	19.7
13	21.5	0	0	0	0	—	56.7	0	21.8
14	27.4	0	0	0	0	—	0	72.6[a]	0
15	13.9	37.0	0	0	0	—	19.5	9.55	20.0
16	25.2	24.8	17.2	12.4	20.4	—	0	0	0
17	32.2	0	5.16	18.0	0	—	26.7	0	17.9
18	0	59.2	0	0	0	—	40.8	0	0
19	28.7	24.6	24.8	21.9	0	—	0	0	0
20	46.5	29.8	23.7	0	0	—	0	0	0
20	34.0	16.4	21.6	0	0	—	28.0[b]	0	0
21	22.4	27.7	17.4	22.4	0	0	10.1	0	0
22	—	42.8	22.6	18.0	0	—	16.6	0	0
23	19.0	27.5	4.89	3.51	0	17.4	—	27.7[c]	—

[a] For n_0.   [b] For n_{12}.   [c] For ω.

pairs in the side-chain. Dupaix and his coworkers (81) have reported values of pk", K_M, k_3, and k_{cat}/K_M for the hydrolysis of N-benzyloxycarbonylamino acid 4-nitrophenyl esters by chymotrypsin in 4.5% aqueous acetonitrile 0.5 M in NaCl (sets AAb 19–22). Once again, all but two of the amino acids studied are nonpolar. The results of the correlations obtained are in fair agreement with those for sets 5, 7, and 8. Significant correlations were obtained for all four sets. In order to establish with certainty the absence of a dependence on an ionic side-chain and the extent of the dependence on hydrogen bonding in the hydrolytic reactions of amino acid derivatives catalyzed by chymotrypsin, it is necessary to study a much wider range of side-chain structure than has been done so far.

Jones and Beck (80) have also reported rate constants for the hydrolysis of N-acetyl- and N-chloroacetylamino acids catalyzed by hog kidney renal acylase. In examining these data sets, it is easily seen that they exhibit bilinear or parabolic behavior (82). We have attempted to treat the problem of bilinear behavior by dividing each set into two subsets. The A subset contains all of the members of the set whose polarizability is less than that of the most active member of the entire set together with the data point for the most active member. The B subset contains all of the set members whose polarizability is greater than that of the most active set member together with the data point for the most active member. The most active set member (designated Aax^0) is that for which the measured quantity reported is a maximum; its data point is part of both subsets. Thus:

Set A contains Aax^0 + all Aax wth $\alpha_X < \alpha_{X^0}$.

Set B contains Aax^0 + all Aax with $\alpha_X > \alpha_{X^0}$.

The results obtained for the N-chloroacetyl amino acids show bilinear behavior very clearly with the coefficients for α, σ_1, n_n and n_1 having opposite signs in sets 9A and 9B. The two subsets differ further in that that former shows a dependence on n_H and i but not on n_2, whereas the latter shows a dependence on n_2 but not on n_H or i.

Bilinear behavior is also observed in the N-acetyl amino acids. Both sets 10A and 10B are a function of n_H with coefficients of opposite sign. It is interesting to note that the coefficients of n_H, n_n and i for the N-chloroacetyl amino acids of set 9A are not significantly different from those of the N-acetyl amino acids of set 10A. The small size of set 10B precludes a meaningful comparison with set 9B.

3. Isoleucyl-tRNA Synthetase

Freist and his coworkers have reported values of k_{cat} and K_M for the amino-acylation of $tRNA^{Ile}$-C—C—A(3'-NH_2)(83) and of $tRNA^{Ile}$-

C—C—A (84) catalyzed by isoleucyl-tRNA synthetase obtained from baker's yeast or from *E. coli* (E_Y and E_B, respectively). The amino acids studied are the PBS. The data sets were correlated with the IMFSB equation. Significant correlations were obtained for all of the k_{cat} data sets (sets AAB 11, 13, 15, and 17). The data point for Ile had to be excluded from sets 11, 15, and 17; that for Cys had to be excluded from set 13. All of these sets were dependent on σ_1 and n_1, sets 13, 15, and 17 on n_3, sets 11 and 15 on polarizability and on the number of lone pairs. Of the K_M sets all but set 14 gave significant results. A dependence on σ_1 is observed for sets 12, 16, and 18 and may also be present in set 14 as well. There is no other common factor.

We may compare the effect of the enzyme source on $\log k_{cat}$ while holding the reactant constant by comparing the results for sets 11 and 15 and for sets 13 and 17. The comparison shows many differences and only one similarity, the steric effect in sets 15 and 17. To carry out the same process for $\log K_M$, we must compare sets 12 and 14, and sets 16 and 18. As the results for set 14 were not significant, no comparison with it is useful. A comparison shows clear-cut differences and no similarities in the nature of the dependence on the IMF parameters. While the collinearities inherent in the PBS make it difficult to draw reliable conclusions, it seems very likely that there are differences in the receptor sites of the two enzymes.

We may also compare the results obtained for k_{cat} using tRNA^{Ile}-C—C—A(3-NH$_2$) as the substrate with those obtained for tRNA^{Ile}-C—C—A by comparing sets 11 and 15 and sets 13 and 17. In the former case, the coefficient of σ_1 has the same sign and magnitude. Both sets depend on α and n_1 but the coefficients are much larger in set 15 than in set 11. The sets differ in that set 11 is a function of n_n while set 15 id not, and set 15 is a function of n_2 and n_3 while set 11 is not. The two sets differ therefore in their dependence on hydrogen bonding and on steric effects. In the latter case, sets 13 and 17 both show a dependence on σ_1, n_1 and n_3 but the coefficients for the first two variables are very different in magnitude. Set 17 also shows a dependence on hydrogen bonding which is not present in set 13. Overall, it seems that there are more similarities when the enzyme source is held constant and the substrate structure is varied than there are when the substrate structure is held constant and the enzyme source is varied. Finally, we note that steric effects make a major contribution to side-chain structural effects on both k_{cat} and K_M.

4. Amino Acid Transport

Radhakrishan (85) has reported values for the percent inhibition of Gly–Leu hydrolase (monkey intestine) by amino acids in cytosol (set AAB 1) and in brush border (set AAB 2). He has also reported percent inhibition of Gly–(^{14}C-Leu) uptake by monkey intestinal strips (set AAB 3). Correlation with the IMFU equation gave good results for sets 1 and 3, and barely significant results for set 2. Other than a dependence on polarizability for all three sets, the results show no consistent pattern.

Ganapathy, Roesel, and Leibach (86) have studied the effect of amino acids on the relative rate of transport of Na^+-dependent-5-(O)-proline. On the exclusion of the data point for Leu, this set gave a good correlation with the IMFU equation. Charged side chains and steric effects determine the activity, with the former being the more important.

5. Other Amino Acid Bioactivities

Kusakabe and coworkers have determined the percent relative activity of the oxidation of amino acids by L-lysine α-oxidase. Correlation with the IMFU equation gave significant results, with σ_1 and i making approximately equal contributions to the description of the activity.

Finally, Trygstad, Foss, and Sletten (88) have determined the mean percent decrease from preinjection levels for amino acids after the injection of human neurophysin II in rabbits (after 2 hours) and in boys (after 3 hours). As only nine amino acids were studied in each organism, we have made use of the Omega method (Appendix 2) to combine the data for the two organisms into a single set. Correlation with the IMFU equation gave excellent results. The data set shows a dependence on $\omega, \alpha, \sigma_1, \upsilon, n_H,$ and n_n in order of their importance. In view of the large difference in the organisms studied, it seems likely that in this case the bioactvitity is determined by transport rather than by receptor binding or chemical reaction.

F. Extension of the Model to Variant Amino Acids

The models described up to this point for the quantization of amino acid properties are applicable to α-amino acids having the structure **3** with $R^1 = R^2 = R^3 = H$. We will refer to amino acids with alternative structures are variant amino acids. While the majority of the amino acids of interest are of this type, a number of others that are well-known or of possible use have R^1 = alkyl, or one or both of R^2 and R^3 are alkyl. A common example of this type is sarcosine. Another amino acid type of importance is exemplified by proline, **4**, and its derivatives in which X and R^2 are joined.

$$\begin{array}{cc}
R^1 & H \\
| & | \\
X-C-CO_2H & CH_2-C-CO_2H \\
| & | \quad | \\
N & CH_2 \quad NH \\
/ \ \backslash & \backslash \quad / \\
R^2 \quad R^3 & CH_2 \\
\mathbf{3} & \mathbf{4}
\end{array}$$

To include Pro and its derivatives in a data set, it is necessary to account for the change in the number of NH bonds on the α-amino group and the change in its σ_1 value. It is also necessary to determine an effective value of v for the Pro system. In order to account for the number of NH bonds, we must introduce a new parameter. It is convenient for this parameter to take the value 0 when the amino acid is not a variant type. We therefore define the variable n_{NC} as the number of NC bonds on the α-amino group minus one. For ordinary amino acids, this will have the value 0, for Pro and its derivatives and for Sar it will have the value 1. To obtain an effective value of v, we have used the data sets AAH 16 and 17. The IMFU equation can be written for Pro and its derivatives in the form

$$Q_X = L\sigma_{1X} + A\alpha_X + H_1 n_{HX} + H_2 n_{nX} + I_{iX} + B_{NC} n_{NCX} + Sv_X + B^0 \quad (46)$$

Using the experimentally determined Q_{Pro} values, the $\sigma_1, \alpha, n_H, n_n$, and i values for $X = (CH_2)_3$, and n_{NC} for Pro and substituting into the regression equations whose coefficients are given in Table 3, we have the relationships

$$0.293 v_{Pro} + B_{NC,16} = 0.0733 \quad (47)$$

and

$$0.709 v_{Pro} + B_{NC,17} = -0.2633 \quad (48)$$

As n_{NC} like n_H is an H donor hydrogen bonding parameter and H_1, the coefficient of n_H has the same value in both set 16 and set 17, we have assumed that $B_{NC,16} = B_{NC,17}$. We then obtain $v_{Pro} = -0.81$. We further assume that Pro derivatives such as Hyp will have the same value of v.

In order to include Sar and other variant amino acids with R^2 and possibly R^3 not equal to H, we require a steric term for NR^2R^3, a parameterization of its polarizability, and a parameterization of its electrical effect. The first two factors are also required for variant amino acids in which R^1 is not H. The polarizability can be accounted for in all three types of variant α-amino acid by replacing α with $\sum\alpha$ where

$$\sum\alpha = \alpha_X + \alpha_{R^1} + \alpha_{R^2} + \alpha_{R^3} \quad (49)$$

The additional steric parameters v_{R^1} and $v_{NR^2R^3}$ must be included in the IMF equation in the general case. To parameterize the electrical effect of the NR^2R^3 group, we make use of the variable σ_1'' which is defined as

$$\sigma_1'' = \sigma_{1X} + (\sigma_{1NHR^2R^3} - \sigma_{1NH2}) \quad (50)$$

For Sar v_{NHMe} is not significantly different from v_{NH2}, and σ_{1NHMs} is not significantly different from σ_{INH2}, so that only the charge in polarizability need be accounted for.

The problem in applying this expanded form of the IMFU equation to data sets which include variant amino acids is that there must be enough of them in the data set to establish reliable values of the coefficients of the additional parameters. Consider, for example, the case in which Pro type amino acids are to be included in the data set. The appropriate form of the IMFU equation is Equation 46. If the data set consists of the PBS and Pro, then of the 20 amino acids in the set, 19 will have $n_{NC} = 0$ and only for Pro will $n_{NC} = 1$. This will not be sufficient to provide a reasonable estimate of B_{NC}. At the very least, two amino acids of the Pro type are necessary if the results obtained with Equation 46 are to be atl all reliable. In any event, it is necessary to correlate the data set of interest with the IMFU equation with the exclusion of the Pro type acids, and with Equation 46 including the Pro type acids in order to compare the coefficients of the variable common to both equations. No significant difference in the coefficients should be observed if the results obtained with Equation 46 are indeed reliable. We have examined 14 data sets for which at least two Pro type data points are available. The additional data points are given in Table 30, results of the correlations in Table 31, zeroth-order partial correlation coefficients in

TABLE 30
Data for Variant Amino Acids

AAH4.	$\log P$ (1–OcOH–H_2O). Pro, -2.50; Hyp, -3.17.
AAS1.	Solubility (mol/kg H_2O). Pro, 14.1; Hyp, 2.75.
AAS2.	$H_2^{\prime 0}-H_2^\prime$, Aax (s) to Aax (aq) at infinite dilution. Pro, 1.3; Hyp, 1.4.
AAS3.	ΔG_{soln}. Pro, -2.24; Hyp, -0.629.
AAS4.	Partial molar heat capacities at infinite dilution. Pro, 170; Hyp, 197.
AAC18.	$100R_F$ (TLC), cellulose, iPrOH–HCO_2H–H_2O (20–1–5). Pro, 35; Hyp, 22; Sar, 29.
AAC19.	$100R_F$ (TLC), cellulose, sBuOH–EtAc–cHxNH–H_2O (10–10–2–5). Pro, 19; Hyp, 19; Sar, 29.
AAC20.	$100R_F$ (TLC), cellulose, PhOH–H_2O (3–1) + 7.5 mg NaCN chamber saturated with NH_3. Pro, 87; Hyp, 66; Sar, 80.
AAC21.	$100R_F$ (TLC), cellulose, iPrOH–EtAc–1 M HCl (60–15–25). Pro, 58; Hyp, 48; Sar, 48.
AAC22.	cellulose, tBuOH–EtAc–MeAc–MeOH–H_2O–conc. aq NH_3 (40–20–20–1–14–5) Pro, 30; Hyp, 17; Sar, 24.
AAO1.	K_P^N, 5′-Up, 5′-Cp, 5′-Ip, 5′-Ap, 5′-Gp. Pro, 1.10; 1.01, 1.41, 1.42, 2.10.
AAO9.	K_h/K_c, poly(A)·poly(U), poly(I)·poly(C). Pro, 0.84, 0.95.
AAO10.	K_h, poly(A)·poly(U), poly(I)·poly(C). Pro, 510, 480.
AAB5.	Rate of Aax transport, Na^+ dependent –5–(O)–Pro. Pro, 75; Hyp, 102.

TABLE 31
Results of Correlations with the IMFU Equation

	AAH4		AAS1		AAS2		AAS3		AAS4	
L	−2.89	−2.88	—	—	—	—	—	—	−1330	—
S_L	1.14	1.16	—	—	—	—	—	—	418	—
A	4.64	4.63	−5.16	−5.13	7.36	7.34	10.3	10.4	1200	1340
S_A	0.626	0.637	0.766	0.768	2.12	1.90	1.68	1.68	126	189
H_1	−0.538	−0.535	0.531	0.530	—	—	−1.23	−1.24	−295	−51.6
S_{H1}	0.0574	0.0589	0.0988	0.0990	—	—	0.248	0.248	92.7	18.2
H_2	—	—	−0.430	−0.420	1.34	1.43	0.911	0.893	183	—
S_{H2}	—	—	0.0651	0.0660	0.197	0.183	0.150	0.151	63.3	—
I	−1.08	−1.08	—	—	—	—	—	—	—	—
S_I	0.129	0.131	—	—	—	—	—	—	—	—
S	0.957	0.960	—	—	−3.50	−3.53	—	—	106	—
S_S	0.244	0.249	—	—	1.33	1.19	—	—	20.1	—
B_{NC}	1.03	—	1.16	—	−6.88	—	−2.01	—	—	—
S_{BNC}	0.420	—	0.196	—	2.11	—	0.512	—	—	—
B^0	−3.501	−3.504	0.536	0.518	3.0	2.9	−1.19	−1.17	51.9	96.6
S_{B^0}	0.169	0.172	0.156	0.157	0.919	0.824	0.360	0.360	20.9	29.5
$100R^2$	94.74	94.72	88.84	82.84	85.30	88.01	86.76	83.20	95.29	86.50
F	84.01	96.85	25.88	19.30	17.41	26.90	21.30	19.81	32.36	28.83
S_{est}	0.289	0.294	0.259	0.260	1.07	0.961	0.673	0.671	30.9	48.2
S^0	0.256	0.254	0.393	0.478	0.456	0.404	0.428	0.473	0.287	0.424
n	35	33	18	16	17	15	18	16	14	12
B_z	—	—	—	—	—	—	—	—	—	—
n_z	—	—	—	—	—	—	—	—	—	—

TABLE 31 (Continued)

	AAC18		AAC19		AAC20		AAC21	
L	—		2.40	2.40	—	—	−1.97	−2.08
S_L	—		0.759	0.727	—	—	0.415	0.435
A	0.777	0.753	0.643	0.640	0.317	0.303	0.640	0.594
S_A	0.138	0.144	0.197	0.194	0.0715	0.0695	0.222	0.233
H_1	—		−0.245	−0.228	0.0289	0.0317	−0.0532	−0.0535
S_{H1}	—		0.0385	0.0398	0.0142	0.0138	0.0206	0.0212
H_2	−0.100	−0.102	—	—	−0.0450	−0.0477	—	—
S_{H2}	0.0149	0.0156	—	—	0.0116	0.0115	—	—
I	−0.543	−0.551	−0.213	−0.267	−0.295	−0.290	−0.571	−0.574
S_I	0.0588	0.0609	0.0923	0.0961	0.0455	0.0446	0.0542	0.0556
S	—		0.156	—	0.191	0.272	—	—
S_S	—		0.0996	—	0.0602	0.0716	—	—
B_{NC}	—		—	—	0.279	—	—	—
S_{BNC}	—		—	—	0.0859	—	—	—
B^0	1.52	1.53	1.42	1.53	1.70	1.64	1.73	1.75
S_{B^0}	0.0408	0.0452	0.0715	0.0661	0.0462	0.0518	0.0453	0.0504
$100R^2$	84.26	84.74	74.97	76.94	87.33	88.67	86.94	87.27
F	60.68	57.39	19.17	23.03	35.62	45.34	59.90	56.54
S_{est}	0.142	0.146	0.203	0.200	0.0698	0.0674	0.125	0.128
S^0	0.419	0.415	0.545	0.519	0.394	0.370	0.386	0.383
n	38	35	38	35	38	35	41	38
B_z	—	—	—	—	—	—	—	—
n_z	—	—	—	—	—	—	—	—

TABLE 31 (Continued)

	AAC22	AAO1		AAP9	AAO10		AAB1	
L	2.83	4.10	4.10	—	—	—	—	—
S_L	0.943	0.719	0.825	—	—	—	—	—
A	1.16	3.09	3.09	—	1.43	1.63	—	—
S_A	0.368	0.284	0.295	−0.870	0.220	0.228	—	—
H_1	0.112	−0.102	−0.102	0.0807	—	—	—	—
S_{H1}	0.0446	0.0388	0.0405	0.131	—	—	—	—
H_2	−0.293	0.123	0.123	0.0306	—	—	—	—
S_{H2}	0.0354	0.0335	0.0349	−0.0609	—	—	—	—
I	−0.573	−0.117	−0.117	0.0186	nd	nd	—	0.319
S_I	0.0830	0.0620	0.0646	nd	nd	nd	0.319	0.0479
S	—	−0.638	−0.638	nd	−0.101	−0.281	0.0487	−0.347
S_S	—	0.125	0.130	−0.0567	0.0487	0.0974	−0.347	0.101
B_{NC}	—	−0.586	—	0.0365	—	—	0.103	—
S_{BNC}	—	0.152	—	—	—	—	−0.299	—
B^0	1.49	−0.456	−0.463	0.0949	2.48	2.57	0.162	1.96
S_{B^0}	0.0763	0.0422	0.0448	0.0222	0.0449	0.0618	1.96	0.0694
$100R^2$	85.77	89.62	94.07	94.14	88.61	69.03	75.36	0.0705	82.29
F	40.97	41.71	69.37	71.10	33.05	21.17	26.00	81.61	25.56
S_{est}	0.209	0.190	0.0786	0.0818	0.0366	0.107	0.101	17.75	0.0803
S^0	0.409	0.359	0.273	0.272	0.404	0.599	0.538	0.0816	0.475
n	40	36	44	39	24	22	22	20	0.495	16	14
B_z	—	—	0.719	0.737	—	—	—	—
n_z	—	—	0.0628	0.0694	—	—	—	—

TABLE 32
Zeroth-Order Correlation Coefficients for Data Sets Containing Variant Amino Acids

	AAH1	AAS1	AAS2	AAS3	AAS4	AAC18	AAC19
σ, α	0.196	0.141	0.009	0.139	0.067	0.254	0.225
σ, n_H	0.044	0.363	0.355	0.515	0.679	0.311	0.353
σ, n_n	0.416	0.720	0.732	0.807	0.733	0.613	0.619
σ, i	0.314	0.541	0.579	0.673	—	0.331	0.339
σ, υ	0.186	0.308	0.288	0.305	0.267	0.312	0.315
σ, n_{NC}	0.226	0.332	0.311	0.329	0.265	0.285	0.288
α, n_H	0.095	0.279	0.266	0.430	0.027	0.353	0.229
α, n_n	0.221	0.131	0.068	0.168	0.010	0.352	0.362
α, i	0.082	0.148	0.022	0.233	—	0.043	0.061
α, υ	0.471	0.274	0.292	0.301	0.204	0.297	0.229
α, n_{NC}	0.130	0.094	0.162	0.145	0.065	0.248	0.200
n_H, n_n	0.682	0.698	0.731	0.766	0.987	0.701	0.721
n_H, i	0.272	0.483	0.571	0.618	—	0.614	0.575
n_H, υ	0.099	0.116	0.099	0.139	0.060	0.145	0.144
n_H, n_{NC}	0.070	0.116	0.096	0.139	0.000	0.202	0.200
n_n, i	0.164	0.598	0.542	0.765	—	0.465	0.431
n_n, υ	0.115	0.089	0.052	0.124	0.009	0.214	0.215
n_n, n_{NC}	0.028	0.079	0.032	0.115	0.049	0.198	0.200
i, υ	0.075	0.195	0.179	0.195	—	0.154	0.146
i, n_{NC}	0.794	0.917	0.917	0.917	—	0.867	0.867
υ, n_{NC}	0.134	0.189	0.169	0.189	0.925	0.187	0.175

	AAC20	AAC21	AAC22	AAO1[z]	AAO9	AAO10[z]	AAB5
σ, α	0.225	0.297	0.298	0.234	0.108	0.173	0.209
σ, n_H	0.315	0.207	0.232	0.197	0.593	0.598	0.241
σ, n_n	0.629	0.575	0.612	0.385	0.732	0.741	0.732
σ, i	0.494	0.185	0.321	0.435	—	—	0.370
σ, υ	0.315	0.339	0.317	0.559	0.172	0.169	0.335
σ, n_{NC}	0.288	0.280	—	0.435	0.258	0.271	0.323
α, n_H	0.221	0.446	0.453	0.364	0.183	0.192	0.350
α, n_n	0.360	0.354	0.362	0.286	0.079	0.089	0.105
α, i	0.005	0.396	0.284	0.089	—	—	0.134
α, υ	0.228	0.326	0.332	0.620	0.350	0.353	0.282
α, n_{NC}	0.200	0.274	—	0.305	0.097	0.105	0.096
n_H, n_n	0.681	0.701	0.694	0.865	0.889	0.885	0.623
n_H, i	0.425	0.561	0.507	0.756	—	—	0.617
n_H, υ	0.140	0.069	0.110	0.365	0.185	0.196	0.156
n_H, n_{NC}	0.197	0.174	—	0.283	0.196	0.219	0.135
n_n, i	0.680	0.248	0.424	0.610	—	—	0.655
n_n, υ	0.214	0.137	0.186	0.346	0.145	0.154	0.114
n_n, n_{NC}	0.198	0.179	—	0.267	0.170	0.188	0.079
i, υ	0.116	0.094	0.153	0.321	—	—	0.250
i, n_{NC}	0.867	0.852	—	0.896	—	—	0.924
υ, n_{NC}	0.127	0.170	—	0.250	0.856	0.858	0.218

[z] r values are less than or equal to 0.100 for zeroth-order correlation coefficients with ζ.

TABLE 33
C_i Values for Data Sets Containing Variant α-Amino Acids

Set	ζ	σ_I	α	n_H	n_n	i	υ	n_{NC}
AAH4	—	5.01	23.1	11.6	0	23.4	14.5	22.3
AAS1	—	0	35.9	16.1	13.0	0	0	35.1
AAS2	—	0	13.7	0	10.8	0	19.8	55.7
AAS3	—	0	36.4	18.9	13.9	0	0	30.8
AAS4	—	11.4	29.5	31.6	19.6	0	7.93	0
AAC18	—	0	21.7	0	12.2	66.1	0	0
AAC19	—	21.2	16.3	27.0	0	23.5	12.0	0
AAC20	—	0	8.53	3.38	5.27	34.5	15.7	32.7
AAC21	—	17.0	15.8	5.73	0	61.5	0	0
AAC22	—	15.4	18.2	7.58	19.9	38.9	0	0
AAO1	11.5	12.0	26.0	3.74	4.50	4.28	16.4	21.5
AAO9	0	0	51.1	33.4	15.5	0	0	0
AAO10	0	0	76.5	0	0	0	23.5	0
AAB5	—	0	0	0	0	37.0	28.2	34.7

Table 32, and C_i values in Table 33. Table 31 includes the results for the same data sets obtained with the exclusion of the Pro type data points.

In all of the data sets studied, coefficients of the regression equations obtained, including and excluding the variant amino acids, show no significant differences. Considering the $100R^2$ statistic, three sets gave better results on the inclusion of the variant amino acids, five sets showed no major difference, and six sets gave better results on the exclusion of the variant amino acids. Considering the S^0 values, three sets gave better results on the inclusion of the variant amino acids, for eight sets the difference in S^0 was insignificant (the difference was less than 0.030), and for three sets the difference in S^0 was meaningful, with inclusion of the variant amino acids giving poorer results. Of the 14 data sets studied, three contained two or more Pro data points, eight contained one Pro and one Hyp data point, and three contained one Pro, one Hyp, and one Sar data point. Overall, our results seem to support the validity of the method. The IMFU equation can indeed be extended to variant amino acids.

IV. PEPTIDES

A. Introduction

In the course of applying the IMF equation to amino acid hydrophobicities, we have studied three data sets in which the substrates (5) are protopeptides in the sense that they contain peptide bonds (sets AAH 16, 17, and 20). The successful modeling of data sets based on amino acid residues in protein (AAH 22, 27, 29, and 30) implies that the IMF equation is applicable to peptide properties and bioactivities.

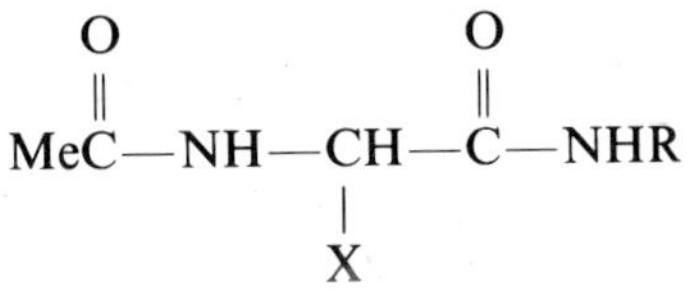

$$\text{MeC(=O)—NH—CH(X)—C(=O)—NHR} \qquad \begin{array}{l} a.\ R = Me \\ b.\ R = H \end{array}$$

5

We have previously reported correlations of peptide chromatographic properties (89) and bioactivities (90–92) with the IMFU and IMFSB equations and relationships derived from them.

While substitution in peptides most frequently involves the replacement of one amino acid residue by another at a given position of the peptide chain (side-chain substitution, designated X^n where n is the position at which substitution occurs), there are other possible types as is shown in **6**. Substitution may involve replacement of the α-amino group of the N-terminal residue (N-substitution designated X^N) or of the OH of the C^1 carboxyl group of the C-terminal residue (C-substitution, designated X^C). It may occur at the N atom in a peptide bond (P-substitution, designated X^P). Finally, there is a special case of interest, that in which side-chain substitution involves variation of only part of a side chain. In that case, the substituent X may be written in the form Z^0Z where Z^0 is constant and Z varies.

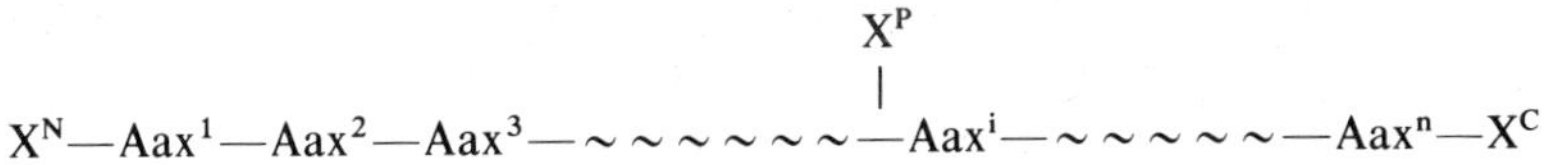

$$X^N—Aax^1—Aax^2—Aax^3—\sim\sim\sim\sim\sim\sim—\overset{X^P}{\underset{|}{Aax^i}}—\sim\sim\sim\sim\sim\sim—Aax^n—X^C$$

6

B. Peptide Properties

We have studied the correlation of 22 data sets of peptide properties with the IMFU and the IMFSB equations. All but one of these data sets involve chromatographic parameters. The data used are set forth in Table 34, results of the correlations are presented in Table 35, zeroth-order partial correlation coefficients are given in Table 36, and C_i values are reported in Table 37.

Meek (92a) has developed a method for the determination of retention coefficients for amino acid residues in peptides. These retention coefficients

TABLE 34
Peptide Properties Data Sets

PPC1, 2. Retention coefficients (HPLC) of amino acid residues in peptides. Initial solutions were 0.1 M NaClO$_4$–0.1% H$_3$PO$_4$ and 0.1 M NaH$_2$PO$_4$–0.1% H$_3$PO$_4$ respectively[a]. Aax, —, —

Trp, 17.1, 15.1; Phe, 13.4, 12.6; Leu, 11.0, 9.6; Ile, 8.5, 7.0; Tyr, 7.4, 6.7; Val, 5.9, 4.6; Met, 5.4, 4.0; Ala, 1.1, 1.0; Glu, 0.7, 1.1; Gly, −0.2, 0.2; Arg, −0.4, 2.0; His, −0.7, −2.2; Asp, −1.6, −0.5; Thr, −1.7, −0.6; Lys, −1.9, −3.0; Gln, −2.9, −2.0; Ser, −3.2, −2.9; Asn, −4.2, −3.0.

PPC3. 100R_F, reversed-phase TLC, His–Aax, SiO$_2$ = m% H–DBS detergent[b]
Aax, m = 0, 1, 2, 3, 4

Gly, 96, 43, 14, 7, 2; Ala^z, 96, 35, 10, 5, 2; Ser, 96, 47, 18, 10, 3; Leu, 90, 4, 2, 2, —; Met, 93, 10, 4, 3, —; Phe, 82, 3, 2, 2, —: Tyr, 93, 13, 5, 3, —.

PPC4. Met–Aax conditions as above[b]

Gly, 95, 54, 40, 30, 17; Val, 80, 25, 16, 12, 7; Leu^z, 62, 13, 7, 5, 4; Met, 78, 23, 16, 13, 8; Phe, 55, 9, 6, 5, 3; Tyr, 77, 30, 22, 17, 11; His, 96, 15, 5, 3, —.

PPC5. Aax–Gly, conditions as above[b]

Ser, 96, 90, 72, 62, 45; Thr, 96, 89, 72, 60, 43; Val, 94, 63, 49, 40.25; Leu^z, 90, 45, 30, 22, 15; Met, 95, 54, 40, 30, 17; Trp, 77, 24, 14, 8, 5; Tyr, 93, 61, 43, 34, 20; His, 96, 43, 14, 7.2^x.

PPC6. 100R_F, reversed-phase TLC, His–Aax, SiO$_2$ + 4% H–DBS, mobile-phase 30% aq MeOH, pH = 1.25, 1.55, 2.75, 3.30, 5.10, 8.15[b]

Gly, 6, 5, 10, 12, 19, 81; Ala^z, 3, 3, 8, 9, 18, 81; Ser, 6, 5, 11, 16, 20, 81; Leu, 1, —, 2, 2, 4, —; Met, 2, 2, 3, 2, 8, —; Phe, —, —, —, —, 3; Tyr, 2, 1, 3, 2, 11, —.

PPC7. Met–Aax, conditions as above[b]

Gly, 34, 31, 37, 32, 41, 68; Val, 14, 12, 17, 12, 30, 70; Leu^z, 7, 6, 7, 5, 13, 46; Met, 13, 11, 13, 10, 24, 58; Phe, 5, 4, 5, 3, 9, 27; His, 2, 2, 3, 4,5, 54; Tyr, 17, 16, 17, 14, 31, 67.

PPC8. Aax–Gly, conditions as above[b]

Ser, 74, 67, 73, 72, 77, 85; Thr, 73, 67, 72, 70, 73, 85; Val, 45, 42, 45, 42, 50, 76; Leu^z, 23, 20, 25, 21, 31, 54; Met, 34, 31, 37, 32, 41, 68; Trp, 11, 9, 10, 7, 12, 30; Tyr, 38, 35, 45, 33, 43, 67; His, 6, 5, 10, 12, 19, 81^x.

PPC9. His–Aax, SiO$_2$ w/wo 4%H–DBS/4% N–DPC aq AcOH/30% aq MeOH[b]

Gly, 3, 13, 96, 97, 96, 81; Ala^z, 2, 11, 96, 97, 96, 78; Ser, 3, 13, 96, 97, 96, 81; Leu, —, 3, 88, 97, 96, —; Met, —, 4, 96, 97, 96, —; Phe, —, 2, 76, 93, 89, —.

PPC10. Met–Aax, conditions as above[b]

Val 10, 37, 88, 92, —, 76; Leu^z, 4, 26, 74, 82, 73, 49; Met, 9, 37, 85, 87, 78, 63; Phe, 4, 25, 58, 77, 56, 24; Tyr, 15, 53, 81, 84, 72, 48; His, —, 9, 96, 97, 95, 70; Gly, 20, 51, 96, 96, 96, 92.

PPC11. Aax–Gly, conditions as above[d]

Ser, 49, 68, 96, 97, 96, 95; Thr, 49, 65, 96, 97, 96, 95; Val, 26, 52, 96, 97, 96, 95; Leu^z, 11^x, 44, 96, 96, 94, 86; Met, 20, 51, 96, 96, 96, 92; Trp, 9, 41, 73, 77, 59, 44; Tyr, 28, 64, 91, 94, 85, 81; His, 3^x, 13^x, 96, 97, 96, 81.

PPC12, 13, 14. 100R_F, TLC, [1-(β-mercapto β,β-pentamethylene-propionic acid)-2-ile-4-Aax] arginine vasopressin, BuOH–AcOH–H$_2$O(4:1:1 v/v), BuOH–AcOH–H$_2$O (4:1:5, v/v), (upper-phase), BuOH–AcOH–H$_2$O–PyH (15:3:3:10, v/v)[c]

Abu, 25, 30, 47; Ile, 29, 32, 51; Thr, 20, 29, 44; Ala, 36, 29, 44; Ser, 27, 28, 42; Nva, 37, 34, 49; Gln, 12, 31, 35; Leu, 30, 33, 50; Lys, 3, 23, 20; Cha, 42, 36, 51; Asn, 22, 28, 35; Orn, 4, 25, 20; Phe, 38, 36, 49.

PPC15, 16, 17. 100R_F, TLC [1-(β-mercapto-β,β-pentamethylene-propionic acid)-2-phe-4-Aax] arginine vasopressin, conditions as above[d]

Ile, 33, 39, 58; Abu, 28, 35, 53; Thr, 22, 34, 48; Ala, 21, 35, 50; Gln, 13, 32, 40; Lys, 2, 28, 19;

TABLE 34 (*Continued*)

Cha, 36, 41, 62; Nle, 34, 39, 58; Nva, 32, 38, 57; Phe, 33, 38, 55; Leu, 34, 39, 59; Gly, 19, 34, 47; Tyr, 31, 37, 53.

PPC18. $100R_F$, TLC, Z–pGlu–Aax–Pro–NH$_2$, mobile-phase EtOAc–(PyH–AcOH–H$_2$O, 20:6:11, v/v), 3:2^e
Phe, 65; Met,56; Leu, 60; Gly, 42; Ala, 47; Val, 54; Ile, 57; Abu, 48; Nva, 58; Nle, 58; Ade, 68; Cha, 63; Ser (tBu), 61; Thr(Bzl), 67.

PPC19, 20. $100R_F$, TLC, pGlu–Aax–Pro–NH$_2$, mobile-phase, EtOAc–(PyH–AcOH–H$_2$O, 20:6:1, v/v) 3:2/CHCl$_3$–MeOH, 4:1, v/v^e
Phe, 40, 33; Met, 32, 20; Leu, 37, 29; Ala, 12, 20; Val, 33, 28, Ile, 39, 32; Abu, 27, 23; Nva, 33, 23; Nle, 40, 24; Ade, 51, 31; Cha, 45, 27; Ser(tBu),40, 34; Thr, 12, 13.

PPC21. $100R_F$, TLC LL or DD- $^+$HAax^1-Aax^2 CF$_3$CO$_2^{-f}$ Aax^1, Aax^2, LL, DD
Ala, Ala, 22, 23; Ala(Cl), Ala, 29, 29; Ala, Ala(Cl), 26, 26; Ala(Cl), 36, —, Met, Ala(Cl), 34, —; Nva, Ala(Cl),34, —; Ala, Pga, 22, 22; Pga, Ala, 25, 25; Pga, Pga, 35, —; Ala(Cl), Pga, 32, 32.

PPO1. Rates of hydrolysis of MeO$_2$C–Phe–OPnNO$_2$-4 and BzlO$_2$C–Phe–OPnNO$_2$-4 catalyzed by BzlO$_2$–Aax–His–R or BzlO$_2$C–Aax–his–R R = H or Me) in 0.02 M phosphate buffer, pH = 7.30, containing cetyltrimethylammonium bromide micelles at 25°^{g}. Aax, R, —, —, —, —.
Ala, H, 139, 32.1, 98,9, 37.9; Ala, Me, 32.6, 15.4, 33.0, 18.9; Val, H, 309, 44.7, 225, 56.7; Leu, H, 645, 52.7, 473, 74.9; Leu, Me, 291, 46.2, 210, 56.4; Phe, H. 541, 74.5, 439, 110; Trp, H, 85.9, 47.2, 103, 67.5.

^{a}Ref. 93. ^{b}Ref. 94. ^{c}Ref. 97. ^{d}Ref. 98. ^{e}Ref. 95. ^{f}Ref. 96. ^{g}Ref. 75.

can be used to predit the retention times of peptides in HPLC chromatography by an additive method. Meek and Rossetti (93) have reported retention coefficients for amino acid residues in peptides. Their results are based on a study of 100 peptides of no more than 20 residues. The mobile phase was composed of mixtures of acetonitrile with either aqueous NaClO$_4$ or aqueous NaH$_2$PO$_4$. Retention times should not properly be correlated directly but rather in the form of their logarithms. As some of the retention coefficients are negative, we have chosen to correlate them directly with the IMFU equation (sets PPC 1 and 2). Alternatively, we could have rescaled them and then used the logarithms of the rescaled values as the dependent variable. Best results for set 2 are obtained when the acidity of the mobile-phase phase is taken into account (see Section III.C). The major difference between sets 1 and 2 is the dependence of the latter on i. The intercept, the coefficient of σ_1 and that of α, is the same for the two sets within a standard deviation. The coefficient of n_H is somewhat smaller for set 2 than it is for set 1.

Lepri, Desideri and Heimler (94) have reported R_F values obtained from reversed-phase TLC for dipeptides on silanized silica gel with and without

TABLE 35
Results for Peptide Properties Data Sets

Set	PPC1	PPC2	PPC3	PPC4	PPC5	PAC6
Z	—	—	1.04	0.866	0.925	0.843
S_Z	—	—	0.0783	0.0531	0.0855	0.0575
L	−26.7	−22.9	—	7.46	—	—
S_L	14.2^k	11.6^k	—	1.77	—	—
A	47.0	42.9	−2.10	−3.01	−0.943	−2.21
S_A	6.42	5.18	0.386	0.339	0.336^g	0.343
H_1	−3.74	−2.36	0.283	−1.51	−0.189	0.244
S_{H1}	0.625	0.619^f	0.0949^g	0.253	0.105^k	0.0530
H_2	—	—	—	0.978	0.168	—
S_{H2}	—	—	—	0.137	0.0557^f	—
I	—	−5.35	—	—	−0.217	—
S_I	—	1.60^g	—	—	0.0809^h	—
S	—	—	—	—	—	−0.236
S_S	—	—	—	—	—	0.121^k
B^0	−0.913	−1.22	0.0156	0.752	0.372	0.326
S_{B^0}	1.43^a	1.15^o	0.112^g	0.0757	0.148^i	0.0787
$100R^2$	87.62	90.77	87.54	92.76	85.68	93.66
F	33.03	31.96	63.24	71.74	39.48	103.5
S_{est}	2.44	1.96	0.238	0.132	0.140	0.138
S^0	0.399	0.357	0.378	0.297	0.411	0.273
n	18	18	31	34	39	33

Set	PPC7	PPC8	PPC9	PPC10	PPC11	PPC12
Z	0.774	0.724	1.08	0.749	0.589	—
S_Z	0.0481	0.102	0.0503	0.0375	0.0364	—
L	8.86	—	—	3.08	—	—
S_L	1.36	—	—	1.26^i	—	—
A	−4.36	−0.838	−0.778	−1.48	−0.304	—
S_A	0.404	0.237^f	0.250^f	0.242	0.167^k	—
H_1	0.574	−0.398	0.106	0.209	−0.137	—
S_{H1}	0.0551	0.0841	0.0609^k	0.0527	0.0524^h	—
H_2	0.317	0.238	—	—	0.0909	−0.0977
S_{H2}	0.0973^f	0.0363	—	—	0.0277^f	0.0234
I	−1.51	−0.445	—	−0.225	0.0750	−0.890
S_I	0.107	0.0522	—	0.100	0.0436^k	0.0739
S	—	−0.596	—	—	—	—
S_S	—	0.200^f	—	—	—	—
B^0	0.825	1.187	−0.0963	0.686	0.848	1.527
S_{B^0}	0.0608	0.214	0.0888^h	0.0695	0.0744	0.0359
$100R^2$	95.85	92.84	93.90	93.42	89.46	94.30
F	130.9	86.40	153.8	93.73	66.21	82.72
S_{est}	0.0947	0.0979	0.160	0.103	0.0761	0.0961
S^0	0.224	0.290	0.283	0.279	0.349	0.272
n	41	47	34	39	45	13

TABLE 35 (Continued)

Set	PPC13	PPC14	PPC15	PPC16	PPC17	PPC18
Z	—	—	—	—	—	—
S_Z	—	—	—	—	—	—
L	—	—	—	—	−0.420	—
S_L	—	—	—	—	0.229^m	—
A	0.348	0.0872	0.926	0.154	0.156	0.369
S_A	0.705	0.0436^k	0.131	0.0480^h	0.0562^j	0.0481
H_1	—	−0.172	−0.465	−0.0800	−0.179	—
S_{H1}	—	0.0230	0.123^g	0.00807	0.0428^g	—
H_2	−0.0133	0.0688	0.183	0.0270	0.0790	—
S_{H2}	0.00457^h	0.0147^f	0.0739^j	0.00593^f	0.0287^j	—
I	−0.128	−0.106	−0.483	—	−0.192	—
S_I	0.0142	0.0325^h	0.180^j	—	0.0606^h	—
S	—	0.0714	—	0.0368	0.0683	0.0735
S_S	—	0.0196^g	—	0.0173^k	0.0201^h	0.0247^h
B^0	1.453	1.616	1.327	1.524	1.669	1.628
S_{B^0}	0.0139	0.0126	0.0250	0.00926	0.0107	0.0160
$100R^2$	92.62	99.80	99.01	95.89	99.54	91.33
F	37.64	700.9	200.7	46.64	214.3	57.97
S_{est}	0.0181	0.00833	0.0416	0.0111	0.0129	0.0195
s^0	0.327	0.0608	0.127	0.258	0.100	0.332
n	13	13	13	13	13	14

Set	PPC19	PPC20	PPC21^u	PC21^u	PPO1
Z	—	—	—	—	1.10
S_Z	—	—	—	—	0.111
L	—	—	0.556	0.885	−8.85
S_L	—	—	0.138^f	0.255^g	3.22^h
A	0.980	0.424	0.501	−0.949	3.79
S_A	0.276^g	0.193^k	0.231^k	0.553^m	0.530
H_1	−0.624	−0.391	—	—	−1.22
S_{H1}	0.155	0.0819	—	—	0.166
H_2	—	—	—	—	—
S_{H2}	—	—	—	—	—
I	—	—	—	—	—
S_I	—	—	—	—	—
S	0.344	—	—	—	—
S_S	0.188^m	—	—	—	—
B^0	1.069	1.336	1.355	—	−0.455
S_{B^0}	0.141	0.0451	0.0391	—	0.219^k
$100R^2$	81.80	69.48	83.18	—	89.74
F	13.48^f	11.38^f	13.60	—	38.47
S_{est}	0.0975	0.0704	0.0370	—	0.159
s^0	0.513	1.630	0.495	—	0.361
n	13	13	16	—	28

^{u}Values of L, S_L, A, and S_A are from X^1 and X^2, respectively.
^{v}Values of B_{Me} and S_{BMe} are −0.348 and 0.0778, respectively.
For footnotes see Table 3.

TABLE 36
Zeroth-Order Partial Correlation Coefficients for Peptide Property Data Sets

Set	PPC1	PPC2	PPC3	PPC4	PPC5	PPC6
σ_1, α	0.231	0.231	0.034	0.518	0.592	0.042
σ_1, n_H	0.267	0.267	0.735	0.637	0.556	0.747
σ_1, n_n	0.730	0.730	—	0.428	0.758	—
σ_1, i	0.381	0.006	—	—	0.300	—
σ_1, v	0.007	0.008	0.062	0.149	0.734	0.094
α, n_H	0.283	0.283	0.119	0.462	0.180	0.257
α, n_n	0.115	0.115	—	0.497	0.368	—
α, i	0.115	0.297	—	—	0.077	—
α, v	0.455	0.455	0.726	0.704	0.166	0.745
n_H, n_n	0.646	0.646	—	0.937	0.734	—
n_H, i	0.517	0.605	—	—	0.267	—
n_H, v	0.023	0.023	0.063	0.086	0.721	0.123
n_n, i	0.515	0.085	—	—	0.046	—
n_n, v	0.044	0.044	—	0.080	0.644	—
i, v	0.046	0.004	—	—	0.092	—

Set	PPC7	PPC8	PPC9	PPC10	PPC11	PPC 12, 13, 14
σ_1, α	0.508	0.590	0.037	0.512	0.616	0.321
σ_1, n_H	0.669	0.557	0.764	0.640	0.534	0.489
σ_1, n_n	0.445	0.757	—	0.427	0.756	0.788
σ_1, i	0.800	0.305	—	0.776	0.274	0.138
σ_1, v	0.149	0.733	0.089	0.155	0.725	0.359
α, n_H	0.461	0.180	0.116	0.457	0.176	0.055
α, n_n	0.500	0.368	—	0.493	0.376	0.253
α, i	0.151	0.078	—	0.135	0.070	0.187
α, v	0.704	0.166	0.734	0.713	0.185	0.631
n_H, n_n	0.930	0.734	0.090	0.935	0.731	0.815
n_H, i	0.645	0.272	—	0.614	0.243	0.589
n_H, v	0.089	0.721	—	0.092	0.711	0.242
n_n, i	0.320	0.047	—	0.295	0.037	0.029
n_n, v	0.083	0.644	—	0.086	0.638	0.251
i, v	0.057	0.093	—	0.056	0.074	0.128

Zeroth-order partial correlation coefficients for ζ with other variables are less than 0.200 for sets PPC3-11 and PPO1. For set PPC21, zeroth-order partial coerrelation coefficients are: $\sigma_1\sigma_2$, 0.248; σ_1, α_1, 0.387; $\sigma_1\alpha_2$, 0.090; $\sigma_1 v_1$, 0.454; $\sigma_1 v_2$, 0.049; $\sigma_2\alpha_1$, 0.086; $\sigma_2\alpha_2$, 0.854; $\sigma_2 v_1$, 0.166; $\sigma_2 v_2$, 0.575; $\alpha_1\alpha_2$, 0.037; α_1, v_1, 0.779; $\alpha_1 v_2$, 0.123; $\alpha_2 v_1$, 0.146; $\alpha_2 v_2$, 0.917; $v_1 v_2$, 0.101.

For set PPO1, the remaining zeroth-order partial correlation coefficients are:
$\sigma_1 n_{Me}$, 0.447; $\sigma_1 n_1$, 0.474; $\sigma_1 n_2$, 0.121; αn_{Me}, 0.368; αn_1, 0.434; αn_2, 0.464; $n_H n_{Me}$, 0.258; $n_H n_1$, 0.091; $n_H n_2$, 0.070; $n_{Me} n_1$, 0.354; $n_{Me} n_2$, 0.108; $n_1 n_2$, 0.230.

TABLE 36 (Continued)

Set	PPC 15, 16, 17	PPC18	PPC 19, 20	PPO1
σ_1, α	0.070	0.370	0.366	0.457
σ_1, n_H	0.548	—	0.542	0.000
σ_1, n_n	0.789	0.909	0.908	—
σ_1, i	0.096	—	—	—
σ_1, v	0.039	0.017	0.167	—
α, n_H	0.194	—	0.446	0.754
α, n_n	0.158	0.384	0.376	—
α, i	0.154	—	—	—
α, v	0.683	0.477	0.195	—
n_H, n_n	0.854	—	0.677	—
n_H, i	0.595	—	—	—
n_H, v	0.011	—	0.071	—
n_n, i	0.111	—	—	—
n_n, v	0.019	0.006	0.163	—
i, v	0.003	—	—	—

TABLE 37
C_i Values for Peptide Properties Data Sets

PPC Set	ζ	σ_1	α	n_H	n_n	i	v
1	—	12.8	64.8	22.4	0	0	0
2	—	9.42	50.8	12.1	0	27.6	0
3	57.4	0	26.1	15.7	—	—	0
4	16.1	13.2	15.4	33.5	21.7	0	0
5	63.4	0	10.0	8.75	7.77	10.0	0
6	46.6	0	29.6	14.2	—	—	9.62
7	16.1	14.8	20.9	12.0	0	31.6	4.63
8	30.0	0	7.99	16.5	9.86	18.4	17.3
9	79.1	0	13.1	7.80	—	—	0
10	42.3	13.9	19.2	11.8	0	12.7	0
11	61.2	0	7.27	14.2	9.45	7.80	0
12	—	0	0	0	9.89	90.1	0
13	—	0	36.1	0	5.99	57.9	0
14	—	0	4.82	41.2	16.5	25.4	12.0
15	—	0	15.9	34.6	13.6	35.9	0
16	—	0	21.1	47.6	16.0	0	15.3
17	—	5.92	6.31	31.5	13.9	33.9	8.42
18	—	0	62.2	—	0	—	37.8
19	—	0	20.7	57.2	0	—	22.1
20	—	0	20.0	80.0	0	—	0

For set PPC21, C_i values are; σ_1, 9.90; σ_2, 15.8; α_1, 25.7; α_2, 48.6; v_1, 0; v_2, 0.
For set PPO1, C_i values are ζ, 25.9; σ_1, 16.7; α, 20.5; n_H, 28.7; n_{Me}, 8.18.

impregnation by detergents. We have studied three types of dipeptide, His–Aax (sets PPC 3, 6, and 9), Met–Aax (sets PPC 4, 7, and 10), and Aax–Gly (sets 5, 8, and 11), making use of the Zeta method to combine into a single data set R_F values obtained with different layers and mobile-phases. The data sets were correlated with the IMFU equation. The results for His–Aax show strong similarities, with sets 3, 6, and 9 all dependent on ζ, α, and n_H. As the data sets included no ionizable side chains and in all of them $2n_H = n_n$, the dependence on these variables could not be determined. The results for Aax–Gly also show strong similarities, with set 5, 8, and 11 all dependent on ζ, α, n_H, n_n and i. These sets differ principally in that 8 is a function of steric effects, whereas 5 and 11 are not. By contrast, the Met–Aax data sets show fewer similarities in behavior. Though they are all dependent on ζ, σ_1, α, and n_H, set 4 is also a function of n_n and v while sets 7 and 10 are not, and set 4 is not a function of i while 7 and 10 are. ζ is the dominant parameter in the His–Aax data sets and in two of the three Aax–Gly data sets but not in the Met–Aax data sets. Good results were obtained for all of these sets. Due to the narrow range of side chains studied, they do not provide a strong test of the validity of the IMF model.

Szirtes, Kisfaludy, and Szporny (95) have reported R_F values obtained from TLC for $BzlO_2C$–pGlu–Aax–Pro–NH_2 (set PPC 18) and pGlu–Aax–Pro–NH_2 (sets PPC 19 and 20). As these sets included no ionizable side chains and only two capable of hydrogen bonding, the terms in i and n_n were excluded from the IMF equation. Correlation with the IMFU equation gave good results for sets 18 and 19, and poor but significant results for set 20.

Cheung and coworkers (96) have determined R_F values obtained from TLC for the trifluoroacetates of the dipeptides Aax^1–Aax^2 and aax^1–aax^2. Only nonpolar amino acids were studied. As the configuration seemed to have no effect on the magnitude of the R_F, value we combined the data for the L, L and the D, D peptides into a single set (PPC 21) and correlated it with the equation

$$\log 100 * R_F = L_1 \sigma_{1X^1} + L_2 \sigma_{1X^2} + A_1 \sigma_{X^1} A_2 \alpha_{X^2} + S_1 v_{X^1} + S_2 v_{X^2} + B^0 \quad (51)$$

The results obtained were interesting as well as significant. There is no dependence on steric effects, but a significant dependence on σ_1 for both X^1 and X^2, and on α for X^1 and probably for X^2 as well. The coefficients of σ_1, L_1 and L_2, are not significantly different, those of α, A_1 and A_2, are different in both magnitude and sign.

Manning and coworkers (97, 98) have measured R_F values obtained from TLC for [1-(β-mercapto-β, β-cyclopentamethylenepropionic acid)-2-aax-4-Aax] arginine vasopresion (7) with the 2-aax either ile (sets PPC 12–14) or phe (sets PPC 15–17).

```
          1   2   3   4   5   6   7   8   9
         CH₂CO-aax-Phe-Aax-Asn-Cys-Pro-Arg-Gly-NH₂
```

$$\text{(structure 7: cyclohexyl-CH}_2\text{CO-aax-Phe-Aax-Asn-Cys-Pro-Arg-Gly-NH}_2 \text{ with S—S bridge)}$$

7

Attempts to apply the Zeta method to sets 12–14 were unsuccessful. All of these data sets gave good results. Every set was a function of n_n, all but one set of α and i, and four of the six of n_H. Sets 15 and 17 were dependent on steric effects, and the latter showed a small dependence on σ_1 as well.

Overall, of the 21 chromatographic data sets studied, all but one showed a dependence on α. Of the 19 sets in which n_H was a variable, 17 were a function of it. In order of their importance as determined by the frequency of their significance as a variable (the number of data sets in which they are significant divided by the number in which they were studied), the chromatographic data sets are a function of α, n_H, i, n_n, v, and σ_1. The results obtained for peptide chromatographic data sets are in general agreement with those observed for amino acid chromatographic data sets.

Finally, we have applied the Zeta method to rate constants determined by Ihara and coworkers (75) for the hydrolysis of 4-nitrophenyl N-benzyloxycarbonyl- and N-methoxycarbonyl-phenylalaninates in 0.02 M phosphate buffer, pH 7.30, at 25° catalyzed by BzlO₂Aax–His–OR (R = H or Me) (set PPO1). Correlation with the IMFU equation gave good results. The hydrolysis rate constants are a function of σ_1, α, n_H, and n_n.

Overall, the correlation of peptide properties with the IMF equation gave very good results. Of the 22 data sets studied, 14 gave $100R^2$ values greater than 90.0, seven gave values between 80 and 89.9, and one gave a value less than 70.0. All of the correlations were significant according to the F statistic.

C. Peptide Bioactivity, an Overview

Having already shown that the IMF model is applicable to amino acid bioactivity, we now consider its applicability to peptide bioactivity. For this purpose, we have examined examples of the various types of peptide substitution.

1. X^n Substitution

Manning and coworkers (97, 98) have determined pA_2 values for the antiantidiuretic, antivasopressor, and in both vivo and vitro the antioxytocin

TABLE 38
Peptide Bioactivity Data Sets

PPB1. pA_2, antiantidiuretic, [1-(β-mercapto-β, β-cyclopentamethylene propionic acid)-2-phe/ile-4-Aax]arginine vasopressin. Aax, phe, ile[a]
Ala, 7.52, 7.76; Abu, 7.96, 8.22; Cha, 7.19, 6.64; Gln, 7.21; 6.96; Gly, 5.85[x], —; Ile, 8.24, 8.04; Leu, 6.07, 6.80; Lys, 7.22, 6.76; Nle, 7.12, —; Nva, 6.99, 7.01; Phe, 6.07, —; Thr, 7.62, 7.91; Tyr, 5.57[x](approx), —; Val, 8.07, 7.98; Orn, 6.50, 6.46; Ser, 7.33, 7.26; Asn, —, 6.51.

PPB2. pA_2, antivasopressin, substrate as above. Aax, phe, ile[a]
Ala, 7.55, 6.03; Abu, 7.70, 6.73; Cha, 6.93, 6.56; Gln, 8.35, 7.79; Gly, 6.79[x], —; Ile, 7.86, 6.42; Leu, 7.70, 6.75; Lys, 6.93, —; Nle, 7.52, —; Nva, 7.73, 6.96; Phe, 6.45, —; Thr, 7.38, 6.83; Tyr, 6.58, —; Val, 8.06, 6.94; Orn, 7.57, 6.76; Ser 7.63, 6.21; Asn, —, 7.26.

PPB3. pA_2, antioxytocin activity in vitro (rat uterus), [1-(β-mercapto-β, β-cyclopentamethylene propionic acid)-2-2phe-4-Aax] arginine vasopressin, no $Mg^{++}/0.5\,mM\,Mg^{++}$. Aax, $0\,Mg^{++}$, $0.5\,mM\,Mg^{++\,a}$
Ala, 7.68, 7.73; Abu, 7.73, 7.76; Cha, 7.78, 7.59; Gln, 8.59, 8.43; Gly, 7.76, 7.33; Ile, 8.04, 8.19; Leu, 7.76, 7.88; Lys, 7.60, 7.47; Nle, 7.54, 7.78; Nva, 8.03, 7.72; Phe, 7.57, 7.67; Thr, 8.04, 8.01; Tyr, 7.68, 7.45[x]; Val, 7.74, 8.29.

PPB4. pA_2, antioxytocin activity in vitro (rat uterus), [1-β-mercapto-β, β-cyclopentamethylene propionic acid)-2-ile-4-Aax] arginine vasopressin, no $Mg^{++}/0.5\,mM\,Mg^{++\,a}$
Ala, 7.50, 7.47; Abu, 7.49, 7.03; Cha, 7.44, 7.09; Gln, 7.62, 7.81; Ile, 7.47[x], 7.82; Leu, 7.84, 7.57; Nva, 7.83, 7.80; Phe, 7.13, 6.92; Ser, 7.28, 6.82; Thr, 8.10, 7.69; Val, 7.98, 7.53; Asn, 7.93, 7.40; Lys, 7.10, 6.60; Orn, 7.57, 6.71.

PPB5. pA_2, antioxytocin activities (rat uterus) in vivo. [1-β-mercapto-β, β-cyclopentamethylene propionic acid)-2-ile/phe-4-Aax] arginine vasopressin, As, phe, ile[a]
Abu, 7.52, 6.65; Ala, 6.91, 6.14; Gln, 7.25, 6.63; Ile, 7.11; 6.90; Ser, 6.70, —; Thr, 7.35, 6.87; Val, 6.92, 6.21; Nva, —, 6.53.

PPB6, 7. PCI (% inhibition) of the monkey intestinal Gly–Leu uptake system, GlyAax, (^{14}C-Gly)–Leu/Gly–(^{14}C-Leu)[b].
Aax, (^{14}C-Gly)–Leu, Gly–(^{14}C-Leu) Gly, 24, 20; Ala, 42, 50; Val, 70, 65; Ile, 80, 78; Leu, 82, 86; Ser, 53, 70; Phe, 51, 73; Trp, 44, 67; His, 45, 63; Met, 70, 86; Asp, 38, 50; Glu, 49, 42; Asn, 50, 63.

PPB8. A_{rel}, γ-glutamyltranspeptidase and Gly–Aax, L, D-γ-glutamyl-4-nitroanilide[c]. Aax, L, D-(γ-glutamyl)-4-nitroanilide
Gly, 100, 100; Met, —, 5; Ala, 65, 83; Ans, 14, —; Asp, 11, —; Thr, 11, 3; Val, 10, —; Lys, —, 60; Leu, —, 7; Tyr, —, 3.

PPB9. A_{rel}, γ-glutamyl transpeptidase and Aax–Gly, D-glutamyl-4-nitroanilide[c]. Aax, A_{rel}
Met, 210[x]; Gln, 182; Ala, 154; Gly, 100; Ser, 99; Cys, 81; Lys, 5.57; Phe, 28; Leu, 19; Val, 14.

PPB10. Relative potency, , C-substituted Lutein-hormone-releasing factor (LRF)[d]. X[c], % (average values)
NH_2, 10.5; NHMe, 90; NHEt, 400; NHPr, 250; $NHCH_2OH$, 125; NMe, 15; pyrrolidinyl, 75; morpholinyl, 25; piperidinyl, 1.5; NHiPr, 150; NHiBu, 3; NHcHx, 0.7; NHBu, 7; OMe, 6; OEt, 11.

PPB11–14. Enkephalin activity [(Met[5])enkephalin = 1], (Z)Try–Gly–Gly–Phe–Met, guinea pig ileum/mouse vas deferens: IC_{50}, n mol dm^{-3} rat brain homogenate, vs[^{3}H]-[ala^2; leu^5]enkephalin; K_D, n mol dm$^{-3\,e}$. Z, A(gpi), A(mvd), IC_{50}, K_D
iPr, 0.013, 0.058, 1100, 21; $PhCH_2CH_2$, 0.15, 0.01, 35, 15.6; $cPrCH_2$, 0.014, 0.0025, 460, 206; Me, 1.45, 0.23, —, —; Et, 0.04, 0.0086, 50, 22.4; Pr, 0.045, 0.0081[x], 56, 25; Bu, 0.20, 0.0078, 58, 26; Pe, 0.67, 0.011, 26, 11.6; Hx, 0.16, 0.027, 10, 4.5; Hp, 0.11, 0.0064, 36, 16.1; Oc, 0.08, 0.0078, 45, 20.

TABLE 38 (Continued)

PPB15. Relative potency, mouse vas deferens, mouse hot plate test, Tyr–ala–Gly–(4-Z)–Phe–MeMet–NH$_2{}^f$. Z, A$_{rel,mvd}$, A$_{rel,mhpt}$

H, 1, 1; F, 15.8, 16.5; Cl, 5.6, 0.54; Br, 5.6, 0.23; I, 0.94, —; CF$_3$, 8.1, 0.84; NO$_2$, 19.1, 2.3; OH, 0.03, —.

PPB16, 17. Relative potency (mouse vas deferens/mouse hot plate test). Tyr–ala–Gly–(CH$_2$Z)–PheNH$_2{}^g$. Z, A$_{mvd}$, A$_{mhp}$

H, 0.10, 0.77; Me, 5.81, 121; Et, 1.11, 2.4; Hp, 0.46, 0.03; Vi, 1.44, 36.3; cPr, 7.53, 30.25; C$_2$H, 2.3, 4.54; CH=CMe$_2$, 1.89, 0.91; CH$_2$SMe, 2.87, 7.3; CO$_2$Me, 0.06, 0.26; CH$_2$OH, 0.04, 1.1; CH$_2$F, 3.2, 27.9.

PPB18. $K_{cat}/K_M \times 10^{-4}$, ArCO–Leu–Gly–Pro–Aax–OR + β-collagenase, water 50 mM in Tricine, 0.4 M in NaCl, 10 mM in CaCl$_2$, pH = 7.5, 25°^{h}. Ar, Aax, R, $10^4 K_{cat}/K_M$

Ph, Gly, H, 7.2; Ph, Gly, Me, 2.7; Ph, Ala, H, 9.4; Ph, Ala, Me, 4.7; Ph, Ser, H, 48; Ph, Ser, Me, 3.5; Ph, Leu, H, 160; Ph, Leu, Me, 24; Ph, Pro, H, 780; Ph, Hyp, H, 570; Ph, Hyp, Me, 4.6; Ph, Arg, H, 500; Ph, Arg, Me, 470; 2-furyl, Gly, H, 7.3; 2-furyl, Ala, H, 61; 2-furyl, Leu, H, 150; 2-furyl, Pro, H, 460.

PPB19. $K_{cat}/K_M \times 10^{-4}$. Hydrolysis of ArCOAax–GlyProAla + β-collagenase, conditions as above^h. Ar, Aax $10^4 K_{cat}/KM$

Ph, Gly, 3.4; Ph, Ala, 64; Ph, Ser, 50; Ph, Glu, 0.42; Ph, Val, 2.2; Ph, Ile, 1.2; Ph, Phe, 900; Ph, (4-NO$_2$)Phe, 4400; Ph, Tyr, 1300; Ph, Pro, 26; Ph, Hyp, 280; Ph, Leu, 94; 2-furyl, Gly, 1.1; 2-furyl, Leu, 61; 2-furyl, Phe, 690.

PPB20. $K_{cat}/K_M \times 10 \text{x}^{-4}$, Hydrolysis of ArCOLeuGlyAax(Pro/Hyp) + β-collagenase, conditions as above^h. Ar, Aax^3, Aax^4, $10^4 K_{cat}/K_M$

Ph, Ala, Pro, 27; Ph, Ser, Pro, 6.0; Ph, Gly, Pro, 1.8; Ph, Leu, Pro, 3.0; Ph, Hyp, Pro, 1.1; Ph, Pro, Pro, 780; Ph, Ala, Hyp, 32; Ph, Ser, Hyp, 2.3; Ph, Glu, Hyp, 2.0; Ph, Leu, Hyp, 3.4; 2-furyl, Pro, Pro, 460; 2-furyl, Ala, Hyp, 24.

PPB21. Relative potency, P$_t$, of 2-substituted thyrotropin-releasing factor (TRF) [pGlu–Aax–Pro–NH$_2$]^i. Aax, P

His(π-Me), 0.04; Ser(tBu), 0.2; Orn, 0.025; Met, 1; Thr, 25.7; Val, 3.1; Arg, 0.05; Lys, 0.02; Leu, 2.7; Phe, 10; Tyr, 0.084; Ile, 0.2; Nle, 3.8; Cha, 45.6; His, 100; His(τ-Me), 800.

[a] Ref. 97, 98. [b] Ref. 85. [c] Ref. 99. [d] Ref. 100. [e] Ref. 103. [f] Ref. 105. [g] Ref. 104. [h] Ref. 106. [i] Ref. 95, 100. [x] Excluded from the correlation.

activities of [1-(β-mercapto-β,β-cyclopentamethylenepropionic acid)-2-aax-4-Aax] arginine vasopression, **7**. The pA$_2$ values are given by the relationship

$$pA_2 = [(-1)*\log ba]/V_d \tag{52}$$

where ba is the substrate bioactivity and V_d the volume of distribution. V_d is set equal to 67 ml/kg. The pA$_2$ values for antiantidiuretic activity of **7a** (aax^2 = ile) and **7b** (aax^2 = phe) were combined into a single data set by means of the Zeta method (set PPB1). Correlation of the data set with the IMFSB equation gave best results on the exclusion of the data point for Gly; a good fit was obtained. No dependence on ζ was observed, showing that the activity does not depend on which of the two D amino acids, phe or ile, is present in the 2 position. Although, as is so often the case, collinearity

TABLE 39
Results of Correlation of Peptide Bioactivity Data Sets

Set:	PPB1	PPB2	PPB3	PPB4
Z	—	0.625	—	11.3
S_Z	—	0.0760	—	3.09^f
L	−25.9	—	—	−6.03
S_L	5.75	—	—	2.29^h
A	—	−11.5	−0.936	−3.23
S_A	—	2.94^f	0.323^g	1.09^g
H_1	−3.87	3.46	1.86	—
S_{H1}	1.26^g	0.651	0.297	—
H_2	2.98	−2.04	−0.993	0.157
S_{H2}	0.928^g	0.416	0.180	0.0742^j
I	0.468	−5.00	−2.92	−0.519
S_I	1.69^h	0.924	0.433	1.38^f
S	—	—	—	—
S_S	—	—	—	—
B_1	0.481	0.763	0.331	0.454
S_{B1}	0.117	0.174	0.0515	0.103
B_2	0.718	0.518	—	0.232
S_{B2}	0.0984	0.170^g	—	0.123^k
B_3	—	0.461	—	—
S_{B3}	—	0.220^k	—	—
B^0	7.31	3.22	7.61	−77.5
S_{B^0}	0.173	0.543	0.0668	23.1
$100R^2$	83.01	85.49	82.99	74.90
F	16.29	12.52	20.49	8.102
S_{est}	0.302	0.282	0.138	0.237
S^0	0.479	0.471	0.468	0.597
n	27	26	27	27

Set:	PPB5	PPB6	PPB7	PPB8	PPB9
Z	0.802	—	—	—	—
S_Z	0.134^j	—	—	—	—
L	−25.9	—	2.01	−1.90	3.41
S_L	6.56^h	—	0.571^g	1.11^m	1.37^k
A	—	—	—	−4.20	—
S_A	—	—	—	0.597	—
H_1	−4.60	—	—	—	0.209
S_{H1}	1.32^j	—	—	—	0.0768^j
H_2	3.61	—	−0.0513	—	—
S_{H2}	0.987^i	—	0.0186^i	—	—
I	nd	−0.107	−0.135	—	—
S_I	nd	0.0436^j	0.0537^j	—	—
S	—	0.513	0.606	—	—
S_S	—	0.0768	0.0667	—	—
B_1	0.230	—	—	−0.281	−0.541
S_{B1}	0.0774^j	—	—	0.0830^g	0.115^g

TABLE 39 (*Continued*)

Set:	PPB5	PPB6	PPB7	PPB8	PPB9
B_2	nd	—	—	—	—
S_{B2}	nd	—	—	—	—
B_3	nd	—	—	—	—
S_{B3}	nd	—	—	—	—
B^0	1.14	1.38	1.36	2.02	2.03
S_{B^0}	0.893^n	0.0559	0.0499	0.0836	0.123
$100R^2$	86.46	82.97	92.31	94.04	85.97
F	10.22^i	24.35	24.01	47.34	10.21^i
S_{est}	0.187	0.0662	0.0573	0.160	0.189
S^0	0.487	0.471	0.253	0.293	0.503
n	14	13	13	13	9

Set:	PPB10	PPB11	PPB12	PPB13	PPB14	PPB15
Z	—	—	—	—	—	0.706
S_Z	—	—	—	—	—	0.282^j
L	−6.35	—	—	—	—	1.01^u
S_L	2.82^k	—	—	—	—	0.608^m
A_1	−5.55	6.51	—	−6.26	−6.25	−9.11
S_{A1}	1.80^h	2.07^j	—	1.82^h	1.82^h	4.20^k
A_2	−1.02	−61.9	—	47.9	47.7	—
H_1	21.5^f	22.6^j	—	19.5^k	19.5^k	—
S_{H1}	—	—	—	—	—	−1.85
H_2	—	—	—	—	—	0.655^j
S_{H2}	—	—	—	—	—	—
I	—	—	—	—	—	—
S_I	—	—	—	—	—	—
S	—	—	—	—	—	—
S_S	—	—	—	—	—	—
B_1	—	—	—	—	—	—
S_{B1}	—	−1.98	−0.299	2.15	0.776	—
B_2	—	0.400^g	0.136	0.424^g	0.425^m	—
S_{B2}	—	1.33	−1.09	1.26	1.26	—
B_3	—	0.441^j	0.135	0.378^j	0.378^j	—
S_{B3}	—	—	—	—	—	—
B^0	3.05	2.631^j	−0.627	−1.918	−0.893	−0.242
S_{B^0}	0.570	0.886	0.188^h	0.815^k	0.815^h	0.284^p
$100R^2$	76.78	82.55	90.83	87.59	74.00	72.25
F	12.13^f	7.096^i	34.67	8.825^i	3.558^k	5.860^i
S_{est}	0.454	0.341	0.193	0.283	0.283	0.507
S^0	0.563	0.566	0.362	0.498	0.721	0.657
n	15	11	10	10	10	14

TABLE 39 (Continued)

Set:	PPB16	PPB17	PPB18	PPB19	PPB20	PPB21
L	—	−3.18	—	—	—	—
S_L	—	1.46^k	—	—	—	—
A_1	−3.22	−15.5	—	7.80	−5.74	7.87
S_{A1}	1.49^k	2.67	—	1.06	2.85^k	2.76^h
A_2	−17.7	—	—	—	—	—
S_{A2}	12.0^m	—	—	—	—	—
H_1	—	−1.29	—	—	−1.44	—
S_{H1}	—	0.571^k	—	—	0.342^f	—
H_2	−0.386	—	—	—	—	-0.692^f
S_{H2}	0.0758^f	—	—	—	—	0.177
I	—	—	1.29	−2.28	0.750	−0.995
S_I	—	—	0.302^f	0.451	0.500^m	0.348^h
S	2.30	5.29	1.23	—	−0.864	—
S_S	0.771^i	1.21^f	0.320^f	—	0.224	—
B_1	—	—	—	-1.78^v	—	—
S_{B1}	—	—	—	0.331	—	—
B_2	—	—	-1.02^w	—	—	3.40^y
S_{B2}	—	—	0.197	—	—	0.418
B_3	—	—	2.42^x	—	—	—
S_{B3}	—	—	0.477	—	—	—
B^0	−1.028	−2.675	1.075	0.722	2.420	−1.32
S_{B^0}	0.553^m	0.722^g	0.211	0.215^g	0.376	0.629^k
$100R^2$	87.46	83.25	85.89	90.65	83.37	87.17
F	12.20^f	8.697^g	18.26	35.55	8.775^g	18.68
S_{est}	0.300	0.531	0.384	0.733	0.783	0.591
S^0	0.677	0.536	0.447	0.357	0.534	0.432
n	12	12	17	15	12	16

[u]Values are for P, S_P. [v]Values are for B_{12}, S_{B12}. [w]Values are for B_{Me}, S_{BMe}.
[x]Values are for B_{NC}, S_{BNC}. [y]Values are for H_3, S_{H3}.

TABLE 40
Zeroth-Order Partial Correlation Coefficients for Peptide Bioactivity Data Sets

Set:	1	2	3	4	5	6, 7	8	9
σ, α	0.347	0.337	0.038	0.306	0.078	0.181	0.094	0.074
σ, n_H	0.494	0.551	0.531	0.482	0.725	0.615	0.452	0.122
σ, n_n	0.805	0.813	0.779	0.781	0.823	0.809	0.871	0.378
σ, i	0.098	0.060	0.089	0.141	—	0.667	0.363	0.249
σ, v	—	—	—	—	—	0.003	—	—
σ, n_1	0.220	0.216	0.380	0.265	0.244	—	0.532	0.221
σ, n_2	0.416	0.413	0.267	0.385	—	—	0.058	—
σ, n_3	0.166	0.152	0.019	0.180	—	—	—	—
α, n_H	0.026	0.027	0.175	0.025	0.290	0.072	0.403	0.255
α, n_n	0.191	0.201	0.108	0.222	0.214	0.262	0.224	0.060
α, i	0.224	0.166	0.179	0.201	—	0.022	0.188	0.328
α, v	—	—	—	—	—	0.509	—	—

TABLE 40 (*Continued*)

Set:	1	2	3	4	5	6, 7	8	9
α, n_1	0.144	0.156	0.319	0.110	0.605	—	0.474	0.529
α, n_2	0.800	0.802	0.753	0.763	—	—	0.677	—
α, n_3	0.786	0.780	0.766	0.806	—	—	—	—
n_H, n_n	0.808	0.835	0.860	0.820	0.987	0.744	0.736	0.845
n_H, i	0.646	0.598	0.614	0.593	—	0.399	0.511	0.614
n_H, υ	—	—	—	—	—	0.052	—	—
n_H, n_1	0.055	0.040	0.099	0.019	0.000	—	0.374	0.131
n_H, n_2	0.000	0.020	0.033	0.027	—	—	0.278	—
n_H, n_3	0.274	0.236	0.297	0.207	—	—	—	—
n_n, i	0.087	0.077	0.142	0.041	—	0.683	0.457	0.112
n_n, υ	—	—	—	—	—	0.047	—	—
n_n, n_1	0.059	0.061	0.190	0.088	0.049	—	0.403	0.124
n_n, n_2	0.169	0.172	0.050	0.127	—	—	0.225	—
n_n, n_3	0.037	0.030	0.200	0.035	—	—	—	—
i, υ	—	—	—	—	—	0.067	—	—
i, n_1	0.118	0.104	0.035	0.093	—	—	0.045	0.069
i, n_2	0.135	0.120	0.092	0.135	—	—	0.310	—
i, n_3	0.369	0.330	0.233	0.369	—	—	—	—
n_1, n_2	0.115	0.112	0.040	0.145	—	—	0.089	—
n_1, n_3	0.200	0.094	0.094	0.158	—	—	—	—
n_2, n_3	0.574	0.572	0.591	0.574	—	—	—	—

Set	10	11, 12	13, 14	15	16, 17
σ, α^*	0.440	—	—	0.448	0.163
σ, α^{*2}	0.109	—	—	0.589	0.346
σ, n_H	0.502	—	—	0.574	0.029
σ, n_n	0.338	—	—	—	0.559
σ, υ	—	—	—	0.420	0.156
σ, n_1	0.032	—	—	—	—
σ, n_2	0.255	—	—	—	—
$\alpha^* \cdot \alpha^{*2}$	0.008	0.195	0.602	0.433	0.007
α^*, n_H	0.198	—	—	0.191	0.189
α^*, n_n	0.059	—	—	—	0.070
α^*, υ	—	—	—	0.593	0.632
α^*, n_1	0.434	0.213	0.319	—	—
α^*, n_2	0.612	0.525	0.472	—	—
α^{*2}, n_H	0.190	—	—	0.048	0.037
α^{*2}, n_n	0.131	—	—	—	0.201
α^{*2}, υ	—	—	—	0.590	0.423
α^{*2}, n_1	0.520	0.497	0.148	—	—
α^{*2}, n_2	0.064	0.371	0.134	—	—
n_H, n_n	0.055	—	—	—	0.380
n_H, υ	—	—	—	0.140	0.017
n_H, n_1	0.737	—	—	—	—
n_H, n_2	0.218	—	—	—	—
n_n, υ	—	—	—	—	0.072
n_n, n_1	0.054	—	—	—	—
n_n, n_2	0.031	—	—	—	—
n, n_2	0.119	0.000	0.630	—	—

TABLE 40 (Continued)

Set	18	19	20	21	Set	18	19	20	21
σ, α	0.035	0.368	0.175	0.125	n_n, n_{Ar}	0.397	0.325	—	0.108^a
σ, n_H	0.460	0.371	0.842	0.012	n_n, n_{Me}	0.220	—	—	0.270^b
σ, n_n	0.661	0.794	—	0.582	n_n, n_{NC}	0.085	0.018	—	—
σ, i	0.256	0.267	0.427	0.282	n_n, υ	0.006	0.035	—	—
σ, n_{Ar}	0.294	0.199	—	0.068^a	i, n_{Ar}	0.203	0.134	—	0.372^a
σ, n_{Me}	0.167	--	—	0.380^b	i, n_{Me}	0.112	—	—	0.000^b
σ, n_{NC}	0.258	0.257	0.371	—	i, n_{NC}	0.203	0.105	0.258	—
σ, υ	0.224	0.247	0.286	—	i, υ	0.264	0.123	0.264	—
α, n_H	0.698	0.000	0.041	0.086	n_{Ar}, n_{Me}	0.410	—	—	$0.641^{a,b}$
α, n_n	0.548	0.320	—	0.207	n_{Ar}, n_{NC}	0.019	0.196	—	—
α, i	0.711	0.034	0.330	0.229	n_{Ar}, υ	0.038	0.038	—	—
α, n_{Ar}	0.150	0.027	—	0.681^a	n_{Me}, n_{NC}	0.119	—	—	—
α, n_{Me}	0.056	—	—	0.519^b	n_{Me}, υ	0.115	—	—	—
α, n_{NC}	0.141	0.098	0.255	—	n_{NC}, υ	0.887	0.870	0.971	—
α, υ	0.213	0.408	0.033	—	$n_\pi H, \sigma$	—	—	—	0.580
n_H, n_n	0.889	0.658	—	0.696	$n_\pi H, \alpha$	—	—	—	0.184
n_H, i	0.945	0.443	0.529	0.571	$n_\pi H, n_H$	—	—	—	0.009
n_H, n_{Ar}	0.308	0.302	—	0.154^a	$n_\pi H, n_n$	—	—	—	0.238
n_H, n_{Me}	0.171	—	—	0.173^b	$n_\pi H, i$	—	—	—	0.289
n_H, n_{NC}	0.090	0.207	0.098	—	$n_\pi H, n_1$	—	—	—	0.179
n_H, υ	0.164	0.162	0.038	—	$n_\pi H, n_2$	—	—	—	0.320
n_n, i	0.691	0.571	—	0.383					

$^a n_1$ in place of n_{Ar}. $^b n_2$ in place of n_{Me}.

Zeroth-order partial correlation coefficients of ζ with other variables are less than 0.200 for sets PPB1, 2, 3, 4, 5. For set PPB15, all but n_H (0.240) are less than 0.200.

makes the separation of the contributions from the different intermolecular forces difficult, it seems highly probable that ionic side chains and hydrogen bonding dominate the structural effects, with branching at the first and second side-chain atoms making small but significant contributions. The data for the antivasopressor activities of **7a** and **7b** were also combined into a single set by the application of the Zeta method (set PPB 2). It was again necessary to exclude the data point for Gly for the best results. Correlation with the IMFSB equation again gave a good fit. In set 2 a significant dependence on both ζ and polarizability was observed, in contrast to the results obtained for set 1. As in set 1, hydrogen bonding capacity and an ionic side-chain were the dominant structural factors, with small but significant steric contributions from branching at the first, second, and third atoms of the side-chain also contributing. The in vitro antioxytocin data for **7b** in the presence and absence of Mg^{++} were combined into a single data set by means of the Zeta method (set PPB 3). The same was done for the in vitro antioxytocin

TABLE 41
C_i Values for Peptide Bioactivity Sets

PPB Set	ζ	σ	α	α^2	n_H	n_n	i	v	n_1	n_2	n_3
1	0	18.3	0	—	3.42	26.3	41.3	—	4.25	6.34	0
2	4.03	0	17.0	—	22.3	13.2	32.2	—	4.92	3.34	2.97
3	0	0	3.41	—	29.4	15.7	46.3	—	5.24	0	0
4	81.4	3.47	5.33	—	0	1.13	3.73	—	3.26	1.66	0
5	7.09	18.3	0	—	40.6	31.9	—	—	2.03	—	—
6	—	0	0	—	0	0	23.0	77.0	—	—	—
7	—	20.8	0	—	0	6.66	17.5	55.0	—	—	—
8	—	10.8	69.1	—	0	0	0	—	20.1	0	—
9	—	26.7	0	—	20.5	0	0	—	52.9	—	—
10	—	33.0	5.77	61.2	0	0	—	—	0	0	—
11	—	—	11.0	7.61	—	—	—	—	48.7	32.0	—
12	—	—	0	0	—	—	—	—	21.6	78.4	—
13	—	—	10.6	5.87	—	—	—	—	52.7	30.9	—
14	—	—	15.9	8.81	—	—	—	—	28.7	46.6	—
15	21.0	11.4[a]	12.5	—	55.1	—	—	—	—	—	—
16	—	0	12.9	7.11	0	31.0	—	49.0	—	—	—
17	—	5.84	25.9	0	21.5	0	—	46.8	—	—	—
18	—	0	0	—	0	0	23.1	15.4	18.2[b]	43.3[c]	—
19	—	0	30.6	—	0	0	38.9	—	30.4	—	—
20	—	0	32.1	—	35.0	0	18.2	14.7	—	—	—
21	—	—	26.3	—	—	10.0	14.4	—	49.3[d]	—	—

[a] Variable is σ_p. [b] Variable is n_{Me}. [c] Variable is n_{NC}. [d] Variable is $n_{\pi H}$.

data for **7a** (set PPB 4). The data were successfully correlated with the IMFSB equation on the exclusion of the Tyr value determined in the absence of Mg^{++} from set 3 and the Ile value determined in its presence from set 4. Sets 3 and 4 give very different results; while 3 is indepenent of ζ, in 4 it is the dominant parameter. In view of the very small difference in the values of ζ in set 4, it seems unlikely that the dependence on it is real. It is much more likely that it is an artifact. Supporting this conclusion is the very large value of Z which is obtained. It is about ten times that normally observed. Sets 3 and 4 show a number of other differences in their behavior aside from their dependence on ζ. Thus, 3 is dependent on n_H and n_2 but not on σ_1, whereas the reverse is true for 4. Though both sets are a function of α, n_n, i, and n_1, the coefficients of the first three variables differ greatly in magnitude. The pA_2 values for in vivo antioxytocin activity determined for **7a** and **7b** were again combined into a single data set by means of the Zeta method (set PPB 5) and the data were correlated with a modification of the IMFSB equation. The i parameter was deleted from the correlation equation as none

TABLE 41A.
Location of OH and NH Bonds in Amino Acid Side Chains

High-Potency Residues

His, (τ—Me)His, Thr

Low-Potency Residues

—CH$_2$CH$_2$CH$_2$CH$_2$$\overset{5}{\text{N}}$HC(NH$_2$)$_2^+$ Arg

—CH$_2$CH$_2$CH$_2$CH$_2$$\overset{5}{\text{N}}H_3^+$ Lys

—CH$_2$CH$_2$CH$_2$$\overset{4}{\text{N}}H_3^+$ Orn

—CH$_2$—⟨phenyl⟩—$\overset{6}{\text{O}}$H   Tyr

(π—Me)His

Proposed High-Potency Residues

—(CH$_2$)$_n$NH$_3^+$ $n = 1$, Dap; $n = 2$, dab

—(CH$_2$)$_n$NHAc $n = 1$, AcDap; $n = 2$, AcDab

—(CH$_2$)$_2$OH   Hse

---

of the data points in the set had an ionizable side-chain. Due to the small size of the set, the effect of branching at the second and third atoms of the side-chain was not investigated. A good fit with the model was obtained. A significant dependence on ζ was observed with Z within the common range of 0.7 to 1.2. The set shows a large dependence on hydrogen bonding and a much smaller but still significant dependence on branching at the first atom of the side-chain. It is interesting to note that the coefficients H_1, H_2, and A of set 5 are in excellent agreement with those of set 1 (as no dependence on α is observed for either set, they are both assumed to have a value of 0 for A). The difference in the B_1 values for sets 1 and 5 is just barely significant. No comparison is possible for B_2, B_3, or I which were not determined for set 5. Radhakrishnan (85) has reported percent inhibition values of the monkey intestinal Gly–Leu system by Gly–Aax determined with

(^{14}C-Gly)–Leu and Gly–(^{14}C-Leu) (sets PPB 6 and 7, respectively). These sets were correlated successfully with the IMFU equation. Both sets show a dependence on i and v with the coefficients I and S having about the same values as would be expected. The two sets also have essentially the same value for the intercept B^0. They differ, however, in that set 7 also is dependent on σ_1 and n_n while set 6 is not. This may be due to the much poorer fit of the latter set.

Meister and coworkers (99) have determined relative activities of γ-glutamyl transpeptidase toward Gly–Aax using GluNHPnNO$_2$-4, **8L**, or its D configuration **8D** as the source of the glutamyl residue. We have combined the data into a single set by assuming that the configuration of the glutamyl anilide has no effect on the measured activity (set PPB 8). These authors also determined relative activities for this enzyme toward Aax–Gly using **8D** as the source of the glutamyl residue (set PPB 9). Correlation with the IMFU equation gave good results for both sets. They differ greatly in the nature of their dependence on intermolecular forces. Though both are a function of σ_1 they differ in the sign of L. Set 8 is dependent on α while set 9 is not; the reverse is true for n_H. They both show a dependence on branching at the first side-chain atom and have the same value of the intercept, however.

We have combined relative potencies of 2-substituted thyrotropin-releasing factor (TRF) determined by Szirtes and coworkers with those reported by Wade in his review (100) into a single data set (PPB 21). The set was correlated with a modification of the IMFSB equation. Inspection of the data points showed that (π-Me)His has a relative potency which is three or four orders less than those of (τ-Me)His or of His itself. The potency of Thr is two to three orders of magnitude larger than that of the other polar amino acids studied, Ser(tBu), Tyr, Arg, Lys, and Orn. It seemed clear that the high-potency amino acid residues must have a structural feature that distinguishes them from the low-potency residues. Examination of the side-chains shown in Table 41 suggests that the difference between high and low potency may depend at least in part on the availability of an OH or NH bond on the second or third atom of the side chain. We have therefore introduced the variable $n_{\pi H}$ which takes the value 1 when the side chain has an OH or NH bond on the second or third side-chain atom and 0 otherwise. The correlation equation used has the form

$$\log P_{\text{rel}} = L\sigma_{1X} + A\alpha_X + H_1 n_{HX} + H_2 n_{nX} + H_3 n_{\pi H} + B_1 n_{1X} + B_2 n_{2X} + B^0 \tag{53}$$

A good fit was obtained. The variables in order of their decreasing contribution to the side-chain structural effects are $n_{\pi H}$, α, i, and n_n. Our results suggest that Dap, Dab, and their acetyl derivatives, and amino acids having an OH bond on the second or third side-chain atom such as Hse should produce large relative potencies when introduced into the 2 position of TRF.

2. X^C Substitution

Wade (100) has reported data for the relative potencies of carbonyl substituted lutein hormone-releasing factor (LRF) (set PPB 10). In order to account for bilinear or parabolic behavior in this and other peptide bioactivity data sets, we have included a term in α^2 in the correlation equation. As any quantity is a highly linear function of its square, we have made use of the method of Goodford and Berntsson (101, 102) for rescaling to avoid collinearity with α^2. We define α^* as

$$\alpha^* = \alpha - (\alpha_{max} + \alpha_{min})/2 \tag{54}$$

where α_{max} and α_{min} are the maximum and minimum values of α in the data set to be studied. As no ionizable substituents were present in the data set, it was correlated with the IMFSB equation in the form

$$\log P_{rel} = L\sigma_{1X} + A_1\alpha_X^* + A_2\alpha_X^{*2} + H_1 n_{HX} + H_2 n_{nX} + B_1 n_{1X} + B_2 n_{2X} + B^0 \tag{55}$$

Values of n_1 and n_2 were obtained by using those of the corresponding alkyl and cycloalkyl groups. Unknown σ_1 values for NHAk group were assumed equal to 0.15 which is the mean of the values for NHMe and NHEt; those for NAk$_2$ and cyclic dialkylamino groups were assumed equal to 0.17, the value for NMe$_2$. A poor but highly significant fit was obtained. It must be noted, however, that the range of structural variation in the substituents was very limited. The major contribution to the structural effect of the substituent was made by α^{*2} with α^* having a small but significant effect. In the calculation of C_i values for this data set, the reference group chosen was NHEt.

3. Z^0Z Substitution

Summer and Hayes have reported values of the enkephalin activity relative to that of (Met^5)enkephalin for the peptides (Z)Tyr–Gly–Gly–Phe–Met, **9**, in guinea pig ileum (set PPB 11) and in mouse vas deferens (set PPB 12). They have also determined values of IC$_{50}$ (nmol/dm^3) and K$_D$ versus [^{3}H]–[ala^2–leu^5]enkephalin with rat brain homogenate (sets PPB 13 and 14, respectively). As Z in these sets is alkyl, cycloalkyl, or 2-phenylethyl, the σ_1 values for the substituents are essentially constant and of course

$n_H = n_n = i = 0$. Thus, the only parameters required other than steric are α^* and α^{*2}. We have therefore correlated these data sets with the equation

$$\log ba_Z = A_1 \alpha_Z^* + A_2 \alpha_Z^{*2} + B_1 n_1 + B_2 n_2 + B^0 \tag{56}$$

The enkephalin activities in guinea pig ileum gave a good fit with the correlation equation, with all four parameters making significant contributions. The best equation for the description of enkephalin activity in mouse vas deferens was obtained on the exclusion of the data point for Pr. Unlike the result obtained for guinea pig ileum, that for mouse vas deferens was dependent only on steric effects. The IC_{50} and the K_D values gave good and barely significant results, respectively. As expected, the coefficients A_1, A_2, and B_2 were almost identical. There was a difference in the values of B_1 but in view of the uncertainty in the B_1 coefficient for the K_D data set, this is not surprising. Gesellchen and coworkers (104) have reported relative potencies of the peptides Tyr–ala–Gly–(4-Z)Phe–MetNH$_2$ in mouse vas deferens (set PPB 15). Although the set is too small for a reliable analysis, we have examined it as an example of the $Z^0 Z$ type of substitution in which Z^0 is an aromatic ring. As under the conditions used for the determination of the bioactivities none of the side chains ionizes, the i parameter was deleted from the correlation equation. The σ_1 parameter used in the IMF equation was replaced by the σ_p parameter as σ_1 values were not available for all of the groups of interest and for 4-substituted benzyl groups the σ_1 value is a linear function of the σ_p value. The correlation equation then takes the form

$$\log P_{\text{rel} Z} = \rho \sigma_{pZ} + A_1 \alpha_Z^* + A_2 \alpha_Z^{*2} + H_1 n_{HZ} + H_2 n_{nZ} + S v_Z + B^0 \tag{57}$$

Though the fit is exceptionally good, this is almost certainly fortuitous. The potencies are a function of the polarizability, electrical effect, hydrogen bonding capability, and steric effect of the substituent Z.

4. X^P Substitution

Gesellchen and coworkers (105) have determined the relative potencies in mouse vas deferens (set PPB 16) and the analgesic activities by the mouse hot plate test (set PPB 17) of the peptides Tyr–ala–Gly–(CH$_2$Z)–PheNH$_2$ in which substitution occurs at the peptide bond nitrogen atom between Gly and Phe (the 3–4 peptide bond). As no ionizable side chain is present in either set 16 or set 17, i has been deleted from the correlation equation which is therefore a modification of the IMFU equation. The relative potencies in mouse vas deferens were dependent on α^*, n_n, v, and possibly on α^{*2} as well, and gave a good fit with the correlation. The analgesic activities in the mouse were a function in order of decreasing importance of α^*, v, σ_1, and n_H. The relative potencies have a very different dependence on intermolecular forces from that observed for the analgesic activities.

5. Multiply Substituted Peptides

We now consider some examples of peptides with more than one type of substitution. Steinbrink, Bond, and Van Werf (106) have measured values of (k_{cat}/K_M) for the hydrolysis of the peptides ArCOLeu–Gly–Pro–AlaOR, **10** (set PPB 18); ArCOAax–Gly–Pro–Ala, **11** (set PPB 19); and ArCOLeu-Gly-Aax-(Pro/Hyp), **12** (set PPB 20) with β-collagenase in water 50 nM in Tricine, 0.4 M in NaCl, 10 mM in $CaCl_2$, at pH 7.5 and 25°. **10** is undergoing limited X_N and X_C substitution as well as X^n substitution at position 4 of the peptide. To account for the X_N substitution, we have introduced the indicator variable n_{Ar} which takes the value 1 when the aryl group is 2-furyl and 0 when it is phenyl. To account for the X_C substitution, we have introduced the indicator variable n_{Me} which takes the value 1 when R is Me and 0 when it is H. Finally, as the data set includes variant amino acids of the proline type, is was necessary to introduce the n_{NC} parameter to account for the chage in hydrogen bonding capacity of the α-amino group. The correlation equation therefore takes the form

$$\log(10^4 * k_{cat}/K_M) = L\sigma_{1X} + A_1\alpha_Z^* + A_2\alpha_Z^{*2} + Ii_X$$
$$+ B_{Ar}n_{Ar} + B_{Me}n_{Me} + B_{NC}n_{NC} + Sv_X + B^0 \qquad (58)$$

Set 18 was well fit by the model; it is a function of i, n_{Me}, n_{NC}, and v but is independent of the nature of the aryl group. As **11** has no X^C substitution, the variable n_{Me} was deleted from the correlation equation. Best results were obtained by replacement of v as the steric parameter with n_{12} which equals the number of second branches on the first atom of the side chain. A good fit was obtained with the resulting correlation equation. Set 19 is a function of α, i, and n_{12}. Thus, the dependence on intermolecular forces of **10** is very different from that of **11**. **12** is also free of X^C substitution so that once again n_{Me} was deleted from the correlation equation. In order to minimize the number of variables, it was assumed that there was no dependence on n_n and that no important variation with the amino acid in the 4 position of **12** occurs. The resulting correlation equation gave a fairly good fit to the data. Set 20 is a function of α, n_H, i, and v. It differs in its dependence on intermolecular forces from both set 19 and set 20. Substitution in **10** is X^N, X^4, and X^C, in **11** it is X^N and X^1, and in **12** it is X^N, X^3, and to some extent X^4. In no case was a dependence on the nature of the X^C aryl group observed.

We may reasonably conclude that the differences observed in the dependence on intermolecular forces reflect differences in the binding to the receptor site of the enzyme at different positions in the peptide. Obviously the bioactivity depends on the site of substitution in the peptide.

Overall, of the 21 peptide bioactivity data sets examined here, four gave $100R^2$ values greater than 90.0, 13 were in the range 80.0 to 89.9, and four were in the range 70.0 to 79.9. All of the data sets gave significant correlations as judged by the F statistic.

D. Peptide Antibiotic Bioactivities

1. Introduction

In order to provide a more extensive range of examples of the correlation of peptide bioactivities with the IMF equation, we now consider the application of the method to peptide antibiotic bioactivities. The data sets studied are given in Table 42, results of the correlations in Table 43, zeroth-order partial correlation coefficients in Table 44, and C_i values in Table 45.

Andruszkiewicz and coworkers (107) have determined minimum inhibitory concentrations (MIC) for Aax-FMDP, 13 (FMDP = N^3-(4′-methoxyfumaryl)-L-2, 3-diamino-propanoic acid) in three species of bacteria and one of pathogenic yeast. The organisms studied were *Candida albicans* ABM 25, the yeast; *Staphylococcus aureus* 209P, a Gram-positive sphere; *Bacillus pumilas* 1697, a Gram-positive rod; and *Shigella sonnei* 433, a Gram-negative rod (sets PPB 32–35, respectively). After conversion to nM concentrations, these sets were correlated with the IMFSB equation using the branching parameters n_0, n_1, and n_2 to account for branching at the zeroth, first, and second atoms of the side chain for all but set 33. As the value for Gly was unavailable for this set only n_1 and n_2 were used as steric parameters. The n_n parameter was deleted from the correlation equation to minimize the number of parameters required. The i parameter was required only for set 34, as the other sets did not have ionizable side chains. All four sets are fit by the model. As is expected from the large differences in the types of organism studied, the four data sets show large differences in the nature of their dependence on intermolecular forces. The only similarity lies in the dependence of sets 32, 33, and 35 on α, with A values that are not significantly different from each other. Sets 34 and 35 are both a function of σ_1 but they have opposite signs for L. Set 32 is a function of n_H while the others are not. Set 34 is a function of i while the others are not. By contrast, there is considerable similarity between the sets insofar as steric effects are

TABLE 42
Peptide Antibiotic Activity Data Sets

PPB32–35. Minimum inhibitory concentration (MIC) of N^3-(4-methoxy-fumaryl)-L-2, 3-diaminopropionic acid dipeptides in *Candida albicans*, AMB 25, *Staphylococcus aureus* 209P, *Bacillus pumilas* 1697, *Shigella sonnei* 433 (nM)[a]. Aax, MIC, C.a., S.a., B.p., S.s.
Ala, 2.64, 165, 5.14, 10.3; Met, 2.20, 1.38, 8.59, 8.59; Gly, 43.2, —, 86.4, 173; Val, 4.71, 37.7, 2.41, 4.71; Leu, 2.32, 36.2, 4.52, 4.52; Phe, 32.9, 264, 16.5, 32.9; Tyr, 150, 300, 37.5, 75.0; Nva, 1.21, 37.7, 1.21, 9.42; Nle, 2.32, 72.4, 1.16, 9.03; Abu, 4.92, 78.8, 2.52, 19.7; Lys, —, —, 277, —.

PPB36–43. MIC (nM) of Aax–Ala(P) in *E. coli* NCIB 8879, *K. aerogenes* 310001, *Enterobacter* 250002, *S. marcescens* ATCC 14756, *S. typhimurium* 538003, *H. influenzae* NCTC 4560, *S. faecalis* 58511, *S. aureus* NCIB 8625[b]. Aax, MIC, E.c., K.a., E, S.m., S.t., H.i., S.f., S.a.
Ala, 11.2, 5.61, 11.2, 89.8, 718, 718, 22.4, 359; Arg, 1.44, —, 5.74, 11.5, 184., 735., 91.8, 91.8; 91.8; Cys, 264., 132., —, —, —, —, —, —; Met, 1.68, 0.804, 6.70, 13.4, 53.6, 107, 6.70, 6.70; Phe, 3.03, 6.05, 12.1, 24.2, 194, 194, 12.1, 24.2; Nva, 0.256, 0.128, 2.13, 4.27, 2.13, 34.1, 8.54, 62.3; Ala(F), 2.33, —, —, —, 299, 74.7, 149, 149.; Arg(NO_2),
1.33, 0.637, 2.28, 9.12, 36.5, 18.2, 292., —, 9.12, 146.; Har,
10.6, 10.6, 10.6, 340., 10.6, 680.; Cys(Me), 0.222, 0.222, 0.222, 3.70, 1.85, —, 14.8, 118.; Cys(Pr), 0.368, 0.194, 0.368, 3.06, 1.53, —, 12.3, 49.0; Met(O), 24.2, 24.2, 194., 194., —, 96.8, 24.2, 48.4; Met(O_2), 88.3, Cys(Me), 0.222, 0.222, 0.222, 3.70, 1.85, —, 14.8, 118.; Cys(Pr), 0.368, 0.194, 0.368, 3.06, 1.53, —, 12.3, 49.0; Met(O), 24.2, 24.2, 194., 194., —, 96.8, 24.2, 48.4; Met(O_2), 88.3, 177., —, —, —, 177., 22.1, 177.; 2-Tha, 2.92, 1.46, —, 23.4, 23.4, —, 11.7, 187.

AAB23. $10^4 * K_i$ for the inhibition of the binding of ^{125}I Ristocetin to *Micrococcus leuteus* cell wall by Acaax.[c] Acaax, K_i
phe, 1.0; trp, 1.6; met, 24; ile, 28; leu, 28; val, 31; ala, 32; hpg, 42; asn 90; Gly, 90.

PPB44. $10^4 * K_i$, conditions as above, Acaax–aax.[c] aax, aax, K_i
phe, gly, 0.098; leu, Gly, 5.9; val, Gly, 3.4; ala, Gly, 1.3; asn, Gly, 7.2; Gly, Gly, 46; ser, Gly, 5.4; asp, Gly, 14; thr, Gly, 47; phe, ala, 0.087; ala, ala, 0.73; Gly, ala, 31; leu, leu, 45; Gly, Leu, 110.

PPB45. $10^4 * K_i$, conditions as above, Acaax–aax–Aax/aax. aax, aax,[c] Aax/aax, k_i
Gly, Gly, Trp, 12; Gly, Gly, Tyr, 13; Gly, Gly, Pro, 16; Gly, Gly, Phe, 16; Gly, Gly, leu, 19; Gly, Gly, Leu, 20; Gly, Gly, Gly, 71; ala, ala, Lys, 0.16, ala, ala, Ala, 0.42.

[a] Ref. 107.   [b] Ref. 108.   [c] Ref. 109.

concerned. Thus, sets 32, 34, and 35 are dependent on n_0 and have B_0 values that are not significantly different from each other (note that the dependence of set 33 on n_0 could not be determined); sets 33, 34, and 35 are a function of n_1 with B_1 values that are essentially the same; and sets 32, 34, and 35 are dependent on n_2 with B_2 values that are essentially the same. An attempt to combine the bacterial data sets, 33, 34, and 35, into a single data set by means of the Omega method was unsuccessful. In view of the large differences between the organisms, this is to be expected.

TABLE 43
Results of Correlation of Peptide Antibiotic Bioactivity Data

Set:	PPB32	PPB33	PPB34	PPB35	PPB36	PPB37	PPB38	PPB39	PPB40	PPB41	PPB42	PPB43
L	—	—	20.3	−10.7	−6.70	−5.04	−11.4	−4.14	—	−3.40	4.43	—
S_L	—	—	5.44[g]	7.35[n]	3.10[k]	3.83[n]	2.81[j]	2.47[m]	—	1.52[k]	1.17[f]	—
A	5.48	4.97	—	8.05	8.66	8.66	6.05	4.51	8.28	4.04	—	13.6
S_A	2.09[j]	0.371	—	2.35[j]	2.74[h]	3.28[k]	2.56[k]	2.08[k]	3.68[k]	1.84[k]	—	5.13[j]
H_1	0.989	—	—	—	−2.02	−1.84	−2.97	−2.09	—	—	—	−3.41
S_{H1}	0.431[k]	—	—	—	0.385[f]	0.448[h]	0.517[h]	0.454[g]	—	—	—	1.16[j]
H_2	—	—	—	—	0.661	0.747	1.25	0.783	—	0.139	—	1.27
S_{H2}	—	—	—	—	0.114[f]	0.124[f]	0.187[g]	0.163[g]	—	0.0744[m]	—	0.391[h]
I	—	—	1.91	—	5.31	3.92	7.78	5.61	—	—	0.588	9.99
S_I	—	—	0.336[f]	—	1.24[g]	1.36[j]	1.51[h]	1.33[h]	—	—	0.194[h]	3.29[i]
B_0	−1.63	—	−1.04	−1.59	—	—	—	—	—	—	—	—
S_{B0}	0.442[h]	—	0.403[j]	0.374[h]	—	—	—	—	—	—	—	—
B_1	—	−0.554	−0.367	−0.441	1.17	1.32	—	—	—	—	—	—
S_{B1}	—	0.0506	0.231[m]	−0.175[k]	0.756[m]	0.800[m]	—	—	—	—	—	—
B_2	−0.451	−0.421	—	−0.589	−3.37	−3.49	−1.06	−1.56	−3.12	−1.38	−0.675	−1.93
S_{B2}	0.224[m]	0.0449	—	0.194[j]	0.597[f]	0.640[g]	0.572[m]	0.496[j]	0.972[h]	0.403[j]	0.191[g]	0.671[j]
B_3	—	—	—	—	—	—	—	—	—	—	—	−1.85
S_{B3}	—	—	—	—	—	—	—	—	—	—	—	0.616[i]
B^0	1.635	1.968	1.937	2.238	0.584	0.300	0.656	1.704	2.293	2.523	1.470	1.752
S_{B0}	0.330[f]	0.0672	0.313	0.231	0.435[n]	0.465[q]	0.349[m]	0.304[f]	0.538[f]	0.267	0.206	0.339[f]
$100R^2$	87.57	97.85	89.87	91.23	90.23	93.97	95.36	89.44	57.20	81.85	79.59	76.68
F	8.804[i]	75.76	13.31[f]	8.321[j]	7.919[i]	8.899[j]	10.28[j]	5.647[k]	5.347[j]	4.510[k]	11.70[f]	3.289[k]
S_{est}	0.330	0.0672	0.317	0.231	0.416	0.435	0.325	0.286	0.724	0.270	0.240	0.351
S^0	0.499	0.197	0.431	0.468	0.477	0.425	0.393	0.539	0.767	0.679	0.543	0.711
n	10	9	11	10	14	12	10	11	11	9	13	13

For footnotes see Table 3.

TABLE 44
Zeroth-Order Partial Correlation Coefficients for Peptide Antibiotic Bioactivity Data Sets

Set:	PPB 32, 35	PPB33	PPB34	PPB36	PPB37	PPB38	PPB39	PPB40	PPB41	PPB 42, 43
σ, α	0.670	0.751	0.637	0.322	0.033	0.181	0.189	0.288	0.310	0.280
σ, n_H	0.419	0.412	0.110	0.348	0.350	0.266	0.275	0.285	0.261	0.328
σ, n_n	—	—	—	0.137	0.018	0.091	0.106	0.253	0.102	0.097
σ, i	—	—	0.094	0.306	0.312	0.247	0.255	0.254	0.261	0.290
σ, n_0	0.104	—	0.094	—	—	—	—	—	—	—
σ, n_1	0.253	0.236	0.246	0.370	0.435	0.414	—	0.345	—	0.363
σ, n_2	0.147	0.118	0.132	0.319	0.053	0.414	0.419	0.305	0.336	0.238
σ, n_3	—	—	—	0.043	0.317	0.343	0.348	0.220	—	0.117
α, n_H	0.499	0.537	0.396	0.566	0.583	0.611	0.568	0.566	0.614	0.554
α, n_n	—	—	—	0.561	0.519	0.662	0.606	0.615	0.564	0.540
α, i	—	—	0.190	0.438	0.410	0.459	0.427	0.432	0.614	0.426
α, n_0	0.578	—	0.584	—	—	—	—	—	—	—
α, n_1	0.512	0.280	0.512	0.498	0.623	0.703	—	0.542	0.840	0.537
α, n_2	0.655	0.585	0.664	0.788	0.708	0.703	0.706	0.813	—	0.798
α, n_3	—	—	—	0.504	0.441	0.612	0.617	0.890	—	0.466
n_H, n_n	—	—	—	0.625	0.593	0.751	0.760	0.818	0.570	0.616
n_H, i	—	—	0.886	0.934	0.887	0.930	0.932	0.932	1.000	0.934
n_H, n_0	0.111	—	0.140	—	—	—	—	—	—	—
n_H, n_1	0.062	0.000	0.078	0.138	0.127	0.207	—	0.184	—	0.150
n_H, n_2	0.156	0.125	0.196	0.260	0.188	0.207	0.184	0.274	0.286	0.222
n_H, n_3	—	—	—	0.085	0.041	0.207	0.184	0.439	—	0.048
n_n, i	—	—	—	0.419	0.312	0.490	0.505	0.553	0.564	0.407
n_n, n_0	—	—	—	—	—	—	—	—	—	—
n_n, n_1	—	—	—	0.194	0.199	0.250	—	0.185	—	0.213
n_n, n_2	—	—	—	0.365	0.294	0.250	0.220	0.276	0.456	0.315
n_n, n_3	—	—	—	0.613	0.609	0.442	0.407	0.443	—	0.593
i, n_0	—	—	0.100	—	—	—	—	—	—	—
i, n_1	—	—	0.056	0.113	0.091	0.167	—	0.149	—	0.123
i, n_2	—	—	0.140	0.213	0.135	0.167	0.149	0.222	0.286	0.182
i, n_3	—	—	—	0.070	0.029	0.167	0.149	0.356	—	0.040
n_0, n_1	0.557	—	0.559	—	—	—	—	—	—	—
n_0, n_2	0.364	—	0.373	—	—	—	—	—	—	—
n_1, n_2	0.203	0.000	0.209	0.531	0.674	1.000	—	0.671	—	0.677
n_1, n_3	—	—	—	0.285	0.321	0.444	—	0.418	—	0.322
n_2, n_3	—	—	—	0.537	0.476	0.444	0.450	0.624	—	0.475

TABLE 45
C_i Values for Peptide Antibiotic Bioactivity Sets

Set:	σ_1	α	n_H	n_n	i	n_o	n_1	n_2	n_3
32	0	29.1	22.8	—	—	37.6	0	10.4	—
33	0	54.0	0	—	—	—	26.2	19.9	—
34	32.9	0	0	—	38.7	21.0	7.44	0	—
35	16.1	34.7	0	—	—	29.9	8.26	11.0	—
36	3.56	13.2	13.4	4.39	35.2	—	7.78	22.4	0
37	2.94	14.5	13.4	5.45	28.6	—	9.60	25.5	0
38	5.96	9.06	19.3	8.16	50.6	—	0	6.89	0
39	2.90	9.10	18.4	6.86	49.2	—	—	13.6	0
40	0	37.9	0	0	0	—	0	62.1	0
41	10.0	34.1	0	5.10	0	—	—	50.8	—
42	21.9	0	0	0	36.4	—	0	41.7	0
43	0	14.5	15.8	5.87	46.3	—	0	8.93	8.60

Atherton and coworkers (108) have determined minimum inhibitory concentrations for Aax–Ala(P), **14**, in eight species of bacteria.

$$\begin{array}{c} O \\ \parallel \\ H_2NCHC-NHCHCO_2H \\ | \quad\quad\quad \backslash \\ X \quad\quad CH_2NHCOCH=CHCO_2Me \end{array}$$

13

$$\begin{array}{c} O \quad\quad\quad OH \\ \parallel \quad\quad\quad / \\ H_2NCHC-NH-CH-P=O \\ | \quad\quad\quad\quad \backslash \\ X \quad\quad\quad\quad OH \end{array}$$

14

The organisms studied are: *E. coli* NCIB 8879, *K. aerogenes* 310001, *Enterobacter* 250002, *S. marcescens* ATCC 14576, *S. typhimuriam* 538003, *H. influenzae* NCTC 4560, *S. faecalis* 58511, and *S. aureus* NC1B 8625. After converting the MIC values to nM concentrations, the data sets were correlated with the IMFSB equation using n_1, n_2, and n_3 as the steric parameters in all but sets 39 and 41. In set 39 n_1 is identical to n_2, while in set 41 n_2 is identical to n_3 and almost identical to n_1. We have therefore deleted n_1 from the correlation equation for set 39 and both n_1 and n_3 from that for set 41. All of the data sets for the eight species of bacteria studied gave significant correlations with the IMF equation.

Six of the eight bacteria species studied are Gram-negative, facultatively anaerobic rods. Of these species, five belong to the family Enterobacteriaceae (sets 36–40) while the sixth (set 41) is of uncertain family. The remaining two species are classified as Gram-positive cocci (sets 42, 43).

Applying the Omega method, we combined into a single data set the five species belonging to the family Enterobacteriaceae. Best results were obtained on the exclusion of the data points for Cys(Pr) (set 38), and Arg and Ala (set 40). Thus,

$$\log \text{MIC}_x = -6.08(\pm 1.53)\sigma_{1X} + 7.90(\pm 1.32)\alpha_X - 1.91(\pm 0.217)n_{HX}$$
$$+ 0.730(\pm 0.0708)n_{nX} + 4.62(\pm 0.657)i_X + 1.65(\pm 0.400)n_1$$
$$- 3.48(\pm 0.340)n_2 + 0.840(\pm 0.0959)\omega + 0.886(\pm 0.217) \quad (59)$$
$$100R^2 = 82.56; \quad F = 27.23; \quad s_{est} = 0.415; \quad s^0 = 0.457; \quad n = 55$$

The reference substrate used to define omega is Nva. Inclusion of the sixth rod species in the combined data set gave:

$$\log \text{MIC}_x = -6.76(\pm 1.51)\sigma_{1X} + 8.10(\pm 1.28)\alpha_X - 1.86(\pm 0.214)n_{HX}$$
$$+ 0.701(\pm 0.0703)n_{nX} + 4.51(\pm 0.654)i_X + 1.50(\pm 0.387)n_1$$
$$- 3.38(\pm 0.339)n_2 + 0.779(\pm 0.0715)\omega + 0.934(\pm 0.202) \quad (60)$$
$$100R^2 = 83.07; \quad F = 32.50; \quad s_{est} = 0.427; \quad s^0 = 0.445; \quad n = 62$$

The best equation was obtained on the exclusion of the data points Cys(Pr) (set 38), Arg and Ala (set 40), and Met(SO$_2$) and Arg (set 41). It is interesting to note that the three points which deviated in the combined set 36–40 also deviated in the combined set 36–41. A comparison of the coefficients of equations 59 and 60 and of the C_i values shows that these equations have no significant differences. All of the rod species may therefore be combined into a single data set. Combination of the two species of cocci into a single data set gives on the exclusion of the data points for Har and Ala(F) [set 42]

$$\log \text{MIC}_x = 8.22(\pm 3.31)\alpha_X - 2.14(\pm 0.740)n_{HX} + 0.778(\pm 0.249)n_{nX}$$
$$+ 6.73(\pm 2.13)i_X - 1.32(\pm 0.437)n_2 - 1.08(\pm 0.393)n_3$$
$$+ 0.760(\pm 0.141)\omega + 0.934(\pm 0.202) \quad (61)$$
$$100R^2 = 80.59; \quad F = 9.492; \quad s_{est} = 0.307; \quad s^0 = 0.540; \quad n = 24$$

An attempt to combine all of the data sets for the eight species studied into a single data set was unsuccessful. The fundamental assumption of the Omega method is that in a set of related species, the receptor sites responsible for the observed bioactivity of a given set of substrates are analogous to a set of related chemical reagents interacting with the substrates by a common mechanism. The successful application of the Omega method enables us to infer a common biomechanism for the interaction of the species studied with the given set of substrates. A failure of the Omega method suggests that not all of the organisms studied interact with the substrate set by the same biomechanism. We may therefore conclude that the five species which are members of the family Enterobacteriaceae, together with the sixth species of Gram-negative, facultatively anaerobic rods, all interact with **14** by the same biomechanism, and that the two species of cocci do the same. We further conclude that there are one or more differences in the biomechanisms of the rods and the cocci. A comparison of coefficients and of C_i values for Equation 60 with those for Equation 61 clearly shows a significant difference in both the composition and the magnitude of the steric effect.

For the rods, branching at C^1 (the first carbon atom of the chain) results in incrementation, at C^2 in decrementation, and at C^3 in no observable effect. For the cocci, branching at C^1 has no detectable effect, while at both C^2 and C^3 it results in decrementation. Steric effects make twice as great a contribution to the structural effect in rods as they do in cocci. Both rods and cocci show about the same dependence on the intermolecular force variables n_H, n_n, and alpha. Though no dependence on σ_1 was observed for cocci, this may have been due to some combination of the small contribution made by this variable, the poorer correlation obtained for the cocci, and the smaller size of the data set. The difference in the magnitude of I for rods and that for cocci is not significant. The overall dependence on intermolecular forces is roughly the same for both rods and cocci. There is also no significant difference in the values of O, the coefficient of omega, and the intercept B^0. The difference between rods and cocci in their sensitivity to the substrates lies in the magnitude and composition of the steric effect. It seems reasonable to ascribe this difference in steric effect to the difference in cell wall structure between Gram-positive and Gram-negative organisms. The similarity in the

TABLE 46
Values of C_i

				Variable					
Eq.	ω	σ_1	α	n_H	n_n	i	n_1	n_2	n_3
60	5.15	3.57	12.3	12.3	4.63	29.8	9.91	22.3	—
61	5.17	—	12.9	14.6	5.29	45.8	—	9.00	7.37

dependence on intermolecular forces makes it tempting to conclude that once the substrate passes through the cell wall, the biomechanism is the same for both types of bacteria. Much further work is required before any conclusion can be reached, however.

We now consider the application of our results to the design of bioactive substrates. In order to design peptides with the **14** structure, we must use Equation 60 and 61 to increase the substrate activity. Considering the coefficients of Equation 60, and remembering that the smaller the MIC value the more active the substrate will be, we must design side chains that maximize those variables which have negative coefficients and minimize those variables which have positive coefficients. The order of importance of the variables is given by the magnitude of their C_i values. Thus, in designing side chains for antibiotics to be used against facultatively anaerobic Gram-negative rods, we wish to maximize n_2, n_H, and σ_1 in that order; and to minimize i, α, n_1, and n_n in that order.

Considering steric effects first, to maximize n_2 we must have three atoms other than hydrogen bonded to each of the second atoms of the side chain. To minimize n_1 we must have no more than one atom other than hydrogen bonded to the first atom of the side chain.

We now turn our attention to the intermolecular forces. We wish to maximize the n_H hydrogen bonding parameter and minimize the n_n hydrogen bonding parameter. In order to acheive this end, it is necessary to recognize that there are very few groups with NH bonds and none with OH bonds that do not also have lone pairs on N or O. If we are to maximize n_H while minimizing n_n, we must use groups for which the ratio n_H/n_n is equal to or greater than 1/2. Such groups include (the n_H/n_n value is given in parentheses) OH (1/2), $CONH_2$ (2/3), NHZ [Z = SMe, Me, CSMe] (1), NH_2 (2), and $CSNH_2$ (2).

To minimize the polarizability, the side chain must have the smallest possible number of atoms other than H, F, and O.

To minimize i we need only use nonionizable side chains. The very small contribution of σ_1 suggests that we can ignore this variable.

We have chosen as side chains those shown in Table 47 and designated A, B, C, and D. Calculated log MIC values for them are given in Table 48. Clearly the proposed side chains should produce very active substrates.

We now consider the design of **14** to be used against the cocci species on the basis of Equation 61. In this case, we wish to maximize n_H, n_2, and n_3 while minimizing i, α, and n_n. As the intermolecular force design requirements are the same for cocci as they are for rods, they have already been discussed above. With regard to steric effects, we must have as many atoms other than hydrogen as possible bonded to the second and third atoms of the side chain. We therefore propose the side chains B, E, F, and G. Again,

TABLE 47
Parameter Values for Proposed Side Chains

	X	σ_I	α	n_H	n_n	n_1	n_2	n_3
A	—CHOH—CHOHMe$_2$	0.10	0.218	2	4	2	3	0
B	—CH$_2$C(CH$_2$OH)$_3$	0.03	0.280	3	6	1	3	3
C	—CH$_2$(CONH$_2$)$_3$	0.10	0.269	4	6	1	2	2
D	—CH$_2$C(OH)(CONH$_2$)$_3$	0.15	0.285	5	8	1	3	2
E	—CH$_2$(CHOHCH$_2$OH)$_3$	—	0.464	6	12	1	3	6
F	—CH—CHOH(—CHOH—)$_5$	—	0.428	6	12	1	3	4
G	—CH(CHOHCONH$_2$)$_2$	—	0.351	6	10	2	4	2

For all side chains, $i = 0$.

TABLE 48
Calculated Values of $(-1)\log$ MIC

	X			
Organism	A	B	C	D
E. coli	−6.493	−7.476	−6.158	−10.565
K. aerogenes	−6.728	−7.710	−6.753	−10.799
Enterobacteriaceae	−5.777	−6.759	−5.802	−9.878
S. marcescens	−5.541	−6.524	−5.566	−9.613
S. typhimurium	−5.777	−6.759	−5.802	−9.878
H. influenzae	−4.838	−5.821	−4.863	−8.909

	χ			
	B	E	F	G
S. faecalis	−5.009	−8.488	−6.624	−7.973
S. aureus	−4.323	−7.802	−5.938	−7.287

the side-chain parameters are given in Table 47 and the calculated log MIC values in Table 48. These side chains are indeed likely to be very active against cocci.

2. Inhibition of Ristocetin Binding

Kim, Martin, Otis, and Mao (109) have reported data for the inhibition of ^{125}I-Ristocetin binding to *Micrococcus euglens* cell wall by Ac–aax, Ac–aax–aax, and Ac–Aax–aax–aax. Their data are shown in Table 1. We have examined the correlation with the IMF equation of the amino acids, dipeptides, and tripeptides separately and have then combined them into a single data set. To facilitate the discussion of the results, the correlation equations and statistics will not be tabulated but presented directly in the text.

In order to model the di- and tripeptides and the amino acids in various correlations, we consider the general structure to have the form

$$\text{Ac} - (\text{Aax})_m - \text{aax} - \text{aax}$$
$$\phantom{\text{Ac} - (\text{Aax})_m -\ } 2' 1' 1$$

with as usual amino acids having the D configuration designated aax, and those having the L configuration designated Aax. The sets studied are the acetylamino acids, Ac–aax (set AAB23), for which there are no residues in positions 1' and 2'; Ac–Gly–aax (set PPB44g) in which the residue 1' is Gly and the residue 2' is not present; Ac–aax–aax (set PPB44) in which the residue 2' is not present and the X in the residue 1' is H, Me, or iBu; set PPB44gU23, a combination of sets PPB44g and AAB23; set PPB44U23, a combination of sets AAB23 and PPB44; set PPB45 which includes only the tripeptides Ac–Aax–aax–aax; and set PPB23U44U45 which includes all of the data (a combination of sets AAB23, PPB44, and PPB45).

All of the compounds studied have in common the structural fragment

$$-\underset{\underset{\text{O}}{\|}}{\text{C}}-\text{NH}-\underset{\text{X}^1}{\text{CH}}\text{CO}_2\text{H}$$

in which only the side chain X varies. This fragment consists of residue 1 of **1** and its attached peptide bond. Thus, only the effect of X need be considered. Applying the IMF equation we have considered the parameters α, σ, n_{H1}, n_{n1}, v_1, and n_{01}, the last being 0 for Gly and 1 for all other side chains. For residue 1 and its peptide bond, the fragment

$$-\underset{\underset{\text{O}}{\|}}{\text{CNH}}-\underset{\text{R}}{\text{CH}}-$$

is replacing Me in the Ac–aax.

The effect of R which may be H, Me, or iBu, is represented by $\alpha_{1'}$ and $n_{01'}$ while the replacement of Me by —O=CNHCH— (equivalent to O=CNH) is represented by n_{r1} which takes the value 1 when replacement occurs and 0 when it does not. The effect of the X group in the fragment

$$\text{MeC}-\text{NH}-\underset{\text{X}}{\text{CH}}-$$
$$\phantom{\text{Me}}\underset{\text{O}}{\|}$$

which in the tripeptides replaces the terminal Me group of the Ac–aax–aax is parameterized by $\alpha_2, \sigma_2, n_{H2}, n_{n2}, n_{02}$, and v_2. The effect of the remainder is parameterized by $n_{r2'}$.

Set AAB 23 was correlated with the IMFU equation after the deletion of the term in i. None of the side chains in the data set is ionizable. The best equation obtained is

$$\log K_{iX} = -5.10(\pm 1.12)\alpha_X + 0.211(\pm 0.122)n_{nX} + 2.20(\pm 0.252) \quad (62)$$

$100R^2 = 77.45;\quad F = 12.02;\quad S_{est} = 0.396;\quad S^0 = 0.568;\quad n = 10;$

$r_{\alpha nn} = 0.027;\quad C = 84.7;\quad C_{nn} = 15.3.$

It seems that the polarizability of the side chain is its dominant characteristic. We have examined the dipeptides in two ways. First, we have studied the subset in which the residue in position 1' is Gly (set PPB 44g). We have then considered all of the dipeptides combined into a single set (set PPB 44). The best equations obtained are for set 44g:

$$\log K_{iX} = -6.19(\pm 2.66)\alpha_X + 0.160(\pm 0.144)n_{nX} + 1.26(\pm 0.437) \quad (63)$$

$100R^2 = 55.60;\quad F = 3.756^k;\quad S_{est} = 0.632;\quad S^0 = 0.816;\quad n = 9;$

$r_{\alpha nn} = 0.139;\quad C = 89.9;\quad C_{nn} = 10.1.$

Although the results are barely significant, they are in accord with those observed for the Ac-aax. As Gly was constant throughout the data set, there was of course no need to consider it in the parameterization. When we study the entire dipeptide data set, we must take account of the residue in position 1'. The problem is simplified by the fact that the only side chains in position 1' are H, Me, and iBu. We may therefore account for the variation in these side chains by the polarizability $\alpha_{1'}$ and the steric parameter $n_{01'}$ which indicates the presence or absence of a branch at the α carbon atom of the amino acid residue. A similar parameter, n_{01}, was introduced to account for the presence or absence of branching at the carbon of residue 1. The best equation obtained for set 44 is:

$$\log K_{iX} = -3.98(\pm 2.38)\alpha_X + 0.223(\pm 0.148)n_{nX} - 0.739(\pm 0.593)n_{01}$$
$$+ 11.1(\pm 4.03)\alpha_{1'} - 1.05x \pm 0.560)n_{01'} + 1.58(\pm 0.443) \quad (64)$$

$100R^2 = 75.95;\quad F = 5.052^k;\quad S_{est} = 0.613;\quad S^0 = 0.644;\quad n = 14;$

$r = \alpha_1 n_n, 0.051;\quad \alpha_1 n_{01}, 0.630;\quad \alpha_1 \alpha_{1'}, 0.094;\quad \alpha_1 n_{01'}, 0.074;$

$n_n, n_{01}, 0.311;\quad n_n \alpha_{1'}, 0.340;\quad n_n n_{01'}, 0.444;\quad n_{01} \alpha_{1'}, 0.335;$

$n_{01} n_{01'}, 0.337;\quad \alpha_{1'} n_{01'}, 0.766;\quad C_{\alpha 1}, 26.6;\quad C_{nn}, 6.50;$

$C_{n01}, 21.4;\quad C_{\alpha 1'}, 14.8;\quad C_{n01'}, 30.6.$

The best equation obtained for the tripeptides (set 45) is

$$\log K_{iX} = -2.26(\pm 0.392)\alpha_{2'} + 0.178(\pm 0.0806)n_{H2'}$$
$$- 2.01(\pm 0.0919)n_{Me} + 1.75(\pm 0.0862) \quad (65)$$

$100R^2 = 99.09$; $F = 182.1$; $S_{est} = 0.104$; $S^0 = 0.128$; $n = 9$;

$r = \alpha_{2'}n_{H2'}, 0.662$; $\alpha_{2'}n_{Me}, 0.102$; $n_{H2'}, n_{Me}, 0.236$;

$C_{\alpha 2}, 19.2$; $C_{nH2'}, 6.57$; $C_{nMe}, 74.2$.

Note that one member of the set has the 2' residue in the D configuration. The n_{Me} parameter accounts for the effect of the side chain in the two data points for which the 1 and 1' residues are ala. The C_i values clearly show that the dominant structural effect is the side chain of the ala residues in positions 1 and 1'.

The combination of sets AAB 23 and PPB 44g gives set PPB 23U44g, for which the best regression equation is

$$\log K_{iX} = -4.35(\pm 1.35)\alpha_1 + 0.223(\pm 0.0912)n_{nX}$$
$$- 0.620(\pm 0.441)n_{01} + 1.09(\pm 0.237)n_{r1'} + 2.61(\pm 0.354) \quad (66)$$

$100R^2 = 74.39$; $F = 10.17$; $S_{est} = 0.472$; $S^0 = 0.590$; $n = 19$;

$r = \alpha_1 n_n, 0.157$; $\alpha_1 n_{01}, 0.511$; $\alpha_1 n_{r1'}, 0.325$;

$n_n, n_{01}, 0.221$; $n_n n_{r1'}, 0.276$; $n_{01} n_{r1'}, 0.018$;

$C_{\alpha 1}, 34.1$; $C_{nn}, 7.61$; $C_{n01}, 21.1$; $C_{nr1'}, 37.2$.

We have also combined sets AAB 23 and PPB 44 into a single set for which the best equation is

$$\log K_{iX} = -4.18(\pm 1.24)\alpha_1 + 0.226(\pm 0.0925)n_{nX} - 0.641(\pm 0.366)n_{01}$$
$$+ 11.2(\pm 3.19)\alpha_{1'} - 1.03(\pm 0.437)n_{01'} - 1.08(\pm 0.242)n_{r1'}$$
$$+ 2.60(\pm 0.300) \quad (67)$$

$100R^2 = 79.36$; $F = 10.90$; $S_{est} = 0.487$; $S^0 = 0.540$; $n = 24$

α_1	n_n	$n_{r1'}$	$\alpha_{1'}$	$n_{01'}$	n_{01}	
1	0.075	0.331	0.174	0.188	0.596	α_1
	1	0.116	0.225	0.282	0.246	n_n
		1	0.346	0.434	0.151	$n_{r1'}$
			1	0.798	0.317	$\alpha_{1'}$
				1	0.321	$n_{01'}$
					1	n_{01}

Finally, we have combined sets AAB 23, PPB 44, and PPB 45 to give set PPB 23υ44υ45 for which the best regression equation is

$$\log K_{ix} = -3.82(\pm 1.01)\alpha_1 + 0.229(\pm 0.0783)n_{nx} - 0.819(\pm 0.251)n_{01}$$
$$+ 11.9(\pm 2.61)\alpha_{1'} - 1.26(\pm 0.310)n_{01'}$$
$$- 1.04(\pm 0.197)n_{r1'} - 1.81(\pm 0.782)\alpha_{2'} + 2.69(\pm 0.223) \quad (68)$$

$100R^2 = 83.04;\quad F = 17.49;\quad S_{est} = 0.416;\quad S^o = 0.473;\quad n = 33$

α_1	n_n	$n_{r1'}$	$\alpha_{1'}$	$n_{01'}$	n_{01}	$\alpha_{2'}$	
1	0.089	0.472	0.063	0.093	0.690	0.446	α_1
	1	0.009	0.180	0.234	0.318	0.225	n_n
		1	0.263	0.342	0.326	0.329	$n_{r1'}$
			1	0.770	0.044	0.101	$\alpha_{1'}$
				1	0.052	0.019	$n_{01'}$
					1	0.498	n_{01}
						1	$\alpha_{2'}$

$C_{\alpha 1}, 16.9;\quad C_{nn}, 4.41;\quad C_{n01}, 15.8;\quad C_{nr1'}, 20.0;\quad C_{\alpha 1'}, 10.6;$
$C_{n01'}, 24.2;\quad C_{\alpha 2'}, 8.04.$

The contributions of structural effects at positions 1, 1′, and 2′ are given by the appropriate sums of the C_i values: $C_1, 37.1;\quad C_{1'}, 54.8;\quad C_{2'}, 8.04$.

In interpreting these results, we must note that our conclusions refer to the effect of structural change on the binding of ristocetin. Williams (110) has reviewed the binding of peptides to this antibiotic. The experimental results available are said to show that the carboxyl group of the C-terminal residue 1 and the peptide bond between it and residue 1′ are strongly involved in hydrogen bonding to ristocetin. As these structural features are present in all members of the combined set 23U44U45, no assessment of their contribution to the binding can be made from Equation 68. What we can conclude is that the side chain of residue 1 contributes 37.1% to the structural effects on binding and that this overall effect of the side chain of the 1 residue is due to approximately equal contributions from its polarizability and the presence or absence of a branch on the alpha carbon atom (zeroth carbon atom) of the residue, with a small but significant contribution from hydrogen bonding of side-chain lone pairs. The presence of residue 1′ and its side chain accounts for the major part of the structural effects, with side-chain polarizability contributing slightly less than half as much as branching at the carbon atom. The polarizability of the side chain of residue 2′ makes a significant contribution as well, but this residue has the least effect on binding.

The C_i values reported are based on the reference substrate Ac–his–ala–his for which $\alpha_1 = 0.230$, $n_n = n_{01} = n_{r1'} = n_{01'} = 1$, $\alpha_{1'} = 0.046$, and $\alpha_{2'} = 0.230$.

Binding is favored by increasing polarizability of side chains on the residues 1 and 2', by a side chain other than hydrogen on residues 1 and 1', and by the fragment CH_2CONH of residue 1'. An increase in the number of lone pairs on O and/or N of residue 1, or an increase in the polarizability of residue 1' causes a decrease in binding.

Kim and coworkers have obtained, for a somewhat different overall combined set, the regression equation

$$\log(1/K_i) = 0.43(\pm 0.08)MR_1 - 0.84(\pm 0.17)POLAR_1 - 1.08(\pm 0.17)I_1$$
$$+ 0.11(\pm 0.04)MR_3 + 0.63(\pm 0.21)ALA_2 + 2.31(\pm 0.31) \quad (69)$$
$$100R^2 = 86.0; \quad F = 31.95; \quad n = 32$$

Their data set did not include the dipeptides leu-Gly and leu-leu but did include the data point for (Cbz)-ala. The parameter POLAR is an indicator variable which takes the value 1 for a polar side chain and 0 for a nonpolar side chain. ALA is another indicator variable which is 1 for a methyl side chain at residue 1' and 0 otherwise. I_1 is a third indicator variable which has the value 1 for acylamino acids (Ac–aax) and 0 for di- and tripeptides. The MR values represent the polarizability of the side chains. In our notation residue 2 is 1', residue 3 is 2'. Our results are in reasonable agreement with Equation 69. In order to make comparisons of the coefficients, we must first change the signs of those in Equation 69 as we used K_i as the dependent variable and they used (i/K_i). On the scale of α values the coefficient of MR_1 is $-4.3(\pm 0.9)$ compared with our value of $-3.8(\pm 1.0)$ for the coefficient of α_1. The coefficient of MR_3 is $-1.1(\pm 0.4)$ compared with our value for that of $\alpha_{2'}$ of $-1.8(\pm 0.8)$. Comparisons of the coefficients of the remaining variables can only be qualitative. Our coefficient for n_n agrees with that of $POLAR_1$ when allowance is made for the difference in definition, and our coefficient for $n_{r1'}$ agrees with that for I' when we allow for the difference in definition of these two parameters. There is no significant difference between the intercepts of the two regression equations.

It is very encouraging to see that using two different approaches to the parameterization of a model, we and Kim and coworkers have been able to arrive at roughly equivalent quantifications of these peptide bioactivities. The advantages of the IMF model are that it is not necessary to treat each new data set as a new problem, that the resulting model can be easily compared with those obtained for similar systems, and that the interpretation of the equation obtained is easier with the IMF variables than it is with indicator variables.

V. PROTEINS

We now briefly review our work on the application of the IMF model to structural effects on protein properties and bioactivities.

A. Protein Conformation

Side-chain structural effects on protein conformation result from:

1. Intermolecular forces between the side chain X and an aqueous phase.
2. Intermolecular forces between amino acid side chains:
   a. in the same region of the protein molecule;
   b. in different regions of the protein molecule.
3. Steric effects
   a. which determine the torsional angles φ and ψ of the amino acid residue;
   b. between residues in different regions of the protein;
   c. on the hydration of the protein backbone.

All of these effects can be accounted for by the IMF model.

Our earliest venture into the modeling of amino acid, peptide, and protein properties and activities involved the quantification of the probability of finding an amino acid residue on the surface of a globular protein as compared to the probability of finding it in the interior (111). Chothia (112) has reported calculated values of f_X, the fraction of the amino acid residue with side chain X which are inaccessible to solvent over 95% of their surface area. Thus, f_X is given by the ratio

$$f_X = n_X/n_{tX} \tag{70}$$

where n_X is the number of residues with side chain X which are inaccessible to solvent over 95% of their surface area and n_{tX} is the total number of these residues present in the protein.

Janin (113) has reported ΔG values for the transfer of residues from the interior to the surface of a globulin. They are defined by the equation

$$\Delta G_{trX} = RT \ln(x_b/x_a) \tag{71}$$

where x_b and x_a are the mole fractions of buried and accessible residues with side chain X, respectively. We originally used the equation

$$Q_X = a_1 n_1 + a_2 I_1 + a_3 I_2 + a_0 \tag{72}$$

to model f_X and ΔG_{trX}. In this relationship, n_1 is the branching parameter which accounts for the steric effect of branching on the first atom of the side chain; I_1 is an indicator variable which takes the value 1 when the side chain has one or more OH or NH bonds and 0 otherwise; and I_2 is an indicator variable which has the value 1 when X is strongly basic (Arg, Lys) and 0 otherwise. Equation 71 is a primitive version of the IMFSB equation. Correlation of the data sets with it was successful. We later showed that the IMFSB equation can be used to describe these data sets.

We next studied (114) the Chou–Fasman (115) conformational parameters P_X for the probability of finding a residue in α-helix, β-sheet, and β-turn, and from the Chou–Fasman data have calculated P_X values for coil. P_X is defined as

$$P_{Xcf} = f_{Xcf} / \bar{f}_{cf} \tag{73}$$

where f_{Xcf} is the frequency of the residue with side chain X in conformation cf, and $\bar{f}_{cf}$ is the average frequency of all residues in that conformation. f_{Xcf} is given by the ratio

$$f_{Xcf} = n_{Xcf} / n_X \tag{74}$$

where n_{Xcf} is the number of residues with side chain X in the conformation cf, and n_X is the total number of these residues in all conformations. Similarly, $\bar{f}_{cf}$ is given by the ratio

$$\bar{f}_{cf} = n_{cf} / n \tag{75}$$

where n_{cf} is the total number of residues in conformation cf and n the total number in all conformations.

These P_X values were correlated successfully with a form of the IMF equation in which parameterization for charge-transfer interactions was included. The steric effect parameterization was of the combined type including both v and the simple branching parameters n_1, n_2, n_3, and the n_b parameter which serves as a measure of side-chain length as well. The correlation equation (the IMFUSB equation) is

$$Q_X = L\sigma_{lX} + A\alpha_X + H_1 n_{HX} + H_2 n_{nX} + I i_X + C_A n_A + C_D n_D$$
$$+ S v_X + B_1 n_{1X} + B_2 n_{2X} + B_3 n_{3X} + B_b n_{bX} + B^0 \tag{76}$$

Significant results were obtained for all four conformations. The C_i values show that β-turn and coil have fairly similar dependencies on intermolecular forces and steric effects. In both conformations, steric effects are the major factor in side-chain structural effects. Steric effects are the major factor in the case of α-helix as well, though this conformation shows some distinct differences in behavior from that of β-turn and coil. By contrast, in the case of β-sheet, intermolecular forces are the major factor in determining side-chain structural effects.

We then turned our attention (116) to the Zimm–Bragg s and σ parameters (117) which were intended to characterize ΔG_X values for the equilibrium between α-helix and coil. These parameters as well as the Δ_X values themselves were well fit by the IMFUSB equation. The ΔG_X values are predominantly a function of steric effects, being dependent on σ_1, v, n_1, n_2, and n_3. In further work (118, 118a) we have studied the detailed structure of β-sheet, including aligned versus orthogonal strands (119, 120) and parallel versus antiparallel strands (121). We have successfully modeled Omega loop (122) as well. Finally, we have modeled the results of Yutani and coworkers (123) who studied the effect of replacing residues in the alpha subunit of *E. coli* tryptophan synthetase on its denaturation. Again the model gave a good fit to the data.

The goodness of fit as measured by $100R^2$, the percent of the variance of the data accounted for by the correlation equation, depends on the number of residues in the data base used to determine the conformational preferences, N_r. Thus

$$100R^2 = a_1 N_r^{1/2} + a_0 \tag{77}$$

Our results validate the assumption that protein conformation is a function of intramolecular forces and steric effects. They aid in understanding the structural requirements which determine conformational preferences and make possible their prediction for amino acid residues not normally found in protein.

B. Protein Bioactivities

In the last decade, a number of reports have appeared concerning the effect on bioactivity of substitution of amino acid residues in proteins. Though in most of this work the number of data points is too small to permit the application of the IMF model, there are some data sets in which this is not the case. Thus, Fersht and coworkers (125) have determined kinetically values of $\Delta\Delta G$ for the effect of replacement of one residue by another on the binding of ATP and of Tyr to tyrosyl-tRNA synthetase. We (126) have combined the data for the two substrates into a single data set by means of an indicator variable. In order to justify this combination, we first showed that the $\Delta\Delta G$ values for ATP binding are a linear function of those for Tyr binding. We have chosen to make use of the IMFSB equation. Applying it to the effect of the side chains initially present, designated X^i, we have

$$\Delta G_{X^i} = L\sigma_{1X^i} + A\alpha_{X^i} + H_1 n_{HX^i} + H_2 n_{nX^i} + Ii_{X^i}$$
$$+ B_1 n_{1X^i} + B_2 n_{2X^i} + B_3 n_{3X^i} + B^0 \tag{78}$$

and for the side chains finally present, designated X^f,

$$\Delta G_{X^f} = L\sigma_{1X^f} + A\alpha_{X^f} + H_1 n_{HX^f} + H_2 n_{nX^f} + Ii_{X^f}$$
$$+ B_1 n_{1X^f} + B_2 n_{2X^f} + B_3 n_{3X^f} + B^0 \tag{79}$$

Then, subtracting Equation 78 from Equation 79 gives

$$\Delta\Delta G_{X^f - X^i} = L\sigma_1^d + A\alpha^d + H_1 n_H^d + H_2 n_n^d + Ii^d$$
$$+ B_1 n_1^d + B_2 n_2^d + B_3 n_3^d \tag{80}$$

where the superscript d indicates that the variable has the form

$$v^d = v_{X^f} - v_{X^i} \tag{81}$$

The data were correlated with a modification of Equation 80 to which an intercept was added. The combined set is well fit by the correlation equation with $100R^2 = 90.83$. The $\Delta\Delta G$ values are a function of polarizability, ionic side chains, and steric effects. It must be noted that the substitution in this data set is at several positions in the protein. Three data points are for substitution at each of the positions 35, 48, and 51; one each is at positions 34 and 195. The success of the model requires that in this protein these positions are equivalent or nearly so with regard to the structural effects of side chains.

Alber and coworkers (127) have reported activities of mutant Phage T4 Lysozyme substituted at position 86. The activity measured is the quantity required to give a clear area of 7.00-mm radius in the Lysoplate assay. Correlation of $\log A_{rel}$ with the IMFU equation gave a good fit. Hydrogen bonding is the dominant factor in determining side-chain structural effects on the activity. Next in importance is polarizability followed by steric effects. As the data set is small in size (10 data points) and the range in $\log A_{rel}$ is less than one order of magnitude, the possibility that the quality of the correlation is fortuitous cannot be excluded. Further work is needed to verify our results.

We expect that as more data accumulates involving structural effects of side-chain substitution on the bioactivities of proteins, the IMF model will prove to be a valuable tool for the quantitative description of these effects.

VI. CONCLUSIONS

We have applied the IMF model to 109 data sets for amino acids and 57 data sets for peptides in this work. We have also referred to results published or presented at meetings involving the application of the IMF model to 32 data sets of protein conformational parameters or properties and two data sets of protein bioactivities. We have in addition some 30 other data sets involving amino acids, peptides, and proteins which have not yet been published. The IMF model has therefore been shown to apply to the structural effects of side chains on the properties and bioactivities of more than 200 data sets. We feel that we can now regard it as a generally reliable model for this purpose. In Appendix 1 we have tabulated IMF parameter values for more than 110 amino acids. We hope that this will make the method generally useful to biochemists, medicinal chemists, and all others interested in the quantitative description of amino acid, peptide, and protein properties and bioactivities.

Our results have also underlined the utility of the Zeta and Omega methods in correlation analysis. The ability to combine data sets into a single set makes possible analyses that otherwise could not be done. It also makes it possible to draw conclusions regarding a whole group of related organisms.

It is useful at this point to once again note that collinearities among the parameters of the IMF model will frequently occur, particularly among the 19 members of the protein basis set. While this in no way interferes with the predictive capacity of the regression equations obtained, it does make difficult separation of the intermolecular forces responsible for the observed structural effects.

Our results, particularly those for amino acid hydrophobicities, solution properties, and chromatographic properties, and for peptide chromatographic properties, support the conclusion that hydrophobicity is not a special, unique property or type of bond. Clearly, it is the manifestation of the difference in intermolecular forces between an initial and a final phase. It further seems likely that there is no advantage in the use of the various kinds of available surface or volume parameters that have been proposed as measures of hydrophobicity. It would appear that they simply represent yet another parameterization of polarizability, a variable which has been overparameterized already. Our results also show that there is no one single hydrophobicity parameter which will serve to model all data sets. That nature is more complicated than we would wish is a problem frequently encountered, but we can ignore the complications only at the cost of false results.

Our bioactivity correlations shed light on the nature of the quantitative description of bioactivities as a function of structural variation in the substrate. The success of models which make use of $\log P$ or other hydrophobicity parameters is often based on the use of other physicochemical parameters which represent electrical or steric effects. Our results imply that the function of these additional parameters is to modify the composition of the hydrophobicity parameters so that the resulting mix will meet the requirements of the system being studied.

Our results provide many additional examples of the utility of the Zeta and Omega methods. They permit the combination of data sets into a large, more comprehensive set whose analysis leads to a more general and reliable conclusion. When the attempted combination of data sets is unsuccessful, we can infer that the sets which were to be combined do not share a common mechanism in their dependence on structural effects.

Finally, we must take note of the fact that quantitative models of structural effects on properties, reactivities, and bioactivities can be obtained by statistical methods such as factor analysis and principal component analysis. These methods are of great use when no theoretical model is available and only an empirical description is required. Until and unless the factors or principal components can be identified, usually by regression analysis against well-understood physicochemical parameters, we have no more understanding of how and why the structural effects are manifested than we did before the analysis. It seems to us that it is the search for how and why that distinguishes science from art or craft. We prefer science. Others may heed a different call.

REFERENCES

1. R. Gabler, *Electrical Interactions in Molecular Biophysics*, Academic Press, New York, 1978.
2. M. Charton, *Prog. Phys. Org. Chem.*, *10*, 81–204 (1973).
3. O. Exner, *Coll. Czech. Chem. Communs.*, *25*, 642 (1960).
4. L. W. Deady, M. Kendall, R. D. Topsom, and R. A. Y. Jones, *J. Chem. Soc. Perkin Trans. II*. 416 (1973).
5. M. Charton and B. I. Charton, *J. Org. Chem.*, *44*, 2284–2288 (1979).
6. M. Charton, *Prog. Phys. Org. Chem.*, *13*, 119–125 (1981).
7. M. Charton, and B. I. Charton, *J. Theoret. Biol.*, *99*, 629–644 (1982).
8. M. Charton, *J. Org. Chem.*, *31*, 2991–2996 (1966).
9. M. Charton, *Prog. Phys. Org. Chem.*, *16*, 287–315 (1987).
10. M. Charton, in *Design of Biopharmaceutical Properties Through Prodrugs and Analogs*, E. B. Roche, Ed., American Pharmaceutical Society, Washington, D.C., 1977, pp. 228–280.
11. M. Charton, *Topics in Current Chem.*, *114*, 57–91 (1983).
12. M. Charton, *J. Org. Chem.*, *43*, 3995–4001 (1978).
13. M. Charton, *Proc. 3rd Cong. Hungarian Pharmacologic Soc.*, Budapest, 1979, Akademiai Kiado, Budapest, 1980, pp. 211–220.

14. M. Charton, *Environ. Health Perspect.*, 61, 229–238 (1983).
15. M. Charton and B. I. Charton, *Abstr., 2nd Noordwijkerhout IUPAC–IUPHAR Symposium: Strategy in Drug Design*, 1981, p. 79.
16. M. Charton, in *Rational Approaches to the Synthesis of Pesticides*, P. S. Magee, J. J. Menn, and G. K. Koan, Eds., American Chemical Society, Washington, D.C., 1984, pp. 247–278.
17. W. Kauzman, *Adv. Protein Chem.*, 14, 1 (1959).
18. C. Tanford, *The Hydrophobic Effect*, 2nd ed., Wiley-Interscience, New York, 1980.
19. A. Ben-Naim, *Hydrophobic Interactions*, Plenum Press, New York, 1980.
20. *Faraday Symposium*, Chemical Society, 17, (1982).
21. C. Tanford, *J. Am. Chem. Soc.*, 84, 4240–4247 (1962).
22. Z. Simon, *Quantum Biochemistry and Specific Interactions*, Abacus Press, Tunbridge Wells, 1976, p. 93.
23. P. Cornea, Licentiate's work, Univ. Timisoara (1969).
24. C. C. Bigelow, *J. Theoret. Biol.*, 16, 187 (1967).
25. D. E. Goldsack and R. C. Chalifoux, *J. Theoret. Biol.*, 39, 645 (1973).
26. J. M. Zimmerman, N. Eliezer, and R. Sinha, *J. Theoret. Biol.*, 21, 170–201 (1968).
27. T. P. Hopp and K. R. Woods, *Proc. Natl. Acad. Sci. USA*, 78, 3824–3828 (1981).
28. Y. Nozaki and C. Tanford, *J. Biol. Chem.*, 246, 2211–2217 (1971).
29. M. J. Sternberg and J. M. Thornton, *J. Mol. Biol.*, 115, 1–17 (1977).
30. M. Levitt, *J. Mol. Biol.*, 104, 59–107 (1976).
31. M. F. Kanehisa and T. Y. Tsong, *Biopolymers*, 19, 1617–1628 (1980).
32. D. D. Jones, *J. Theoret. Biol.*, 50, 167 (1975).
33. E. Q. Lawson, A. J. Sadler, D. Harmatz, D. T. Brandau, R. Micomovic, R. D. MacElroy, and C. R. Middaugh, *J. Biol. Chem.*, 259, 2910–2912 (1984).
34. V. Pliska, M. Schmidt and J.-L. Fauchere, *J. Chromatog.*, 216, 79–92 (1981).
35. J.-L. Fauchere, K. W. Do, P. Y. Jow, and C. Hansch, *Experientia*, 36, 1203–1204 (1980).
36. C. Hansch and A. Leo, *Substituent Constants for Correlation Analysis in Chemistry and Biology*, Wiley-Interscience, New York, 1979.
37. R. A. Klein, M. J. Moore, and M. W. Smith, *Biochim. Biophys. Acta*, 233, 420 (1971).
38. A. England and E. J. Cohn, *J. Am. Chem. Soc.*, 57, 634 (1935).
39. S. Moore and W. H. Stein, *Ann. N.Y. Acad. Sci.*, 49, 265 (1948).
40. L. M. Yunger and R. D. Cramer, *Mol. Pharmacol.*, 20, 602–608 (1981).
41. R. F. Rekker, *The Hydrophobic Fragmental Constant*, Elsevier, Amsterdam, 1977.
42. R. V. Wolfenden, P. M. Cullis, and C. C. Southgate, *Science*, 206, 575–577 (1979).
43. R. Wolfenden, L. Anderson, P. M. Cullis, and C. C. Southgate, *Biochem.*, 20, 849–855 (1981).
44. H. B. Bull and K. Breese, *Arch. Biochem. Biophys.*, 161, 665–670 (1974).
45. A. A. Aboderin, *Int. J. Biochem.*, 2, 537 (1971).
46. B. Y. Zaslavsky, N. M. Mestchkina, L. M. Miheeva, and S. V. Rogozhin, *J. Chromatog.*, 240, 21–28 (1982).
47. D. Eisenberg, R. M. Weiss, T. C. Terwilliger, and W. Wilcox, *Faraday Symposium, Chem. Soc.*, 17, 109–120 (1982).
48. C. Frommel, *J. Theor. Biol.*, 111, 247 (1984).

49. J. Kyte and R. F. Doolittle, *J. Mol. Biol.*, *157*, 105–132 (1982).
50. H. J. Wynne, K. H. Van Buuren, and W. Wakelkamp, *Experientia*, *38*, 655–656 (1982).
51. J.-L. Fauchere and V. Pliska, *Eur. J. Med. Chem.*, *18*, 369–375 (1983).
52. R. M. Sweet and D. Eisenberg, *J. Mol. Biol.*, *171*, 479–488 (1983).
53. D. M. Engelman, T. A. Steitz, and A. Goldman, *Ann. Rev. Biophys. Biophys. Chem.*, *15*, 330 (1986).
54. P. Manavalan and P. K. Pannuswamy, *Nature*, *275*, 673–674 (1976).
55. G. van Heijine and C. Blomberg, *Eur. J. Biochem.*, *97*, 175–181 (1979).
56. A. K. Mishra and S. C. Ahluwalia, *J. Phys. Chem.*, *88*, 86–92 (1984).
57. G. D. Rose, A. R. Gelelowitz, G. J. Lesser, R. H. Lee, and M. H. Zehfus, *Science*, *229*, 834–838 (1985).
58. M. Charton and B. I. Charton, in *Quantitative Approaches to Drug Design*, J. Dearden, Ed., Elsevier, Amsterdam, 1983, pp. 260–261.
59. P. M. Hardy, in *Chemistry of the Amino Acids*, G. C. Barrett, Ed., Chapman and Hall, London, 1985, pp. 6–24.
60. J. O. Hutchins, in *CRC Handbook of Biochemistry*, Chemical Rubber Co., Cleveland, 1968, p. B-13.
61. I. D. Kuntz, *J. Am. Chem. Soc.*, *93*, 514–516 (1971).
62. A. J. Hopfinger, *Intermolecular Interactions and Biomolecular Organization*, Wiley, New York, 1977, p. 132.
63. T. H. Lelley, in *Chemistry of the Amino Acids*, G. C. Barrett, Ed., Chapman and Hall, London, 1985, pp. 591–624.
64. B. F. Sleckman and J. Sharma, *J. Liq. Chromatog.*, *5*, 1051–1068 (1982).
65. S. Wold, L. Eriksson, S. Hellberg, J. Jansson, M. Sjostrom, B. Skagerberg, and C. Wikstrom, *Can. J. Chem.*, *65*, 1814–1820 (1987).
66. G. Zweig and J. Sharma, Eds., *CRC Handbook of Chromatography: General Data and Principles*, Vol. 1, CRC Press, Boca Raton, FL. pp. 448–451.
67. M. Brenner, A. Niederwieser, and G. Patak, in *Thin Layer Chromatography*, 2nd ed., E. Stahl, Ed., Springer-Verlag, Heidelberg, 1971, p. 730–786.
68. J. C. Touchstone, E. J. Levin, and S. G. Lee, *J. Liq. Chromatogr.*, *7*, 2719–2793 (1984).
69. N. Seiler and B. Knodgen, *J. Chromatogr. Biomed. Appl.*, *341*, 11–21 (1985).
70. E. S. Hemden and J. Porath, *J. Chromatogr.*, *323*, 255–264 (1985).
71. E. Grushka, S. Levin, and C. Gilon, *J. Chromatogr.*, *235*, 401–409 (1982).
72. Z. Deyl, in *Protein Structure and Function*, J. L. Fox, Z. Deyl, and A. Blazej, Eds., Dekker, New York, 1976, pp. 23–68.
73. N. Saringa, *Nature New Biol.*, *234*, 172–174 (1976).
74. N. Nimura, T. Suzuke, Y. Kusahara, and T. Kinoshita, *Ann. Chem.*, *53*, 1380 (1981).
75. Y. Ihara, N. Kunikiyo, T. Kunimasa, Y. Kimura, M. Nango, and N. Kuroki, *J. Chem. Soc. Perkin Trans. II*, 1741 (1983).
76. P. A. Bash, V. C. Singer, R. Langridge, and P. S. Kollman, *Science*, *236*, 564–568 (1987).
76a. D. Porschke, *J. Mol. Evol.*, *21*, 192–198 (1985).
77. M. Charton, in *QSAR in Drug Design and Toxicology*, D. Hadzi and B. Jerman-Blazic, Eds., Elsevier, Amsterdam, 1987, pp. 285–290.
78. J. W. McFarland, *Prog. Drug Res.*, *15*, 173 (1971).

79. J. R. Knowles, *J. Theoret. Biol.*, 9, 213 (1965).
80. J. B. Jones and J. F. Beck, in *Application of Biochemical Systems in Organic Chemistry, Part 1*, J. B. Jones, C. J. Sih, and D. Perlman, Eds., Wiley-Interscience, New York, 1976, pp. 107–401.
81. J. J. Bechet, A. Dupaix, and C. Coucous, *Biochem.*, 12, 2559–2565; 2566–2572 (1973).
82. H. Kubinyi, *Prog. Drug Research*, 23, 97–198 (1979).
83. W. Freist, H. Sterbach, and F. Cramer, *Eur. J. Biochem.*, 169, 33–39 (1987).
84. W. Freist, H. Sternbach, and F. Cramer, *Eur. J. Biochem.*, 173, 27–34 (1988).
85. A. N. Radhakrishnan, in *Peptide Transport and Hydrolysis*, CIBA Foundation Symposium 50 (New Series), K. Elliott and M. O'Connor, Eds., Elsevier, Amsterdam, 1977, pp. 37–77.
86. V. Ganapathy, R. A. Roesel, and F. H. Leibach, *Biochem. Biophys. Res. Commun.*, 105, 28–35 (1982).
87. H. Kusakabe, K. Kodama, A. Kuninaka, H. Yoshino, H. Misono, and K. Soda, *J. Biol. Chem.*, 255, 976–981 (1980).
88. O. Trygstad, I. Foss, and K. Sletten, *Ann. N. Y. Acad. Sci.*, 248, 304–316 (1975).
89. M. Charton and B. I. Charton, *Abstr., 3rd Intl. EUCHEM Conf. on Correlation Analysis in Organic Chem.* (CAOC), Louvain-la Neuve, 1985, p. B-16.
90. M. Charton, in *Strategies in Drug Design*, J. K. Seydel, Ed., Verlag-Chemie, Heidelberg, 1985, pp. 260–263.
91. M. Charton and B. I. Charton, in *QSAR in Drug Design and Toxicology*, D. Hadzi and B. Jerman-Blazic, Eds., Elsevier, Amsterdam, 1987, pp. 285–290.
92. M. Charton, in *Quantitative Structure Activity Relationships in Drug Design*, J.-L. Fauchere, Ed., Alan R. Liss, New York, 1989, pp. 307–312.
92a. J. L. Meek, *Proc. Natl. Acad. Sci. USA*, 77, 1632–1636 (1980).
93. J. L. Meek and Z. L. Rossetti, *J. Chromatogr.*, 211, 15–28 (1981).
94. L. Lepri, P. G. Desideri, and D. Heimler, *J. Chromatogr.*, 207, 412–420 (1981).
95. T. Szirtes, L. Kisfaludy and L. Szporny, *J. Med. Chem.* 27, 741–745 (1984).
96. K. S. Cheung, S. A. Wasserman, E. Dudek, S. A. Lerner, and M. Johnston, *J. Med. Chem.*, 26, 1741–1746 (1983).
97. M. Manning, E. Nawrocka, B. Misicka, A. Olma, W. A. Klis, J. Seto, and W. H. Sawyer, *J. Med. Chem.*, 27, 423–429 (1984).
98. M. Manning, A. Olma, W. A. Klis, J. Seto, and W. H. Sawyer, *J. Med. Chem.*, 26, 1607 (1983).
99. A. Meister, S. S. Tate, and G. A. Thompson, in *Peptide Transport and Hydrolysis*, CIBA Foundation Symposium 50 (New Series), K. Elliott and M. O'Connor, Eds., Elsevier, Amsterdam, 1977, pp. 123–143.
100. R. Wade, in *Amino Acids, Peptides, and Related Compounds*, H. N. Rydon, Ed., *Intl. Rev. Org. Chem.*, 6, Ser. 2, Butterworths, London, 1978, pp. 480–502.
101. P. J. Goodford, *Adv. Pharmacol. Chemother.*, 11, 51–97 (1973).
102. P. Berntsson, *Acta Pharm. Suec.*, 17, 199–208 (1980).
103. M. C. Summers and R. J. Hayes, *J. Biol. Chem.*, 256, 4951–4956 (1981).
104. P. D. Gesellchen, R. C. A. Frederickson, S. Tafur, and D. Smiley, *Pept. Synth. Struct. Func. Proc., Am. Peptide Symp.*, 7, 621–624 (1981).
105. R. T. Shuman, P. D. Gesellchen, E. L. Smithwick, and R. C. A. Frederickson, *Pept. Synth. Struct. Func., Proc. Am. Peptide Symp.*, 7, 617–620 (1981).

106. D. R. Steinbrink, M. D. Bond, and H. E. Van Werf, *J. Biol. Chem.*, 260, 2771–2776 (1985).
107. R. Andruszkiewicz, H. Chmara, S. Milewski, and E. Borowski, *J. Med. Chem.*, 30, 1715–1719 (1987).
108. F. R. Atherton, C. H. Hassall, and R. W. Lambert, *J. Med. Chem.*, 29, 29–40 (1986).
109. K. H. Kim, Y. C. Martin, E. R. Otis, and J. C. Mao, *J. Med. Chem.*, 32, 84–93 (1989).
110. D. H. Williams, *Acc. Chem. Res.*, 17, 364–369 (1984).
111. M. Charton, *J. Theoret. Biol.*, 91, 115–123 (1981).
112. C. Chothia, *J. Mol. Biol.*, 105, 1–14 (1976).
113. J. Janin, *Nature*, 277, 491 (1979).
114. M. Charton and B. I. Charton, *J. Theoret. Biol.*, 102, 121–134 (1983).
115. P. Y. Chou and G. D. Fasman, *Ann. Rev. Biochem.*, 47, 251–276 (1978).
116. M. Charton, *Abstr., 18th MARM Am. Chem. Soc.*, 1984, p. 39 #14.
117. B. H. Zimm and J. K. Bragg, *J. Chem. Phys.*, 31, 526–535 (1959).
118. M. Charton and B. I. Charton, *Abstr, 3rd Noordwijkerhout Symposium*, Noordwijkerhout, 1985, p. 10.
118a. M. Charton, *Abstr, 4th Meeting on Stereochemistry*, Liblice, zechoslovakia, April 1987, pp. 14a–14b.
119. C. Chothia and J. Janin, *Proc. Natl. Acad. Sci. USA*, 78, 4146–4150 (1981).
120. C. Chothia and J. Janin, *Biochem.*, 21, 3955–3965 (1982).
121. S. Lifson and C. Sander, *Nature*, 282, 109–111 (1979).
122. J. F. Lesczczynski and G. D. Rose, *Science*, 234, 849–855 (1986).
123. K. Yutani, K. Ogasahara, T. Kakuno, and Y. Sugino, *J. Mol. Biol.*, 160, 387–390 (1982).
124. K. Yutani, K. Ogasahara, T. Kakuno, and Y. Sugino, *J. Biol. Chem.*, 259, 14076–14081 (1984).
125. A. R. Fersht, J.-P. Shi, J. Knill-James, D. M. Lowe, A. J. Wilkinson, D. M. Blow, P. Brick, P. Carter, M. M. Y. Waye, and G. Winter, *Nature*, 314, 235–238 (1985).
126. M. Charton, *Intl. J. Peptide Protein Res.*, 28, 201–207 (1986).
127. T. Alber, J. A. Bell, S. Dao-Pin, H. Nicholson, J. A. Wozniak, S. Cook, and B. W. Matthews, *Science*, 239, 631–635 (1988).
128. M. Charton, unpublished results.

APPENDIX 1. AMINO ACID SIDE-CHAIN PARAMETER TABLES

List of Nonstandard Abbreviations

Ad	Adamantyl
Baa	Bromoacetylamino
Bzl	Benzyl
c	Cyclo (e.g., cHx = cyclohexyl)
DNP	2,4-dinitrophenyl
Fu	Furyl
Gu	Guanidino
Hbz	4-hydroxybenzyl
Hp	Heptyl
Hx	Hexyl
Imz	Imidazolyl
Ind	Indolyl
Nh	Naphthyl
No	Nonyl
Oc	Octyl
Ocb	Octhocarboranyl
Pe	Pentyl
Pl	Pyrrolyl
Pn	Phenylene
Pra	Propionylamino
Pym	Pyridylmethyl
Pz	3-pyrazinyl
Tn	Thienyl
Vi	Vinyl

TABLE A
Intermolecular Force Parameters[a]

Aax	X	α	σ_I	n_H	n_n	i	n_D	n_A
Aad	$(CH_2)_3CO_2H$	0.197	0.04	1	4	1	0	1
Abu	Et	0.093	−0.01	0	0	0	0	0
Ada	1-$AdCH_2$	0.442	−0.01	0	0	0	0	0
Ade	Oc	0.370	−0.01	0	0	0	0	0
Adg	1-Ad	0.396	−0.01	0	0	0	0	0
Ahp	Pe	0.232	−0.01	0	0	0	0	0
Ahx(6-OH)	$(CH_2)_4OH$	0.200	0.01	1	2	0	0	0
Ala	Me	0.046	−0.01	0	0	0	0	0
Ala(F)	CH_2F	0.044	0.21	0	0	0	0	0
Ala(NH_2)	CH_2NH_2	0.090	0.06	2	1	0	0	0
Alg	CH_2Vi	0.135	0.02	0	0	0	0	0
Ano	Hp	0.324	−0.01	0	0	0	0	0
Anp	1-$NhCH_2$	0.452	0.08	0	0	0	1	0
Aoc	Hx	0.278	−0.01	0	0	0	0	0
Ape(5-OH)	$(CH_2)_3OH$	0.154	0.02	1	2	0	0	0
Arg	$(CH_2)_3Gu$	0.291	0.04	4	3	1	0	0

TABLE A (Continued)

Aax	X	α	σ_I	n_H	n_n	i	n_D	n_A
Arg(NO₂)	(CH₂)₄NHC=NNO₂(NH₂)	0.344	0.04	2	5	0	1	0
Asn	CH₂CONH₂	0.134	0.06	2	3	0	0	1
Asn(Ph)	CH₂CONHPh	0.377	0.10	1	3	0	1	1
Asn(Me)	CH₂CONHMe	0.180	0.06	1	3	0	0	1
Asp	CH₂CO₂H	0.105	0.15	1	4	1	0	1
Aud	No	0.416	−0.01	0	0	0	0	0
Bnp	2-NhCH₂	0.452	0.04	0	0	0	1	0
Bua	tBuCH₂	0.232	−0.01	0	0	0	0	0
Bug	tBu	0.186	−0.01	0	0	0	0	0
Bza	(CH₂)₂Ph	0.336	0.02	0	0	0	1	0
Can	CH₂CH₂ONH₂	0.143	0.01	2	3	1	0	0
Car	OcbCH₂	0.573	0.06	0	0	0	0	0
Ceg	cHx(CH₂)₂	0.349	−0.01	0	0	0	0	0
Cha	cHxCH₂	0.303	−0.01	0	0	0	0	0
Chg	cHx	0.257	0.00	0	0	0	0	0
Cit	(CH₂)₃NHCONH₂	0.251	0.00	3	4	0	0	0
Cna	CH₂CN	0.099	0.20	0	1	0	0	0
Cpg	cPe	0.214	−0.01	0	0	0	0	0
Cys	CH₂SH	0.128	0.12	0	0	0	1	0
$\overline{Cys}$	CH₂S—	0.128	0.16	0	0	0	1	0
$\overline{CysCys}$	CysSCH₂	0.394	0.16	3	5	1	1	1
Cys(Me)	CH₂SMe	0.174	0.12	0	0	0	1	0
Cys(Me,O)	CH₂SOMe	0.172	0.21	0	2	0	0	0
Cys(Pr)	CH₂SPr	0.266	0.12	0	0	0	0	0
Cys(Bzl)	CH₂SCH₂Ph	0.418	0.08	0	0	0	1	0
Dab	(CH₂)₂NH₂	0.127	0.03	2	1	1	1	0
Dbt	3,5-Br₂-Hbz	0.456	0.06	1	2	0	1	0
Dhp	3-OH-HBz	0.305	0.04	2	4	0	1	0
Djk	CH₂SCH₂Cys	0.440	0.13	3	5	1	1	1
Dpm	(CH₂)₃CHNH₂CO₂H	0.289	0.04	3	5	1	1	1
Fra	2-FuCH₂	0.215	0.05	0	1	0	1	0
Frg	2-Fu	0.169	0.17	0	1	0	1	0
Gln	(CH₂)₂CONH₂	0.180	0.05	2	3	0	0	1
Glu	(CH₂)₂CO₂H	0.151	0.07	1	4	1	0	1
Glu(3-OH)	CHOHCH₂CO₂H	0.166	0.10	2	6	1	0	1
Gly	H	0.000	0.00	0	0	0	0	0
Har	(CH)₄Gu	0.337	0.01	4	3	1	0	0
Hcy	(CH₂)₂SH	0.174	0.04	0	0	0	1	0
Hcy(Et)	(CH₂)₂SEt	0.267	0.05	0	0	0	1	0
Hcy(Bzl)	(CH₂)₂SCH₂Ph	0.465	0.05	0	0	0	1	0
His	4-ImzCH₂	0.230	0.08	1	1	1	1	0
His(π-Me)	3-Me-4-ImzCH₂	0.276	0.08	0	1	1	1	1
His(τ-Me)	1-Me-4-ImzCH₂	0.276	0.08	0	1	1	1	1
His(τ-DNP)	1-DNP-4-ImzCH₂	0.598	0.08	0	9	0	1	1
Hpg	PnOH-4	0.252	0.10	1	2	0	0	0

TABLE A (Continued)

Aax	X	α	σ_I	n_H	n_n	i	n_D	n_A
Hph	CH_2CH_2Ph	0.336	0.02	0	0	0	1	0
Hse	$(CH_2)_2OH$	0.108	0.06	1	2	0	0	0
Hyp	4-hydroxyproline	0.146	0.00	1	2	0	0	0
Ile	sBu	0.186	−0.01	0	0	0	0	0
Kyn	$CH_2COPnNH_2$-2	0.390	0.15	2	3	0	1	1
Lan	$CH_2SCH_2CHNH_2CO_2H$	0.324	0.12	3	5	1	0	1
Leu	iBu	0.186	−0.01	0	0	0	0	0
Lys	$(CH_2)_4NH_2$	0.219	0.00	2	1	1	1	0
Lys(CHO)	$(CH_2)_4NHCHO$	0.274	0.01	1	1	1	0	1
Lys(Me)	$(CH_2)_4NHMe$	0.265	0.00	1	1	1	1	0
Lys(Ac)	$(CH_2)_4NHAc$	0.323	0.01	1	3	0	0	1
Lys(DNP)	$(CH_2)_4NHDNP$	0.587	0.00	1	9	0	0	1
Met	$(CH_2)_2SMe$	0.221	0.04	0	0	0	0	0
Met(O)	$(CH_2)_2SOMe$	0.219	0.10	0	2	0	0	0
Met(O_2)	$(CH_2)_2SO_2Me$	0.217	0.11	0	4	0	0	0
Nip	CH_2PnNO_2-4	0.350	0.08	0	4	0	0	1
Nle	Bu	0.186	−0.01	0	0	0	0	0
Nva	Pr	0.139	−0.01	0	0	0	0	0
Orn	$(CH_2)_3NH_2$	0.173	0.02	2	1	1	1	0
Paa	CH_2Pz	0.263	0.10	0	2	0	0	1
Pea	$(CH_2)_3Ph$	0.382	0.01	0	0	0	1	0
Pen	CMe_2SH	0.221	0.09	0	0	0	1	0
Phe	CH_2Ph	0.290	0.03	0	0	0	1	0
Phe(4-Baa)	$BrCH_2CONHBzl$	0.503	0.04	1	3	0	1	0
Phe(Et)	4-EtBzl	0.383	0.03	0	0	0	1	0
Phe(F_5)	$CH_2C_6F_5$	0.285	0.12	0	0	0	1	0
Phe(4-Me)	4-MeBzl	0.336	0.03	0	0	0	1	0
Phe(4-NH_2)	4-NH_2Bzl	0.322	0.03	2	1	0	1	0
Phe(4-Pra)	4-EtCONHBzl	0.470	0.04	1	3	0	1	0
Phg	Ph	0.244	0.12	0	0	0	1	0
Pla	2-PlCH$_2$	0.247	0.06	1	0	0	1	0
Plg	2-Pl	0.201	0.17	1	0	0	1	0
Ppa	$(CH_2)_4Ph$	0.428	0.00	0	0	0	1	0
Prg	$CH_2C{\equiv}CH$	0.131	0.14	0	0	0	0	0
Pro	proline	0.140	−0.01	0	0	0	0	0
Ser	CH_2OH	0.062	0.11	1	2	0	0	0
Ser(Bzl)	CH_2OBzl	0.352	0.11	0	2	0	1	0
Ser(Et)	CH_2OEt	0.155	0.11	0	2	0	0	0
Ser(Pym)	CH_2OCH_2Py-3	0.328	0.11	0	3	0	0	1
Ser(tBu)	CH_2OtBu	0.248	0.11	0	2	0	0	0
β-PhSer	CHPhOH	0.260	0.10	1	2	0	1	0
Tha	2-TnCH$_2$	0.276	0.06	0	0	0	1	0
Thg	2-Tn	0.230	0.19	0	0	0	1	0
Thr	CHMeOH	0.108	0.09	1	2	0	0	0
Thr(Bzl)	CHMeOBzl	0.398	0.09	0	2	0	1	0
Thr(Me)	CHMe(OMe)	0.154	0.09	0	2	0	0	0
Trp	3-IndCH$_2$	0.409	0.00	1	0	0	1	0

TABLE A (*Continued*)

Aax	X	α	σ_I	n_H	n_n	i	n_D	n_A
Trx	$CH_2(C_6H_2I_2O)_2H$	1.066	0.07	1	4	0	1	0
Tyr	CH_2PnOH-4	0.298	0.03	1	2	0	1	0
Tyr(Bu)	CH_2PnOBu-4	0.484	0.03	0	2	0	1	0
Tyr(Bzl)	$CH_2PnOBzl$-4	0.541	0.03	0	2	0	1	0
Tyr(DNP)	$CH_2PnODNP$-4	0.666	0.03	0	10	0	1	1
Tyr(Et)	CH_2PnOEt-4	0.391	0.03	0	2	0	1	0
Tyr(PnOH-4′)	$CH_2(PnO)_2H$	0.550	0.04	1	4	0	1	0
Tyr(3-I)	3-I-Hbz	0.427	0.04	1	2	0	1	0
Tyr(3,5-I_2)	3,5-I_2-Hbz	0.556	0.06	1	2	0	1	0
Tyr(Me)	CH_2PnOMe-4	0.344	0.03	0	2	0	1	0
Tyr(3-Me)	3-MeHbz	0.344	0.03	1	2	0	1	0
Tyr(3-NO_2)	3-NO_2-Hbz	0.360	0.06	1	6	0	1	0
Tyr(iPr)	$CH_2PnOiPr$	0.436	0.03	0	2	0	1	0
Val	iPr	0.140	0.01	0	0	0	0	0

[a] For amino acids in italics, the value of the parameter n_Z which represents the number of NH bonds on the alpha amino groups is 1; for all other amino acids, it has the value 2.

TABLE B
Steric Parameters[a]

Aax	v	n_1	n_2	n_3	n_b
Aad	0.68	1	1	1	4
Abu	0.56	1	0	0	1
Ada	1.4	1	1	1	4
Ade	0.68	1	1	1	7
Adg	1.33	1	1	1	3
Ahp	0.73	1	1	1	4
Ahx(6-OH)	0.68	1	1	1	4
Ala	0.52	0	0	0	0
Ala(F)	0.62	1	0	0	1
Ala(NH_2)	0.52	1	0	0	1
Alg	0.69	1	1	0	2
Ano	0.73	1	1	1	6
Anp	0.70	1	1	1	5
Aoc	0.73	1	1	1	5
Ape(5-OH)	0.68	1	1	1	3
Arg	0.68	1	1	1	5
Arg(NO_2)	0.68	1	1	1	7
Asn	0.76	1	1	0	2
Asn(Ph)	0.76	1	1	1	6
Asn(Me)	0.76	1	1	1	3
Asp	0.76	1	1	0	2
Aud	0.68	1	1	1	8
Bnp	0.70	1	1	1	6
Bua	1.34	1	3	0	2

TABLE B (*Continued*)

Aax	v	n_1	n_2	n_3	n_b
Bug	1.24	3	0	0	1
Bza	0.70	1	1	1	5
Can	0.89	1	1	1	3
Car	1.7	1	1	1	—
Ceg	0.70	1	1	2	5
Cha	0.97	1	2	2	4
Chg	0.87	1.5	0.75	0	3
Cit	0.68	1	1	1	5
Cna	0.89	1	1	0	2
Cpg	0.71	1.86	0.24	0	2
Cys	0.62	1	0	0	1
$\overline{\text{Cys}}$	0.70	1	1	1	—
$\overline{\text{Cys}}\,\overline{\text{Cys}}$	0.70	1	1	1	6
Cys(Me)	0.70	1	1	0	2
Cys(Me, O)	0.86	1	2	0	2
Cys(Pr)	0.71	1	1	1	4
Cys(Bzl)	0.70	1	1	1	6
Dab	0.68	1	1	0	2
Dbt	0.70	1	1	1	5
Dhp	0.70	1	1	1	5
Djk	0.70	1	1	1	7
Dpm	0.68	1	1	1	5
Fra	0.70	1	1	1	3
Frg	0.57–2.0	1	1	0	2
Gln	0.68	1	1	1	3
Glu	0.68	1	1	1	3
Glu(3-OH)	0.68	1	1	1	3
Gly	0.00	0	0	0	0
Har	0.68	1	1	1	6
Hcy	0.78	1	1	0	2
Hcy(Et)	0.78	1	1	1	3
Hcy(Bzl)	0.78	1	1	1	6
His	0.70	1	1	1	3
His(π-Me)	0.70	1	1	1	3
His(τ-Me)	0.70	1	1	1	4
His(τ-DNP)	0.70	1	1	1	9
Hpg	0.57	1	1	1	4
Hph	0.70	1	1	1	5
Hse	0.77	1	1	0	2
Hyp	−0.81	—	—	—	—
Ile	1.02	2	1	0	2
Kyn	0.76	1	1	1	5
Lan	0.70	1	1	1	5
Leu	0.98	1	2	0	2
Lys	0.68	1	1	1	4

TABLE B (*Continued*)

Aax	v	n_1	n_2	n_3	n_b
Lys(CHO)	0.68	1	1	1	6
Lys(Me)	0.68	1	1	1	5
Lys(Ac)	0.68	1	1	1	6
Lys(DNP)	0.68	1	1	1	10
Met	0.78	1	1	1	3
Met(O)	0.98	1	1	2	3
Met(O_2)	1.01	1	1	3	3
Nip	0.70	1	1	1	6
Nle	0.68	1	1	1	3
Nva	0.68	1	1	0	2
Orn	0.68	1	1	1	3
Paa	0.70	1	1	1	4
Pea	0.70	1	1	1	6
Pen	1.24	3	0	0	1
Phe	0.70	1	1	1	4
Phe(4-Bu)	0.70	1	1	1	8
Phe(4-Et)	0.70	1	1	1	6
Phe(F_5)	0.70	1	1	1	5
Phe(4-Me)	0.70	1	1	1	5
Phe(4-NH_2)	0.70	1	1	1	5
Phe(4-Pra)	0.70	1	1	1	8
Phg	0.57–2.15	1	1	1	3
Pla	0.70	1	1	1	3
Plg	0.57–2.0	1	1	0	2
Ppa	0.70	1	1	1	7
Prg	0.89	1	1	0	2
Pro	−0.81	—	—	—	—
Ser	0.53	1	0	0	1
Ser(Bzl)	0.62	1	1	1	6
Ser(Et)	0.61	1	1	1	3
Ser(tBu)	0.67	1	1	3	3
Ser(Pym)	0.62	1	1	1	6
β-PhSer	0.69	2	1	1	3
Tha	0.70	1	1	1	3
Thg	0.57–2.2	1	1	0	2
Thr	0.70	2	0	0	1
Thr(Bzl)	0.71	2	1	1	6
Thr(Me)	0.70	2	1	0	6
Trp	0.70	1	1	1.5	5
Trx	0.70	1	1	1	10
Tyr	0.70	1	1	1	5
Tyr(Bu)	0.70	1	1	1	9
Tyr(Bzl)	0.70	1	1	1	10
Tyr(DNP)	0.70	1	1	1	11
Tyr(Et)	0.70	1	1	1	7
Tyr(PhOH-4′)	0.70	1	1	1	10
Tyr(3-I)	0.70	1	1	1	5

TABLE B (Continued)

Aax	v	n_1	n_2	n_3	n_b
Tyr(3,5-I$_2$)	0.70	1	1	1	5
Tyr(Me)	0.70	1	1	1	6
Tyr(3-Me)	0.70	1	1	1	5
Try(3-NO$_2$)	0.70	1	1	1	5
Tyr(iPr)	0.70	1	1	1	6
Val	0.76	2	0	0	1

[a] For amino acids in italics, the value of the parameter n_z which represents the number of NH bonds on the alpha amino groups is 1; for all other amino acids, it has the value 2.

APPENDIX 2. THE ZETA AND OMEGA METHODS

A. The Zeta Method

The majority of data sets investigated by correlation analysis have the form XGY, where:

X is a variable substituent
Y is the active site, an atom or group of atoms at which a measurable phenomenon takes place
G is a skeletal group to which X and Y are attached.

The quantity Q_X which is to be correlated is determined under a given set of experimental conditions, and constant structural features. The former include temperature, pressure, reagent, catalyst, and medium. The medium includes the solvent if any, ionic strength, and acidity. The experimental conditions are kept constant throughout the data set. The structural conditions may include the active site Y, the skeletal group G, and constant substituents.

Correlation analysis assumes that the measured quantities Q_X are a linear function of physicochemical parameters which characterize all of the structural and experimental variables. Thus:

$$Q_X = a_1 p_{1X} + a_2 p_{2X} + a_3 p_{3X} + \cdots + a_0 \qquad (1)$$

or

$$Q_X = \sum_{i=1}^{m} a_i p_{iX} + a_0 \qquad (2)$$

If we wish to study the effect of varying the structure of X on the measured property Q, we can factor the sum of the right side of Equation 2 into two terms, one of which includes all the parameters that characterize X, the other of which includes all the parameters that characterize the remaining variables. Then

$$Q_X = \sum_{i=1}^{m} a_i p_{iX} + \sum_{j=1}^{m'} a_j p_{j0} + a_0 \qquad (3)$$

As only X varies and all other variables which do not represent structural effects of X are held constant, we have

$$Q_X = \sum_{i=1}^{m} a_i p_{iX} + a'_0 \qquad (4)$$

We now consider a series of data sets in which X varies, and in which one of the other variables (variables which have nothing to do with X), while constant within each data set, varies from one data set to another. If this variable is extracted from the sum $\sum a_j p_j$ by writing

$$\sum_{j=1}^{m'} a_j p_{j0} = \sum_{j=1}^{m'-1} a_j p_{j0} + a_m p_{m0} \qquad (5)$$

then substituting Equation 5 into Equation 3 gives

$$Q_X = \sum_{i=1}^{m} a_i p_{i0} + \sum_{j=1}^{m'-1} a_j p_{j0} + a_m p_{m0} + a_0 \qquad (6)$$

$$Q_X = \sum_{i=1}^{m} a_i p_{i0} + a_m p_{m0} + a'_0 \qquad (7)$$

If a suitable parameter $p_{m'0}$ is variable, we may then combine all of the data sets into a single set using Equation 7 as the correlation equation. Frequently, however, we do not have a suitable parameter available. In that case, we make use of the Zeta method. We choose as a reference substrate that bears the substituent X^0. For each of the data sets to be combined, a value of Q_{X^0} has been determined. As X^0 is constant in all of these Q_{X^0} values, we obtain on substituting into Equation 7

$$Q_{X^0} = \sum_{i=1}^{m} a_i p_{iX^0} + a_{m'_0} + a'_0 \qquad (8)$$

$$= a_{m'} p_{m'} + C \qquad (9)$$

where

$$C = \sum_{i=1}^{m} a_i p_{iX^0} + a'_0 \tag{10}$$

We define

$$Q_{X^0} \equiv \zeta \tag{11}$$

ζ can now be used as the parameter $p_{m'O}$.

If X^0 is a member of the data sets that are to be combined, this process is designated internal parameterization. If it is not included in the data sets to be combined, it is designated external parameterization.

If it is desirable for the intercept to be identical to $Q_{H,\text{calc}}$ we must modify the definition of ζ by choosing one of the original data sets as the reference set O^0. Then ζ is defined as $\zeta = Q_{X^0O} + Q_{X^0O^0}$.

The combined data set is now correlated with the equation that would be used for any one of the component sets modified by the addition of a term in $Z\zeta$.

B. The Omega Method

The biological activity of a given substrate generally varies within a group of related species. If (1) the biomechanism is the same for all of the species of interest, and (2) the bioactivity is due to receptor–substrate interaction, then it is possible to combine the data sets for the individual species into a single set.

In order to understand this, we must first recognize that the receptor sites in the different species will differ from each other to some extent. The differences between receptor sites cannot be large, however, or the biomechanism could not remain the same. It follows then that bioactivities determined for a set of substrates in a number of different but related organisms are analogous to chemical reactivities determined for a set of substrates with a number of different but related reagents. Just as the Zeta method can be used to combine the chemical reactivities into a single data set, the Omega method can be used to combine bioactivities determined in different but related organisms into a single set. In the Omega method, we are parameterizing the organisms. To do this, we choose a substrate X^0 for which the bioactivity is available for each of the organisms of interest. The organism parameter, ω, is then defined as

$$\omega_{\text{org}} = BA_{X^0_{\text{org}}} \tag{13}$$

The term $O\omega$ is then introduced into the correlation equation that would be used to model the individual organism data sets. Alternatively, in order to have the intercept equal to $Q_{\text{H, calc}}$ we may define ω as

$$\omega_{\text{org}} = BA_{X^0{_\text{org}}} - BA_{X^0{_{\text{org}_0}}}$$

where org_0 is the chosen reference organism.

Author Index

Numbers in parentheses are reference numbers and indicate that the author's work is referred to although his name is not mentioned in the text. Numbers in *italics* show the pages on which the complete references are listed.

Aboderin, A.A., 180 (45), 186 (45), *271*
Abraham, R.J., 155 (41), *161*
Adcock, W., 152 (36), *161*
Agranat, I., 2 (5), *60*
Ahluqalia, S.C., 188 (56), 189 (56), *272*
Aihara, J., 2 (7), 15 (35), 18 (35), 23 (35), 24 (35), *60*, *61*
Albano, C., 88 (32), 89 (32), 112 (117), 113 (117), 120 (143), *124*, *127*, *128*
Alber, T., 268 (127), *274*
Alberts, V., 152 (36), *161*
Alluni, S., 116 (125), *127*
Alperovich, A.M., 27 (82), *63*
Althoff, H., 25 (71), *63*
Ambet, R.W., 252 (108), 255 (108), *274*
Anderson, L., 180 (43), 185 (43), *271*
Andersson, T.W., 78 (7), *123*
Andruszkiewicz, R., 251 (107), 252 (107), *274*
Angus, Hearman, R., 110 (104), *127*
Anulewicz, R., 26 (76), 49 (76b), 52 (76b), 53 (76b), *63*
Ariêns, E.J., 78 (14), *124*
Arnett, E.M., 156 (43), 157 (43), *161*
Arnett, M.M., 116 (127), *128*
Arnold, Z., 25 (64), *62*
Atherton, F.R., 252 (108), 255 (108), *274*
Aue, D.H., 156 (42), 157 (42), 158 (42), *161*

Bacon, N., 25 (65), *62*
Bagno, A., 117 (136), 118 (136), *128*
Bailey, W.F., 90 (50, 51), 91 (50, 51), *125*
Baird, N.C., 33 (99), *64*
Bar, R., 17 (41), *61*
Barron, P., 152 (36), *161*
Bash, P.A., 205 (76), 208 (76), *272*

Bayliss, N.S., 31 (91), *63*
Bechet, J.J., 213 (81), 219 (81), *273*
Beck, J.F., 211 (80), 213 (80), 219 (80), *273*
Beebe, K.R., 78 (12), 120 (12), *124*
Bell, J.A., 268 (127), *274*
Bell, L., 156 (43), 157 (43), *161*
Benson, H.G., 31 (94), *64*
Bergmann, D., 33 (99), *64*
Berntsson, P., 248 (102), *273*
Berrand, D., 108 (85, 87), *126*
Berstein, J.I., 66 (4), *75*
Bigelow, C.C., 180 (24), 188 (24), 189 (24), 190 (24), *271*
Biroš, J., 137 (22), 155 (22), *160*
Blackburn, B.J., 90 (49), 91 (49), *125*
Blomberg, C., 180 (55), 186 (55), *272*
Blow, D.M., 267 (125), *274*
Blunt, W.J., 92 (60), *125*
Bock, H., 13 (31), 33 (31), *61*
Boegel, H., 25 (67), *62*
Bohle, M., 111 (113), *127*
Bond, M.D., 239 (106), 250 (106), *274*
Bornowski, B., 25 (71), 27 (80), 29 (88), *63*
Borowski, E., 251 (107), 252 (107) *274*
Bosman, T., 109 (93), *126*
Botrel, A., 31 (94), *64*
Bowers, M.T., 156 (42), 157 (42), 158 (42), *161*
Boyd, R., 116 (128), *128*
Bozek, J.D., 90 (49), 91 (49), *125*
Bragg, J.K., 267 (117), *274*
Brandau, D.T., 180 (33), 183 (33), *271*
Breese, K., 180 (44), 185 (44), *271*
Brehme, R., 25 (64), *62*
Brenner, M., 197 (67), *272*
Bretschneider, E., 155 (41), *161*
Brick, P., 267 (125), *274*
Bromilow, J., 2 (2), 6 (13), 32 (2, 13), 41 (2), 57 (13a), *59*, *60*

Brown, H.C., 4 (11), *60*, 133 (15), 142 (15), 152 (15), 153 (15), *160*
Brownlee, R.T.C., 2 (2), 3 (12), 4 (10a), 6 (13, 14), 20 (10, 14), 32 (2, 12, 13), 33 (97), 41 (2), 57 (13a), *59, 60, 64*
Buděšínský, M., 91 (53), *125*, 152 (37), *161*
Bull, H.B., 180 (44), 185 (44), *271*
Bulmer, J.F., 107 (79, 80), *126*
Bunnett, J.F., 117 (132), *128*
Bureš, M., 133 (17), 140 (17), 144 (17), 145 (17), 146 (17), 151 (17), *160*
Burgard, D.R., 110 (108), *127*
Butler, A.R., 25 (65), *62*
Butt, G., 55 (104), *64*
Buttgereit, G., 25 (68), *62*

Carey, W.P., 78 (12), 120 (12), *124*
Carlson, R., 112 (117), 113 (117), *127*
Carretto, J., 112 (120), 113 (120), *127*
Carroll, J.B., 86 (25), *124*
Carter, P., 267 (125), *274*
Cartwright, H.M., 80 (18), *124*
Caspentier, J.M., 117 (131), *128*
Cen-Naim, A., 177 (19), *271*
Chakrzwarty, T., 110 (106), *127*
Chalioux, R.C., 180 (25), *271*
Chanon, M., 112 (120), 113 (120c), *127*
Chapman, N.B., 78 (15), *124*
Charma, J., 197 (64), *272*
Charton, B.I., 167 (5), 170 (7), 173 (15, 16), 192 (7, 58), 209 (7, 58), 229 (89, 91), 236 (7), 265 (114), 267 (118), 229 (89, 91), *270, 272, 273, 274*
Charton, M., 2 (3), 51 (103), *59, 64*, 167 (2, 5, 6), 116 (122), *127*, 170 (6, 7, 8, 9), 171 (10, 11), 209 (7, 16, 58, 77), 229 (89, 90, 91, 92), 236 (7), 265 (111, 114), 267 (116, 118, 118a, 126), 229 (89, 91), *270*(12, 13, 14), *272, 273, 274*(128)
Chastrette, M., 112 (120), 113 (120c), *127*
Chawla, B., 156 (43), 157 (43), *161*
Chentli-Benchikha, F., 21 (50b), 23 (50b), 26 (50), *62*
Cheung, K.S., 231 (96), 236 (96), *273*
Chmara, H., 251 (107), 252 (107), *274*
Chothia, C., 265 (112), 267 (119, 120), *274*
Chou, P.Y., 265 (115), *274*
Chretien, J.R., 109 (98, 99, 100, 101), *127*
Christoph, B., 123 (147), *128*
Cioff, E.A., 90 (52), 91 (52), *125*

Clementi, S., 116 (125), *127*
Cline Lowe, L.J., 108 (92), 109 (92), *126*
Cohn, E.J., 180 (38), 183 (38), *271*
Cook, S., 268 (127), *274*
Cornea, P., 180 (23), *271*
Coucous, C., 213 (81), 219 (81), *273*
Coulson, C.A., 26 (76), *63*
Cox, J.D., 144 (27), 145 (27), 146 (27), 150 (27), *160*
Cox, R.A., 65 (1, 2), 67 (9), 68 (13), 74 (13), *75*, 119 (139, 140, 141), *128*
Craik, D.J., 6 (13, 14), 20 (14), 32 (2, 13), 33 (97), 41 (2), 57 (13a), *59, 60, 64*
Cramer, F., 213 (84), 219 (82), 220 (84), *273*
Cramer, R.D., 180 (40), 183 (40), *271*
Cramer, R.D., III, 112 (119), *127*
Crawford, C.B., 86 (24), *124*
Cullis, P.M., 180 (42, 43), 185 (42, 43), *271*

Daehne, S., 2 (9), 7 (9a, 17, 20), 8 (22), 11 (9c, 20), 13 (20c, 27), 15 (33a, 33c, 35), 16 (32), 17 (20a, 38, 41), 18 (35), 20 (43a, 47), 21 (49, 50c), 23 (9c, 20b, 33a, 35, 48, 54, 55, 57b), 24 (33a, 33c, 35, 58b, 59, 62), 25 (66, 67, 69, 71), 26 (48, 50c, 54b, 73), 27 (9, 78, 79, 80, 82), 28 (49), 29 (49, 83, 87, 88), 30 (48, 59, 62, 83, 87, 90), 31 (94), 32 (9c, 17), 33 (9c, 27), 34 (90, 100b), 37 (9c), 38 (9c), 39 (62, 90), 43 (9c), 45 (100), 46 (100a, 100b), *60, 61, 62, 63*(84), *64*
Dais, P., 2 (2), 3 (2b), 5 (2b), 21 (2b), 32 (2), 35 (2b), 36 (2b), 41 (2), *59*
Dale, H., 21 (50a), 26 (50), *62*
Dale, J., 21 (50b), 23 (50b, 55c), 26 (50), *62*
Dalling, D.K., 89 (40), 92 (59), *125*
Daltrozzo, E., 25 (68), *62*
Dao-Pin, S., 268 (127), *274*
Davis, C.T., 66 (6), 67 (6), *75*
Davis, J.C., 88 (30), *124*
Davis, J.M., 108 (90), *126*
Deady, L.W., 167 (4), *270*
Declercq, F., 21 (50b), 23 (50b), 26 (50), *62*
De Jongh, R.O., 25 (72), *63*
de Ligny, C.L., 89 (33), 109 (102), 116 (123, 124), *125, 127*
Dekel'baum, A.B., 157 (47), *161*

Derevjanko, N.A., 20 (44), *61*
Derks, W., 109 (93), *126*
Desideri, P.G., 231 (94), *273*
Devaux, M.F., 108 (87), *126*
Dewar, M.J.S., 33 (25), 11 (25b), *60*
Deyl, Z., 197 (72), 204 (72), *272*
Deyrup, A.J., 116 (126), *128*
Dillon, T.W., 78 (9), *123*
Ditchfield, R., 131 (8), *160*
Do, K.W., 180 (35), 183 (35), *271*
Doddrell, D., 152 (36), *161*
Doerr, F., 23 (56), *62*
Doherty, R.M., 7 (16), 59 (16a), *60*
Domenicano, A., 26 (76), *63*
Dominique, P., 117 (131), *128*
Doolittle, R.F., 180 (49), 186 (49), *272*
Dougherty, R.C., 33 (25), 11 (25b), *60*
Dreux, M., 109 (98, 99, 100, 101), *127*
Dudek, E., 231 (96), 236 (96), *273*
Dunn, G.E., 102 (70), 103 (70), *126*
Dunn, W.J., III, 88 (32), 89 (32), 90 (47), 92 (54), *124*, *125*
Dupaix, A., 213 (81), 219 (81), *273*
Dyadyusha, G.G., 15 (33b, 36), 42 (101), 58 (101), *61*, *64*

Eastment, H., 81 (20), *124*
Ebber, A., 74 (16), 75, 87 (29), *124*
Ebisawa, Y., 17 (41), *61*
Edlund, U., 88 (32), 89 (32), 90 (44, 45, 47), 92 (45, 54), 116 (125), *124*, *125*, *127*
Edward, J.T., 67 (7), 75, 102 (72), 103 (72), 117 (135), *126*, *128*
Ehrenson, S.J., 4 (10a), 20 (10), *60*
Eisenberg, D., 180 (47, 52), 186 (47, 52), *271*, *272*
Elbl, K., 21 (51), *62*
Elguero, J., 111, (115), *127*
Eliasson, B., 90 (45), 92 (45), *125*
Eliezer, N., 180 (26), 188 (26), 189 (26), 205 (26), 208 (26), *271*
Engelhardt, G., 24 (59), 30 (59), *62*
Engelman, D.M., 180 (53), 186 (53), 187 (53), *272*
England, A., 180 (38), 183 (38), *271*
Erbensen, K., 78 (10), 82 (10), *124*
Eriksen, O.I., 21 (50a), 26 (50), *62*
Erikson, B.C., 78 (12), 120 (12), *124*
Eriksson, L., 197 (65), *272*
Esbensen, K., 88 (32), 89 (32, 33), *124*, *125*

Essers, R., 109 (93), *126*
Exner, O., 91 (53), *125*, 129 (1, 7), 131 (1, 7), 132, (7, 14), 133 (18), 135 (18), 136 (1), 137 (22), 141 (18), 143 (7, 14, 24, 25), 144 (26), 145 (28), 146 (28, 29), 147 (29), 152 (1, 37), 155 (1, 22, 39, 40), 157 (14), 158 (7), 159 (26), *160*, *161*, 167 (3), *270*

Fabian, H., 16 (32), *61*
Fabian, J., 13 (28, 30), 14 (30), 15 (34), 16 (16), 17 (28), *61*
Famini, G.R., 7 (16), 59 (16a), *60*
Faraday Symposium, 177 (20), *271*
Fargin, E., 112 (118), *127*
Farley, H.A., 80 (18), *124*
Farncombe, M.J., 80 (17), 110 (104), *124*, *127*
Farral, L., 136 (21), 148 (21), 149 (21), 150 (21), 151 (21), *160*
Fasman, G.D., 265 (115), *274*
Fauchere, J.-L., 180 (34, 35, 51), 183 (34, 35), 186 (51), *271*, *272*
Fedorov, G.F., 157 (47), *161*
Fergusson, G.A., 86 (24), *124*
Fersht, A.R., 267 (125), *274*
Fiske, P.R., 6 (13), 32 (13), 57 (13a), *60*
Flury, B., 78 (9), *124*
Foerster, T., 9 (23), *60*
Ford, G.P., 31 (92), *63*
Foss, I., 213 (88), 221 (88), *273*
Fowler, F.W., 111 (111), *127*
Frans, S.D., 109 (95), *126*
Frederickson, R.C.A., 239 (104, 105), 248 (104, 105), *273*
Freist, W., 213 (84), 219 (82), 220 (84), *273*
Friedl, Z., 136 (20), 137 (22), 155 (22), *160*
Frommel, C., 180 (48), 186 (48), 188 (48), 189 (48), 209 (48), *271*
Fruchier, A., 111 (115), *127*
Futrell, J.H., 108 (91), 109 (91), 110 (91b), *126*

Gabler, R., 166 (1), *270*
Gal, J.-F., 57 (105), *64*, 112 (118), *127*
Ganapathy, V., 213 (86), 221 (86), *273*
Gangadharan, S., 92 (55), 94 (55), *125*
Gann, L., 123 (147), *128*
Gasteiger, J., 123 (146, 147), *128*
Gavilanes Largo, S., 136 (21), 148 (21), 149 (21), 150 (21), 151 (21), *160*

Geiseler, G., 20 (47), *61*
Geissman, T.A., 66 (6), 67 (6), *75*
Geladi, P., 78 (10), 82 (10), 88 (32), 89 (32, 33), *124*, *125*
Gelelowits, A.R., 188 (57), 189 (57), *272*
Geltz, Z., 84 (23), 85 (23), 99 (23), 103 (74), 117 (137), 121 (23), *124*, *126*, *128*
Gemperline, P.J., 109 (94), *126*
George, W.O., 25 (65), *62*
Geribaldi, S., 57 (105), *64*, 104 (78), *126*
Germain, G., 21 (50b), 23 (50b), 26 (50), *62*
Gerstein, R.A., 42 (101), 58 (101), *64*
Gesellchen, P.D., 239 (104, 105), 248 (104, 105), *273*
Gey, E., 15 (33a, 33c), 23 (33a), 24 (33a, 33c), *61*
Gidaspov, B.V., 121 (145), *128*
Giddings, J.C., 108 (90), *126*
Giesbrecht, F.G., 108 (86), *126*
Gillette, P.C., 103 (76), 107 (76), *126*
Gilon, C., 197 (71), 204 (71), *272*
Gimarc, B.M., 15 (37), 17 (37a), 39 (37), *61*
Goldman, A., 180 (53), 186 (53), 187 (53), *272*
Goldsack, D.E., 180 (25), *271*
Goldstein, M., 78 (9), *123*
Golebiewski, A., 32 (96), *64*
Golub, G., 78 (4), *123*
Gomes, A., 6 (13), 32 (13), *60*, 92 (54), *125*
Gompper, R., 29 (85), *63*
Goodford, P.J., 248 (101), *273*
Gordon, M., ed., 27 (81), *63*
Gowenlock, B.G., 55 (104), *64*
Gramaccioni, P., 21 (50a), 26 (50), *62*
Granger, M.R., 66 (5), 67 (5), *75*
Grant, D.M., 89 (38, 40), 92 (59), *125*
Griffiths, J., 7 (21), *60*
Grigoras, S., 90 (47), 92 (54), *125*
Grimm, B., 27 (80), *63*
Gronowitz, S., 90 (46), 92 (46), *125*
Grosse, D., 23 (55), 25 (65), *62*
Groth, P., 21 (50a, 50c, 51), 26 (50c), *62*
Grunwald, E., 129 (6), 131 (6), 132 (6), 155 (6), *160*
Grushka, E., 197 (71), 204 (71), *272*
Gschwendter, W., 92 (60), 98 (60b), 99 (60b), 101 (60b), *125*
Guertler, O., 17 (41), *61*
Gutner, N.M., 148 (30), 150 (30), *160*

Haefelinger, G., 22 (53), 26 (76), 49 (76b), 52 (76b), 53 (76b), *62*, *63*
Halberg, S., 116 (125), *127*
Haldna, Ü., 67 (10, 11), 68 (15), 74 (16), 75, 87 (28, 29), 119 (140, 141), *124*, *128*
Halket, J., 109 (96), *126*
Hamer, G.K., 6 (13), 32 (13), *60*
Hamid, A., 108 (86), *126*
Hammett, L.P., 66 (3), *75*, 78 (13), 114 (13), 116 (126), *124*, *128*, 129 (2, 3), 130 (2), 131 (3), 152 (3), 155 (3), 159 (2), *160*
Hansch C., 78 (14), *124*, 180 (35, 36), 183 (35, 36), *271*
Hardy, P., 192 (59), *272*
Harmatz, D., 180 (33), 183 (33), *271*
Harris, J.M., 109 (95) *126*
Hartmann, H., 8 (22), 13 (28, 30), 14 (30), 16 (30b), 17 (28), *60*, *61*
Hassall, C.H., 252 (108), 255 (108), *274*
Havinga, E., 25 (72), *63*
Hayes, R.J., 239 (103), 248 (103), *273*
Hehre, W.J., 32 (95), *64*, 131 (8), 156 (43), 157 (43), *160*, *161*
Heilbronner, B., 13 (31), 33 (31), *61*
Heimler, D., 231 (94), *273*
Hellberg, S., 88 (32), 89 (32), *124*, 197 (65), *272*
Hemden, E.S., 197 (70), 204 (70), *272*
Hendrickson, A.E., 86 (25), *124*
Hepler, L.G., 132 (9, 10), 155 (9, 10), 157 (10, 44), *160*, *161*
Herman, R.A., 80 (17), *124*
Hess, B.A., 13 (29), *61*
Hibety, P.C., 2 (6), 17 (6), *60*
Hidgins, T., 96 (62), *125*
Hiller, C., 123 (147), *128*
Hinze, J., 33 (98, 99), *64*
Hirsch, B., 25 (69), *62*
Hoefnagel, A.J., 116 (122), *127*
Hohlneicher, G., 29 (85), *63*
Holub, R., 133 (17), 140 (17), 144 (17), 145 (17), 146 (17), 151 (17), *160*
Hoovery, D.G., 109 (97), *126*
Hopfinger, A.J., 192 (62), 194 (62), *272*
Hopp, T.P., 180 (27), *271*
Hoppen, V., 92 (59a), 98 (59a), *125*
Hotelling, H., 78 (2), 104 (2), *123*
Houdard, J., 25 (70), *62*
Howery, D.G., 68 (14), 70 (14), 72 (14), *75*, 78 (11), 80 (11), 103 (11), *124*

AUTHOR INDEX

Huheey, J.E., 94 (61), *125*
Hunter, J.A., 55 (104), *64*
Hutchins, J.O., 192 (60), *272*
Huthings, M.S., 123 (147), *128*
Hutton, H.M., 90 (49), 91 (49), *125*

Ichikawa, H., 15 (35), 17 (41), 18 (35), 23 (35), 24 (35), *61*
Idler, K.L., 119 (140), *128*
Ihara, Y., 205 (75), 208 (75), 231 (75), 237 (75), *272*
Inamoto, N., 152 (35), *161*
Inger, V.C., 205 (76), 208 (76), *272*
Inoshita, T., 204 (74), 205 (74), 208 (74), *272*
Isenhour, T.L., 108 (91), 109 (91), 110 (91), *126*
Istomin, B.I., 151 (31), *161*
Ives, D.J.G., *160* (11)

Jacques, P., 31 (93, 94), *63*, *64*
Jaffé, H.H., 33 (98), *64*, 152 (33), *161*
Jakob, E., 110 (105), *127*
Janata, J., 118 (138), *128*
Janecke, 92 (59d), *125*
Janin, J., 265 (113), 267 (119, 120), *274*
Jankowski, W.C., 92 (59), *125*
Jansson, J., 197 (65), *272*
Jauer, E.A., 25 (69), *62*
Jennings, K.R., 80 (17), 110 (104), *124*, *127*
Jensen, S.A., 108 (84), *126*
Jesse, D., 108 (83), *126*
Johansson, E., 88 (32), 89 (32), *124*
Johnels, D., 90 (48), 91 (48), 92 (48), 116 (125), *125*, *127*
Johnson, B.B., 21 (50a), 26 (50), *62*
Johnson, F., 92 (59), *125*
Johnston, M., 231 (96), 236 (96), *273*
Jones, D.D., 180 (32), *271*
Jones, J.B., 211 (80), 213 (80), 219 (80), *273*
Jones, R.A.Y., 24 (63), *62*, 167 (4), *270*
Jones, R.N., ed., 107 (82), *126*
Joussot-Dubien, J., 25 (70), *62*
Jow, P.Y., 180 (35), 183 (35), *271*
Jutz, C., 23 (55, 58a), 24 (58a), 25 (65), 29 (85), *62*, *63*

Kachkovskii, A.D., 15 (33b, 36), 20 (44), *61*

Kakuno, T., 267 (123), *274* (124)
Kaminski, J.L., 66 (4), *75*
Kamlet, M.J., 7 (16), 59 (16a), *60*
Kanehisa, M.F., 180 (30), *271*
Kanev, I., 17 (40), *61*
Karhunen, K., 78 (5), *123*
Karplus, M., 89 (37), *125*
Kateman, G., 109 (93), *126*
Katritzky, A.R., 24 (63), 31 (92), *62*, *63*, 111 (111), *127*
Kausen, H., 23 (56), *62*
Kauzman, W., 177 (17), 204 (17), *271*
Kebarle, P., 132 (13), 156 (13), 157 (13), *160*
Kendall, M.G., 88 (32), 89 (32), *123*
Kendall, R.D., 167 (4), *270*
Kim, K.H., 252 (109), 259 (109), 264 (109), *274*
Kimura, Y., 205 (75), 208 (75), 231 (75), 237 (75), *272*
Kinimasa, T., 205 (75), 208 (75), 231 (75), 237 (75), *272*
Kirkwood, J.G., 97 (64), *126*
Kisfaludy, L., 231 (95), 236 (95), *273*
Kitching, W., 152 (36), *161*
Klein, J., 15 (37), 39 (37), *61*
Klein, R.A., 180 (37), 183 (37), *271*
Klis, W.A., 231, (98), 236 (98), 237 (98), 239 (98), *273*
Knill-James, J., 267 (125), *274*
Knodgen, B., 197 (69), 204 (69), *272*
Knorr, F.J., 108 (91), 109 (91), 110 (91b), *126*
Knowles, J.R., 211 (79), 213 (79), *273*
Ko, H.C., 157 (44), *161*
Kobischke, H., 27 (80), *63*
Kodama, K., 213 (87), *273*
Koenig, J.L., 103, (76), 107 (76), *126*
Koenig, W., 7 (18), *60*
Kokocińska, H., 103 (74), 113 (121), 114 (121), *126*, *127*
Kollecker, W., 111 (113), *127*
Kollman, P.S., 205 (76), 208 (76), *272*
Koppel, I.A., 111 (112), *127*
Kotschy, J., 23 (56), *62*
Kowalski, B.R., 78 (12), 120 (12), *124*
Kramer, H.E.A., 25 (65), 29 (85), *62*, *63*
Krieger, C., 21 (51), *62*
Kronenberg, M.E., 25 (72), *63*
Krueger, S., 21 (50b), 23 (50b), 26 (50), *62*
Krygowski, M., 2 (8), 26 (76), 49 (76b), 52 (76b), 53 (76b), *60*, *63*

Krygowski, T.M., *63* (76b), 78 (15), 89 (15b), 96 (15b), 103 (74), 111 (15b), *124, 126*
Krzanowski, W., 81 (20), *124*
Kuban, R.J., 21 (50c), 26 (50c), *62*
Kubinyi, H., 219 (82), *273*
Kučera, J., 25 (64), *62*
Kudinova, M.A., 20 (44), *61*
Kulpe, S., 15 (33c), 21 (50c), 23 (48, 57b), 24 (33c), 26 (48, 50c, 73, 74), 27 (78, 80), 30 (48, 89), *61, 62, 63*
Kumar, G., 90 (47), 92 (54), *125*
Kunikiyo, N., 205 (75), 208 (75), 231 (75), 237 (75), *272*
Kuninaka, A., 213 (87), *273*
Kunts, K.R., 90 (49), 91 (49), *125*
Kuntz, I.D., 192 (61), 194 (61), *272*
Kurkina, L.G., 20 (44), *61*
Kuroki, N., 205 (75), 208 (75), 231 (75), 237 (75), *272*
Kusahara, Y., 204 (74), 205 (74), 208 (74), *272*
Kusakabe, H., 213 (87), *273*
Kuura, H., 67 (10), *75*
Kuus, H., 67 (11), *75*
Kyte, J., 180, (49), 186 (49), *272*

Lafosse, M., 109 (98, 100, 101), *127*
Lando, J.B., 103 (76), 107 (76), *126*
Langridge, R., 205, (76), 208 (76), *272*
Lawson, E.Q., 180 (33), 183 (33), *271*
Lawtone, W.H., 103 (75), *126*
Lebart, L., 78 (9), *124*
Le Beuze, A., 31 (94), *64*
Lebedeva, N.D., 148 (30), 150 (30), *160*
Lee, R.H., 188 (57), 189 (57), *272*
Lee, S.G., 197 (68), 204 (68), *272*
Leffler, J.E., 129 (6), 131 (6), 132 (6), 155 (6), *160*
Lefour, J.M., 2 (6), 17 (6), *60*
Leibach, F.H., 213 (86), 221 (86), *273*
Leitner, J., 133 (17), 140 (17), 144 (17), 145 (17), 146 (17), 151 (17), *160*
Lelley, T.H., 192 (63), *272*
Leo, A., 78 (14), *124*, 180 (36), 183 (36), *271*
Lepri, L., 231 (94), *273*
Lerner, S.A., 231 (96), 236 (96), *273*
Le Roy, F., 92 (59), *125*
Lesczczynski, J.F., 267 (122), *274*
Lesser, G.J., 188 (57), 189 (57), *272*

Leupold, D., 7 (20), 11 (20), 13 (13, 27), 17 (20a), 23 (20b), 27 (80), 29 (87), 30 (87), 33 (27), *60, 61, 63*
Levin, E.J., 197 (68), 204 (68), *272*
Levin, S., 197 (71), 204 (71), *272*
Levitt, M., 180 (30), *271*
Levkojev, I.I., 27 (82), *63*
Librovick, N.B., 119 (142), *128*
Lichtenthaler, R.G., 23 (55c), *62*
Lifshits, E.B., 42 (101), 58 (101), *64*
Lifson, S., 267 (121), *274*
Lins, C., 90 (47), 92 (54), *125*
Lippmaa, E., 24 (59), 30 (59), *62*
Liptay, W., 31 (91), *63*
Liu, K.T., 4 (11), *60*
Lloyd, D., 20 (46), 25 (64, 65), *61, 62*
Loeve, M., 78 (5), *123*
Loew, P., 123 (146, 147), *128*
Lohmöller, J., 89 (33), *125*
Lowe, D.M., 267 (125), *274*
Lowry, S.R., 108 (91), 109 (91), 110 (91), *126*
Lueck, R., 15 (33c), 24 (33c), *61*
Lundstedt, T., 112 (117), 113 (117), *127*
Lupton, E.C., 86 (26), 114 (126), *124*
Lynbich, M.S., 42 (101), 58 (101), *64*

McBee, E.T., 96 (62), *125*
McBryde, W.A.E., *160* (12)
McClelland, R.A., 117 (130), *128*
McClure, W.F., 108 (86), *126*
McConnell, M.L., 109 (95), *126*
MacElroy, R.D., 180 (33), 183 (33), *271*
McFarland, J.W., 209 (78), *272*
Mach, G.W., 116 (127), *128*
Maciel, G.E., 89 (38, 39), *125*
MacIntyre, D.W., 2 (2), 3 (2b), 5 (2b), 6 (13), 21 (2b), 32 (2, 13), 35 (2b), 36 (2b), 41 (2), *59*
McIntyre, D.W., 92 (54), *125*
McMahon, T.B., 132 (13), 156 (13), 157 (13), *160*
McNab, H., 25 (64), *62*
Macnaughtan, D., Jr., 103 (77), 108 (77), *126*
McRae, E.G., 31 (91), *63*
Mager, P.P., 78 (9), *124*
Maiorov, V.D., 119 (142), *128*
Malinowski, E.R., 68 (14), 70 (14), 72 (14), *75*, 78 (11), 80 (11, 16), 103 (11), 108 (91), 109 (91), 110 (91c), 119 (141), *124, 126, 128*

Manavalan, P., 180 (54), 186 (54), *272*
Manning, M., 231 (97, 98), 236 (97, 98), 237 (97, 98), 239 (97, 98), *273*
Mao, J.C., 252 (109), 259 (109), 264 (109), *274*
Maria, P.C., 112 (118), *127*
Maron, A., 6 (13), 32 (13), *60*, 92 (54), *125*
Marriott, S., 2 (2), 3 (2b, 12), 5 (2b), 6 (12d, 13), 21 (2b), 24 (61), 27 (77), 31 (77), 32 (2, 12, 13), 35 (2b), 36 (2b), 41 (2), *59*, *60*, *62*, *63*
Marsden, P.D., *160* (11)
Marshall, D.R., 20 (46), 25 (65), *61*, *62*
Marsili, M., 123 (147), *128*
Martens, H., 108 (84), *126*
Martin, D., 111 (113), *127*
Martin, Y.C., 78 (14), *124*, 252 (109), 259 (109), 264 (109), *274*
Marziano, N.C., 117 (134), *128*
Masuda, S., 152 (35), *161*
Mather, P.M., 78 (8), *123*
Matsui, T., 157 (44), *161*
Matthews, B.W., 21 (50b), 23 (50b), 26 (50), *62*, 268 (127), *274*
Maunder, R.G., 6 (13), 32 (13), *60*
Maurin, J., 26 (76), 49 (76b), 52 (76b), 53 (76b), *63*
Mazzeo, P., 26 (76), *63*
Meek, J.L., 229 (92a), 231 (93), *273*
Mehlhorn, A., 15 (34), *61*
Meister, A., 239 (99), 247 (99), *273*
Melue, M., 108 (91), 109 (91), 110 (91c), *126*
Menger, F.O., 78 (14), *124*
Menzelaar, H.L.C., 110 (105, 106, 107), *127*
Mestchkina, N.M., 180 (46), 186 (46), *271*
Micomoivc, R., 180 (33), 183 (33), *271*
Middaugh, C.R., 180 (33), 183 (33), *271*
Miheeva, L.M., 180 (46), 186 (46), *271*
Milewski, S., 251 (107), 252 (107), *274*
Mishra, A.K., 188 (56), 189 (56), *272*
Misono, H., 213 (87), *273*
Mitzinger, L., 23 (54), 26 (54a, 54b), *62*
Modro, T.A., 118 (138), *128*
Moldenhauer, F., 2 (9), 27 (9), 11 (9c), 23 (9c), 27 (9), 32 (9c), 33 (9c), 37 (9c), 38 (9c), 43 (9c), *60*
Monimara, T., 90 (47), 92 (54), *125*
Moore, S., 180 (39), 183 (39), *271*
Moorem, M.J., 180 (37), 183 (37), *271*

Morat, J.L., 108 (87), *126*
More O'Ferrall, R.A., 117 (136), 118 (136), *128*
Morin-Allory, L., 109 (98, 99, 100, 101), *127*
Morineau, A., 78 (9), *124*
Morris, D.G., 57 (105), *64*
Morrison, D.F., *124* (31)
Mukherjee, B.N., 78 (9), *124*
Murrell, J.N., 31, (94), *64*
Murshak, A., 67 (10), 68 (15), 74 (16), 75, 87 (29), *124*
Musumarra, G., 90 (46), 92 (46), *125*

Naether, M., 27 (80), *63*
Nango, M., 205 (75), 208 (75), 231 (75), 237 (75), *272*
Nelsen, C.L., 89 (34), *125*
Neykov, G., 17 (40), *61*
Nicholson, H., 268 (127), *274*
Niderwieser, A., 197 (67), *272*
Nieboer, E., *160* (12)
Nieuwdrup, G.H.E., 116 (123), *127*
Nikolajewski, H.E., 25 (64, 69), 29 (87), 30 (87), *62*, *63*
Nimura, N., 204 (74), 205 (74), 208 (74), *272*
Nolte, K.D., 13 (27), 15 (33a), 17 (39), 23 (33a), 24 (33a, 59, 62), 26 (73), 27 (78, 79), 30 (59, 62, 90), 33 (27), 34 (90), 39 (62, 90), *61*, *62*, *63* (84)
Nowakowski, J., 32 (96), *64*
Nozaki, Y., 180 (28), *271*
Nuñez, L., 136 (21), 148 (21), 149 (21), 150 (21), 151 (21), *160*

Ogasahara, K., 267 (123), *274* (124)
Ohanessian, G., 2 (6), 17 (6), *60*
Öhman, J., 89 (33), *125*
Olma, A., 231 (98), 236 (98), 237 (98), 239 (98), *273*
Olsen, F.P., 117 (132), *128*
Oosterbeck, W., 116 (122), *127*
Otis, E.R., 252 (109), 259 (109), 264 (109), *274*
Ott, J.J., 15 (37), 39 (37), *61*

Palm, V.A., 111 (112), *127*, 129 (4), 151 (31), 155 (4), *160*, *161*
Palm, V.A., ed., 151 (32), 152 (32), *161*
Pannuswamy, P.K., 180 (54), 186 (54), *272*
Parker, A.J., 111 (114), *127*

AUTHOR INDEX

Passerini, R.C., 117 (134), *128*
Passet, B.V., 157 (47), *161*
Patak, G., 197 (67), *272*
Pearson, K., 77 (1), *123*
Peat, I.R., 6 (13), 32 (13), *60*
Pehk, T., 24 (59), 30 (59), *62*
Periasamy, M., 4 (11), *60*
Perone, S.P., 110 (108), *127*
Petersen, E.M., 110 (109), *127*
Pfister-Guillouzo, G., 57 (105), *64*
Pilcher, G., 136 (21), 144 (27), 145 (27), 146 (27), 148 (21), 149 (21), 150 (21, 27), 151 (21), *160*
Pivovarov, S.A., 121 (145), *128*
Pliska, V., 180 (34, 51), 183 (34, 51), *271*
Pollack, S., 49 (102), *64*
Pople, J.A., 89 (37), *125*, 131 (8), *160*
Porath, J., 197 (70), 204 (70), *272*
Porschke, D., 205 (76a), 209 (76a), *272*
Pratt, W.E., 90 (50), 91 (50), *125*
Press, S.J., 78 (8), *123*
Priest, R.G., 108 (89), *126*
Pross, A., 24 (60), *62*, 153 (38), *161*
Purcell, K.F., 112 (120), 113 (120c), *127*
Putelenz, B.U., 107 (81), *126*

Radeglia, R., 15 (33a, 33c), 21 (49), 23 (33a), 24 (33a, 33c, 59), 25 (67), 26 (59b, 75), 27 (78, 82), 28 (49), 29 (49, 83, 86, 87, 88), 30 (59, 83, 87), *61*, *62*, *63*
Radhakrishnan, A.N., 213 (85), 220 (85), 239 (85), 246 (85), *273*
Radom, L., 24 (60), *62*, 131 (8), 153 (38), *160*, *161*
Rajzamann, M., 112 (120), 113 (120c), *127*
Ramana Rao, G., 108 (88), *126*
Ramos, L.S., 78 (12), 120 (12), *124*
Ranft, J., 23 (55), 24 (58b), *62*
Rauh, H.J., 20 (47), *61*
Reat, I.A., 92 (54), *125*
Reeves, R.L., 102 (71), 103 (71), *126*
Regelmann, C., 26 (76), 49 (76b), 52 (76b), 53 (76b), *63*
Reichardt, C., 7 (21), 29 (21c), 31 (21c), *60*, 110 (110), 111 (110), *127*
Reichardt, Ch., 113 (121), 114 (121), *127*
Reijnen, J., 109 (93), *126*
Rekker, R.F., 180 (41), 185 (41), *271*
Repyakh, I.V., 15 (33b), *61*

Reynolds, W., 2 (2), 3 (2b), 5 (2b), 6 (13), 21 (2b), 32 (2, 13), 35 (2b), 36 (2b), 41 (2), *59*, *60*
Reynolds, W.F., 24 (61), *62*, 92 (54), 97 (65, 66), 116 (122), *125*, *126*, *127*
Richards, J.M., 110 (105), *127*
Ritschl, I., 34 (100b), 45 (100), 46 (100b), *64*
Ritter, G., 26 (76), 49 (76b), 52 (76b), 53 (76b), *63*
Ritter, G.L., 108 (91), 109 (91), 110 (91), *126*
Robert, P., 108 (85, 87), *126*
Roemming, C., 21 (50b), 23 (50b), 26 (50), *62*
Roesel, R.A., 213, (86), 221 (86), *273*
Rogers, L.B., 103 (77), 108 (77), *126*
Rogoshin, S.V., 180, (46), 186 (46), *271*
Rose, G.D., 188 (57), 189 (57), 267 (122), *272*, *274*
Rosenquist, N.R., 89 (43), 114 (43), *125*
Rossetti, Z.L., 231 (93), *273*
Rowe, J.E., 6 (13), 32 (13), 57 (13a), *60*
Rozett, R.W., 110 (109), *127*
Ruthenberg, M., 15 (33c), 24 (33c), *61*
Rutherford, R.J.D., 111 (111), *127*
Ryadnenko, V.L., 148 (30), 150 (30), *160*
Rytela, O., 111 (116), 113 (116), *127*

Sadek, M., 2 (2), 6 (13), 32 (2, 13), 33 (97), 41 (2), 57 (13a), *59*, *60*, *64*
Sadler, A.J., 180 (33), 183 (33), *271*
Saika, A., 89 (35), *125*
Salewski, R.I., 67 (12), *75*
Saller, H., 123 (146, 147), *128*
Sander, C., 267 (121), *274*
Sano, N., 23 (54), 26 (54c), 27 (54c, 80), *62*, *63*
Saringa, N., 204 (73), 205 (73), *272*
Sarkice, A.Y., 117 (137), *128*
Sauchez, E., 78 (12), 120 (12), *124*
Sawyer, W.H., 231 (98), 236 (98), 237 (98), 239 (98), *273*
Schaad, L.J., 13 (29), *61*
Scheibe, G., 23 (55, 57a, 58a), 24 (58a), 25 (65, 68), 29 (85), *62*, *63*
Schmidt, M., 180 (34), 183 (34), *271*
Schnabel, R., 20 (47), *61*
Schneider, H.J., 92 (59a, 60), 98 (59a, 60b, 69), 99 (60b), 101 (60b), *125*, *126*
Schob, F., *63* (84)

AUTHOR INDEX

Schuele, J., 26 (76), 49 (76b), 52 (76b), 53 (76b), *63*
Schulz, B., 21 (50c), 26 (50c), 27 (80), 30 (89), *62, 63*
Schwarzenbach, G., 20 (45), *61*
Scorrano, G., 117 (136), 118 (136), *128*
Screevans, J.H., 80 (17), 110 (104), *124, 127*
Seidel, I., 26 (74), *63*
Seiffert, W., 23 (55, 58a), 24 (58a), 25 (65), 29 (85), *62, 63*
Seiler, N., 197 (69), 204 (69), *272*
Selianov, V.P., 121 (145), *128*
Selzer, J.O., 21 (50b), 23 (50b), 26 (50), *62*
Serfaty, I., 96 (62), *125*
Sergeyev, W.M., 92 (59), *125*
Seto, J., 231 (98), 236 (98), 237 (98), 239 (98), *273*
Shaik, S.S., 2 (6), 17 (6, 41), *60, 61*
Shapiro, B.I., 20 (44), *61*
Sharif, M.R., 103 (74), *126*
Sharma, J., ed., 197 (66), *272*
Shi, J.-P., 267 (125), *274*
Shih, L.B., 108 (89), *126*
Shilov, A.E., 152 (34), *161*
Shimizu, M., 23 (54), 26 (54c), 27 (54c, 80), *62, 63*
Shorter, J., 1 (1), 2 (1), *59*, 78 (15), 89 (41), 114 (41), *124, 125*
Shuman, R.T., 239 (105), 248 (105), *273*
Shurwell, H.F., 107 (79, 80, 81), 108 (83), *126*
Signalaufzeichnungsmater, J., 27 (80), *63*
Silvestro, A., 3 (12), 6 (12d), 32 (12), *60*
Simon, Z., 180 (22), 204 (22), *271*
Simonds, J.L., 78 (6), 102 (6), *123*
Simonds, J.T., 67 (8), 68 (8), *75*
Simpson, W.T., 21 (50a), 26 (50), *62*
Sinha, R., 180 (26), 188 (26), 189 (26), 205 (26), 208 (26), *271*
Sinke, G.C., 133 (16), 140 (16), 144 (16), 145 (16), 146 (16), 151 (16), *160*
Sjoestroem, M., 6 (15), 59 (15), *60*
Sjöström, M., 88 (32), 89 (32), 116 (125), 117 (135), *124, 127, 128*, 142 (23), 145 (28), *160*, 197 (65), *272*
Skagerberg, B., 197 (65), *272*
Skancke, A., 2 (5), *60*
Slanina, Z., 133 (18, 19), 135 (18, 19), 136 (19), 141 (18), *160*
Sleckman, B.F., 197 (64), *272*

Sletten, K., 213 (88), 221 (88), *273*
Slichter, C.P., 89 (35), *125*
Smiley, D., 239 (104), 248 (104), *273*
Smith, C.R.T., 67 (9), *75*
Smith, M.W., 180 (37), 183 (37), *271*
Smithwick, E.L., 239 (105), 248 (105), *273*
Snyder, L.R., 109 (103), *127*
Soda, K., 213 (87), *273*
Soroka, J.M., 109 (97), *126*
Southgate, C.C., 180 (43), 185 (43), *271*
Spanjer, M.C., 109 (102), *127*
Springer, B.H., 25 (65), *62*
Springer, H.J., 29 (85), *63*
Staab, H.A., 7 (19), 8 (19), 21 (51), *60, 62*
Steiger, T., 15 (33c), 24 (33c), *61*
Stein, W.H., 180 (39), 183 (39), *271*
Steinbrink, D.R., 239 (106), 250 (106), *274*
Steitz, T.A., 180 (53), 186 (53), 187 (53), *272*
Sterbach, H., 219 (83), *273*
Sternbach, H., 213 (84), 220 (84), *273*
Sternberg, M.J., 180 (29), *271*
Stewart, R., 66 (5), 67 (5), *75*
Stierl, M., 15 (33c), 24 (33c), *61*
Stock, L.M., 133 (15), 142 (15), 152 (15), 153 (15), *160*
Stothers, J.B., 89 (36), 92 (60), *125*
Strub, H., 31 (94), *64*
Stull, D.R., 133 (16), 140 (16), 144 (16), 145 (16), 146 (16), 151 (16), *160*
Subbotin, O.A., 92 (59), *125*
Sugino, Y., 267 (123), *274*
Summers, M.C., 239 (103), 248 (103), *273*
Suzuki, T., 204 (74), 205 (74), 208 (74), *272*
Suzuki, H., 11 (24), *60*
Svoboda, P., 111 (116), 113 (116), *127*
Swain, C.G., 86 (26, 43), 114 (43, 126), *124, 125*
Swain, M.S., 89 (43), 114 (43), *125*
Sweet, R.M., 180 (52), 186 (52), *272*
Sylvestre, E.A., 103 (75), *126*
Szirtes, T., 231 (95), 236 (95), *273*
Szprony, L., 231 (95), 236 (95), *273*

Taagepera, M., 156 (43), 157 (43), *161*
Taft, R.W., 2 (2, 4), 3 (2b, 12), 4 (10a), 5 (2b), 7 (16), 20 (10), 21 (2b), 35 (2b), 32 (2, 12, 95), 36 (2b), 41 (2), 49 (102), 59 (16a), *59, 60, 64*, 89 (42), 96 (63), 97 (63), 116 (63), *125, 126*, 129

Taft, R.W. (*Continued*)
    (5), 130 (5), 131 (5), 155 (5), 156 (43), 157 (45, 46), 158 (45, 46), *160*, *161*
Tafur, S., 239 (104), 248 (104), *273*
Takeuchi, K., 78 (9), *124*
Tanaka, J., 23 (54), 26 (54c), 27 (54c, 80), *62*, *63*
Tanford, C., 177 (18), 180 (21, 28), 204 (21), *271*
Tanin, A., 6 (13), 32 (13), *60*, 92 (54), *125*
Tasumi, M., 23 (54), 26 (54c), 27 (54c), *62*
Tate, S.S., 239 (99), 247 (99), *273*
Teien, G., 23 (55c), *62*
Terwilliger, T.C., 180 (47), 186 (47), *271*
Thompson, G.A., 239 (99), 247 (99), *273*
Thornton, J.M., 180 (29), *271*
Thurstone, L.L., 78 (3), *123*
Tichards, J.M., 110 (106), *127*
Tokita, S., 7 (21), *60*
Tokumaru, K., 152 (35), *161*
Tolmachev, A.I., 20 (44), *61*
Topsom, R.D., 2 (2), 3 (2b, 12), 5 (2b), 6 (12d, 13), 21 (2b), 24 (61), 27 (77), 31 (77, 92), 32 (2, 12, 13, 95), 35 (2b), 36 (2b), 41 (2), 49 (102), 55 (104), *59*, *60*, *62*, *63*, *64*, 96 (63), 97 (63, 67), 116 (63, 67, 122), *126*, *127*, 157 (47), *161*, 167 (4), *270*
Tori, K., 152 (35), *161*
Touchstone, J.C., 197 (68), 204 (68), *272*
Traverso, P.G., 117 (134), *128*
Trifan, D.S., 120 (144), *128*
Trygstad, O., 213 (88), 221 (88), *273*
Tsong, T.Y., 180 (30), *271*
Tway, P.C., 108 (92), 109 (92), *126*

Ugino, Y., *274* (124)
Unger, N.R., 89 (43), 114 (43), *125*
Urbaniak, Z., 92 (57, 58), *125*
Ushomirskii, M.N., 42 (101), 58 (101), *64*

Vaciago, A., 26 (76), *63*
Van Buuren, K.H., 180 (50), 186 (50), *272*
Vanderginste, B., 109 (93), *126*
van Heijine, G., 180 (55), 186 (55), *272*
Van Houvelingen, H.C., 116 (124), *127*
Van Houwelingen, J.C., 109 (102), *127*
VanLoan, C., 78 (4), *123*
Van Meerssche, M., 21 (50b), 23 (50b), 26 (50), *62*
van Rijckevorsel, J.L.A., ed., 78 (9), *124*

Van Werf, H.E., 239 (106), 250 (106), *274*
Vecera, M., 111 (116), 113 (116), *127*
Venkataram, U.V., 78 (14), *124*
Venkataraman, H.S., 152 (34), *161*
Voňka, P., 133 (17), 140 (17), 144 (17), 145 (17), 146 (17), 151 (17), *160*
von Nagy-Felsobuki, E., 2 (2), 3 (2b), 5 (2b), 21 (2b), 32 (2), 35 (2b), 36 (2b), 41 (2), 49 (102), *59*, *64*

Wade, R., 239 (100), 247 (100), 248 (100), *273*
Waehnert, M., 25 (67), 27 (82), 31 (94), *63*, *64*
Wakelkamp, W., 180 (50), 186 (50), *272*
Walczak, B., 109 (98, 100, 101), *127*
Wangen, L.E., 78 (12), 120 (12), *124*
Ware, W.R., ed., 27 (81), *63*
Warwick, K.M., 78 (9), *124*
Wasserman, S.A., 231 (96), 236 (96), *273*
Waye, M.M.Y., 267 (125), *274*
Webb, H.M., 156 (42), 157 (42), 158 (42), *161*
Webster, B.M., 116 (122), *127*
Weeks, W.W., 108 (86), *126*
Weesie, H.M., 109 (102), *127*
Weigand, E.P., 98 (69), *126*
Weiner, P., 86 (27), *124*
Weiss, R.M., 180 (47), 186 (47), *271*
Wells, P.R., 33 (99), *64*
Wengenmayr, H., 23 (58a), 24 (58a), *62*
Werner, H., 92 (59), *125*
Wernimont, G., 103 (77), 108 (77), *126*
Westheimer, F.H., 97 (64), *126*
Westrum, E.F., 133 (16), 140 (16), 144 (16), 145 (16), 146 (16), 151 (16), *160*
White, P.O., 86 (25), *124*
Whitehead, M.A., 33 (99), *64*
Wiberg, K.B., 90 (50, 51, 52), 91 (50, 51, 52), *125*
Wickham, G., 152 (36), *161*
Wiebers, J.L., 110 (108), *127*
Wikstrom, C., 197 (65), *272*
Wilcox, W., 180 (47), 186 (47), *271*
Wilkinson, A.J., 267 (125), *274*
Williams, D.H., 263 (110), *274*
Williams, E.A., 89 (34), *125*
Wilson, B.E., 78 (12), 120 (12), *124*
Windig, W., 110, (105, 106, 107), *127*
Winstein, S., 120 (144), *128*
Winter, G., 267 (125), *274*

Wold, H., 82 (22), *124*
Wold, S., 6 (15), 59 (15), *60*, 78 (10), 82 (10, 19), 88 (32), 89 (32, 33), 90 (44, 46, 47), 92 (46, 54), 99 (19), 116 (125), 117 (135), 120 (143), *124* (21), *125*, *127*, *128*, 142 (23), 145 (28), *160*, *272*
Wolfenden, R., 180 (43), 185 (43), *271*
Wolfenden, R.V., 180 (42), 185 (42), *271*
Wolff, R., 29 (88), *63*
Wong, H.E., 6 (13), 32 (13), *60*
Wong, S.C., 67 (7), *75*, 102 (72), 103 (72), *126*
Woodruff, H.B., 108 (92), 109 (92), *126*
Woods, D.R., 180 (27), *271*
Wozniak, J.A., 268 (127), *274*
Wynne, H.J., 180 (50), 186 (50), *272*

Yanai, H., 78 (9), *124*
Yates, K., 65 (1, 2), 67 (9), 68 (13), 74 (13), *75*, 117 (129, 130, 133), 118 (138), 119 (140), *128*
Yoshida, M., 152 (35), *161*
Yoshima, H., 213 (87), *273*

Yoshimura, Y., 152 (35), *161*
Yuki, K., 123 (147), *128*
Yunger, L.M., 180 (40), 183 (40), *271*
Yutani, K., 267 (123), *274* (124)

Zalewski, R.I., 78 (15), 84 (23), 85 (23), 89 (15b), 92 (55, 56, 57, 58), 94 (55, 56), 96 (15b), 99 (23), 102 (70), 103 (70, 73, 74), 104 (78), 111 (15b), 113 (121), 114 (121), 117 (137), 121 (23), *124*, *125*, *126*, *127*, *128*
Zaslavsky, B.Y., 180 (46), 186 (46), *271*
Zedler, A., 26 (73), *63*
Zefirow, N.S., 98 (68), *126*
Zehfus, M.H., 188 (57), 189 (57), *272*
Zerbi, G., 108 (88), *126*
Ziemer, B., 27 (80), *63*
Zimm, B.H., 267 (117), *274*
Zimmerman, J.M., 180 (26), 188 (26), 189 (26), 205 (26), 208 (26), *271*
Zincke, T., 25 (70), *62*
Zweig, G., ed., 197 (66), *272*

Subject Index

Absorption bands:
  and medium effects, 66, 74
  shifts with increasing acidity, 66
Absorption intensity, of polymethines, 21
Acetophenones, *para*-substituted, 57
Acetylacetonates, 21
Acidity functions, based on PCA analysis, 117, 118
Aggression, of polymethines, 27
Amino acids, 177, 186, 194
  bioactivities of, 209–221
  chromatographic properties of, 194–195, 204
  hydrophobicity parameters for, 177
  selectivity coefficients, 204, 208
  side chain parameters for, 177
  solution properties of, 192, 194
  transport of, 220–221
  variant, 221
  volume and surface area of, 188
Anilines, *para*-acceptor-substituted, 54
Aromaticity, 2
Aromatic pericycles, 13, 14
Aromatic state, ideal, 3, 57

Benzaldehyde:
  para-amino-substituted, 57
  para-hydroxy, anion of, 7
Benzamides, 74
  pK_{BH^+} and m^* values for, 74, 117
Benzenes, *para*-donor–acceptor-substituted, 6–10, 41, 43–45
Benzophenones, *para*-substituted, 57
Benzoquinones, 2,5-donor-substituted, 21, 23
Bioactivity model, 209
Bond angle alteration, in polymethines, 26
Bond length alternation, in respect to quinoid character, 49
Bond lengths:
  of π-electron systems, 37
  of polymethines, 22, 23

Branching equations, 172–173
Butadienes, 1,4-donor–acceptor-substituted, relation to pentamethines, 15

Charge alteration, 12, 23. *See also* π-electron density
  on light excitation, CA, 34
Charge resonance on light excitation, CR, 34, 45
Charge stabilization rule, 17, 39
Charge transfer, 11
Charge transfer on light excitation, CT, 34, 46
Charge transfer transition, 11
Chemical bonds, classified by PCA, 123
Chou–Fasman conformational parameters, 266
Chromostate, 7
C^{13}NMR substituent chemical shifts, 89
  for alicyclic systems, 92, 94
  for axial and equatorial substituents, 94
  correlation by σ_I and σ_R parameters, 89–91
  intensity of transmission, 96
  principal components of, 90
  relation to charge density, 90
  relation to electronegativity, 94, 99
  relation to polar effects, 91, 92
  relation to σ_m parameter, 91
  for unsaturated and aromatic systems, 90–92
Conjugation, strength of, 2, 52
Cyanines, 18, 58. *See also* Polymethine dyes
Cycloaddition reactions, of polymethines, 25

Data dimensionality, 80
  cross-validation of data matrix, 80, 90, 112
  submatrix analysis for, 80
Data pretreatment, 81

297

Davis–Geissman method, 67
Delocalization, see π-electron delocalization
Diphenyl methane dyes, 25
Dipole moment:
    changes on light excitation, 46
    of *para*-donor–acceptor-substitutes benzenes, 49
    of polymethines, 24, 39, 49
        influence of solvent, 28, 31
Double substituent parameter (DSP) equation, 2

Electrical effect parameters, 170
Electron demand parameter, of substituents, 2, 3, 40
Electronegativity, of substituents, 9, 33, 34, 40, 44, 53, 57
Energy:
    activation free of hindered rotation, in polymethines, 23, 29
    delocalization, 13
    of frontier orbitals:
        of *para*-donor–acceptor-substitutes benzenes, 43
        of polymethines, 37
    kinetic, of π-electron systems, 18
    localization, of polymethines, 25
    potential, of π-electron systems, 18
    resonance, 13, 17, 20, 21
Enthalpy, 130, 132
    of combustion, 140
    correlation of reaction enthalpy, 144
    internal and external, 132, 155, 156
    for nuclear substitution, 134
    solvation enthalpy, 130, 137
    of vaporization, 137, 140
Entropy, 130, 132
    internal and external, 132, 155, 156
    solvations entropy, 130
Excimers, 27
Exciplexes, 27
Excited-state properties, 45
Extended selectivity treatment, 153

Factor analysis, in basicity studies, 67
Force field calculations, of polymethines, 23, 26
Formamides, vinylogous, 18
Formamidinium ions, 18

Franck–Condon principle, 21
Frost circle, 13, 14

Gibbs energy, 130, 132
    for nuclear substitution, 134
Glutaconaldehyde sodium, 5-hydroxy, 18

Hammett equation, 1, 2
Hammett linear free energy relationships:
    precondition of, 129
    test of, 153–155
Heat of combustion, of polymethines, 20
Heat content function, 130
Hybridization, changes in polymethines, 25, 26
Hydrocarbon ions, 11
Hydrogen-bonding, 169
Hydrolytic enzymes, 211, 219
Hydrophobicities parameters, 177
    for amino acids, 177

IMF correlation equation, 171, 174, 266
    protein basis sets (PBS) for, 173, 175
Intermolecular forces, 165
    charge transfer, 169–170
    hydrogen-bonding, 169
    ion–dipole, 168
    ion–induced dipole, 168
    steric effects, 170–172
    van der Waals interactions, 166–167
Isodesmic reaction, 131, 133, 136, 159
    thermodynamics of, 142
    at 0° K, 139
Isokinetic relationship, 131, 143, 156, 159
    range of validity, 144
Isokinetic temperature, 156, 157
Isoleucyl-tRNA synthetase, 219–220
Isomerization, *cis–trans*, of polymethines, 23

LDR equation, 169
Lewis–Calvin rule, 21, 28

Merocyanines, 18, 21. See also Polymethine dyes
Mesomerism, 8, 58
Methylene blue, 25
Molecular rotational symmetry numbers, 136, 159
    correlation for, 142

SUBJECT INDEX

Omega method of correlation, 221, 256
Orthogonality, 74
  of protonation process and medium effects, 74
Oscillator strength, of polymethines, 29
Oxonols, 18. *See also* Polymethine dyes

Parker's classification of solvents, 113
Partition functions, from statistical mechanics, 141
Pentadienylium cation, 6
Pentamethine dyes, relation to 1,4-donor–acceptor-substituted butadienes, 9
π-bond order:
  changes on light excitation, 13, 15, 46
  definition, 34
  of polymethines, 39, 40
  in relation to bond length, 22, 23, 37
π-bond order alteration:
  definition, 33
  in *para*-donor–acceptor-substitutes benzenes, 43
  in polymethines, 37, 39
π-bond order equalization:
  in aromatic compounds, 11
  definition, 34
  in *para*-donor–acceptor-substituted benzenes, 44, 58
  in polymethines, 11, 21, 23, 40, 58
π-electron delocalization, 2, 3, 5, 6, 9, 10, 21, 39, 40, 52, 58, 59
π-electron charge transfer parameter, $\Sigma\Delta q_\pi$, 3, 35, 37, 58
π-electron density:
  changes on light excitation, 13, 15, 25
  definition, 34
  dependence on electronegativity of substituents, 46–50
  influence of solvent, 29
π-electron density alternation, 12, 15, 23, 37, 39, 49, 58
  definition, 34
π-polarization effect, 6, 48, 49, 58
π-saturation effect, 3, 5, 58
Peptides, 228
  antibiotic bioactivities of, 251
  bioactivities of, 237
  chromatographic parameters for, 229
  intermolecular forces for, 258
  multiply-subtracted, 250
  properties of, 229
  substitution in, 229
  X^c substitution in, 248
  X^n substitution in, 237
  X^p substitution in, 249
  Z^0Z substitution in, 248
Phenoxide ion, 7, 54
Polar effects of substituents:
  in linear free energy relations (LFER), 153
  significance of, 159
Polarizability, π-electron, of polymethines, 17, 27, 57
Polarizability parameter, 167
  non-directional α values, 167
Polenes, resonance energy, 20
Polene state, ideal, 3, 57
Polyacenes, resonance energy, 20
Polyenals:
  ω-amino, 18
  ω-hydroxy, 18
Polymethine dyes, 13, 15
Polymethine radicals, 13, 15
Polymethines, analytical solutions of molecular parameters, 16
Polymethine state, ideal, 3, 7, 8, 18, 32, 57–59
Polymethinium ions, 18
Potential energy, 130
  calculation of, 141
  for nuclear substitution, 134
  polar effects in, 133, 137
  significance of, 141, 159
Potentials, electrochemical half-wave, of polymethines, 19, 20
Principal component analysis (PCA), 67, 77, 78
  "abstraction" solution from, 83, 116
  in basicity studies, 67, 101, 102, 104
  data matrix for, 78, 79
  and eigenvectors, 68
  of enthalpies of formation, 145–150
  loading for, 39
  mathematical and geometric foundation of, 87
  model for, 78
  "real" solution from, 84, 86, 116
  for solvent classification, 110–113
  transforms of PCA matrices, 68, 71

Proteins, 265
  bioactivities of, 267
  conformation of, 265
  intermolecular forces for, 265
Protonation, enthalpy of, 20
Pyrylium cyanines, 20

Quantum chemical calculations:
  ab initio, 2, 15, 18, 23, 24, 32, 35
  CNDO/2, 15, 24, 30, 39
  Hückel MO (HMO) theory, 11, 32
    analytical solutions of, 16
  Hückel MO (HMO-β_{SC}), 23, 35, 39
    advantages of, 32, 33
  MINDO/3, 15
  Pariser–Parr–Pople (PPP), 15, 23, 24, 39
  perturbational MO (PMO), 11
  valence bond (VB) theory, 9
  VESCF, 30

Refractivity, molar, of polymethines, 27
Resonance, 3, 6, 8, 39, 57, 58
Resonance stabilization, of polymethines, 9, 17, 20
Rotational barrier, in polymethines, 23

Simulation models, 72
Solute selectivity:
  chemical steric factors for, 109
  hydrophobic factors for, 109
Solvation energy relationship, linear, 7, 59
Solvatochromism:
  microstructural model of, 30, 31, 39
  of polymethines, 28, 31, 39, 46
Solvent polarity, 112–113
  influence on polymethines, 27
Solvent polarizability, 112
Solvolysis reaction rates, 120–122
  analyzed by a PCA model, 120
  charge delocalized transition states for, 121, 122
  transition states for, 120
Sorbinaldehyde sodium, 7-hydroxy, 18
Spectra:
  IR, of polymethines, 23, 26
    solvent influence on, 29
  NMR, of polymethines, 23, 24, 26
    solvent influence on, 28–31
  RAMAN, of polymethines, 23, 26

UV-visible, of 1,5-bis(dimethylamino)pentamethinium ion, 8, 10
UV-visible, of *para*-donor–acceptor-substitutes benzenes, 8, 10
UV-visible, of polymethines, 22
Stewart–Granger method, 67
Streptocyanines, 11, 18. *See also* Polymethine dyes
Sulfuric acid, 119
  chemical species in, 119
Substituents:
  basicity of, 15
  effective length of, 21
  electron-donor ability, 15
Substituent constants, σ_R, 3, 6
Substituent effects:
  through-bond, 33
  through-space, 33
  transmission of, 2, 33
  transmissivity of, 51–55, 58
Substituent parameter equations, 2
Substituent reactions, of polymethines, 24, 25,
Symmetry deviation, Σ:
  definition, 9, 11, 34
  of para-donor–acceptor-substituted benzenes, 44, 45
  of polymethines, 37, 39, 41

Target testing (TT), 68
  the key-set, 97
  predicted vector, 71
  target transformation, 86, 109
  test vector for, 69
Third Law of Thermodynamics, 141
  and translational, rotational, and vibrational energies, 141
Topological asymmetry, 43, 53
Topological deviation, 43, 58
Topology, influence on substituent effects, 57
Transition energy, E_T:
  of *para*-donor–acceptor-substituted benzenes, 43
  of polymethines, 20, 37
Transition moment:
  of *para*-donor–acceptor-substituted benzenes, 43
  of polymethines, 37

Triad theory, 2, 57, 59
Triphenyl methane dyes, 20, 25
Tripolonate anion, 7-hydroxy-, 21

Vinylene shift, of polymethines, 21

X-ray structural analyses, of polymethines, 21, 23, 25, 30

Zero-point vibration energy, 130
Zeta method of correlation, 204

Cumulative Index, Volumes 1–18

	VOL.	PAGE
Acetals, Hydrolysis of, Mechanism and Catalysis for (Cordes)	4	1
Acetonitrile, Ionic Reactions in (Coetzee)	4	45
Acidity and Basicity, Correlation Analysis of, From the Solution to the Gas Phase (Gal and Maria)	17	159
Active Sites of Enzymes, Probing with Conformationally Restricted Substrate Analogs (Kenyon and Fee)	10	381
Activity Coefficient Behavior of Organic Molecules and Ions in Aqueous Acid Solutions (Yates and McClelland)	11	323
Alkyl Inductive Effect, The, Calculation of Inductive Substituent Parameters (Levitt and Widing)	12	119
Allenes and Cumulenes, Substituent Effects in (Runge)	13	315
Amines, Thermodynamics of Ionization and Solution of Aliphatic, in Water (Jones and Arnett)	11	263
Amino Acid, Peptide and Protein Properties and Bioactivities, The Quantitative Description of (Charton)	18	163
Aromatic Nitration, A Classic Mechanism for (Stock)	12	21
Barriers, to Internal Rotation about Single Bonds (Lowe)	6	1
Basicity Constants of Weak Bases by the Target Testing Method of Factor Analysis, Estimation of the (Haldna)	18	65
Benzenes, A Theoretical Approach to Substituent Interactions in Substituted (Pross and Radom)	13	1
Benzene Series, Generalized Treatment of Substituent Effects in the, A Statistical Analysis by the Dual Substituent Parameter Equation (Ehrenson, Brownlee, and Taft)	10	1
^{13}C *NMR, Electronic Structure and* (Nelson and Williams)	12	229
Carbonium Ions (Deno)	2	129
Carbonyl Group Reactions, Simple, Mechanism and Catalysis of (Jencks)	2	63
Catalysis, for Hydrolysis of Acetals, Ketals, and Ortho Esters (Cordes)	4	1
Charge Distributions in Monosubstituted Benzenes and in Meta- and Para-Substituted Fluorobenzenes, 4b Initio Calculations of: Comparison with ^1H, ^{13}C, *and* ^{19}F *Nmr Substituent Shifts* (Hehre, Taft, and Topsom)	12	159
Charge-Transfer Complexes, Reactions through (Kosower)	3	81
Chemical Process Systemization by Electron Count in Transition Matrix (Chu and Lee)	15	131
Collage of S_N2 *Reactivity Patterns: A State Correlation Diagram Model, The* (Shaik)	15	197

	VOL.	PAGE
Conformation, as Studied by Electron Spin Resonance of Spectroscopy (Geske)	4	125
Correlation Analysis in Organic Crystal Chemistry (Krygowski)	17	239
Delocalization Effects, Polar and Pi, an Analysis of (Wells, Ehrenson, and Taft)	6	147
Deuterium Compounds, Optically Active (Verbict)	7	51
Electrical Effects, A General Treatment of (Charton)	16	287
Electrolytic Reductive Coupling: Synthetic and Mechanistic Aspects (Baizer and Petrovich)	7	189
Electronic Effects, The Nature and Analysis of Substituent (Taft and Topsom)	16	1
Electronic Structure and ^{13}C NMR (Nelson and Williams)	12	229
Electronic Substituent Effects, in Molecular Spectroscopy (Topsom)	16	193
Electronic Substituent Effects, Some Theoretical Studies in Organic Chemistry (Topsom)	16	125
Electron Spin Resonance, of Nitrenes (Wasserman)	8	319
Electron Spin Spectroscopy, Study of Conformation and Structure by (Geske)	4	125
Electrophilic Substitutions at Alkanes and in Alkylcarbonium Ions (Brouwer and Hogeveen)	9	179
Enthalpy-Entropy Relationship (Exner)	10	411
Fluorine Hyperconjugation (Holtz)	8	1
Gas-Phase Reactions, Properties and Reactivity of Methylene from (Bell)	2	1
Ground-State Molecular Structures, Substituent Effects on and Charge Distributions (Topsom)	16	85
Group Electronegativities (Wells)	6	111
Hammett and Derivative Structure-Reactivity Relationships, Theoretical Interpretations of (Ehrenson)	2	195
Hammett Memorial Lecture (Shorter)	17	1
Heats of Hydrogenation: A Brief Summary (Jenson)	12	189
Hydration, Theoretical Studies of the Effects of, on Organic Equilibria (Topsom)	17	107
Hydrocarbons, Acidity of (Streitwieser and Hammons)	3	41
Hydrocarbons, Pyrolysis of (Badger)	3	1
Hydrolysis, of Acetals, Ketals, and Ortho Esters, Mechanism and Catalysis for (Cordes)	4	1
Internal Rotation, Barriers to, about Single Bonds (Lowe)	6	1
Ionic Reactions, in Acetonitrile (Coetzee)	4	45
Ionization and Dissociation Equilibria, in Solution, in Liquid Sulfur Dioxide (Lichtin)	1	75
Ionization Potentials, in Organic Chemistry (Streitwieser)	1	1
Isotope Effects, Secondary (Halevi)	1	109

	VOL.	PAGE
Ketals, Hydrolysis of, Mechanism and Catalysis for (Cordes)	4	1
Kinetics of Reactions, in Solutions under Pressure (le Noble)	5	207
Linear Free-Energy Relationships, Physicochemical Preconditions of (Exner)	18	129
Linear Solvation Energy Relationships, An Examination of (Abboud, Kamlet, and Taft)	13	485
Methylene, Properties and Reactivity of, from Gas-Phase Reactions (Bell)	2	1
Molecular Orbital Structures for Small Organic Molecules and Cations (Lathan, Curtiss, Hehre, Lisle, and Pople)	11	175
Naphthalene Series, Substituent Effects in the (Wells, Ehrenson, and Taft)	6	147
Neutral Hydrocarbon Isomer, The Systematic Prediction of the Most Stable (Godleski, Schleyer, Osawa, and Wipke)	13	63
Nitrenes, Electron Spin Resonance of (Wasserman)	8	319
Non-Aromatic Unsaturated Systems, Substituent Effects in (Charton)	10	81
Nucleophilic Displacements, on Peroxide Oxygen (Behrman and Edwards)	4	93
Nucleophilic Substitution, at Sulfur (Ciuffarin and Fava)	6	81
Optically Active Deuterium Compounds (Verbict)	7	51
Organic Bases, Weak, Quantitative Comparisons of (Arnett)	1	223
Organic Polarography, Mechanisms of (Perrin)	3	165
Ortho Effect, The Analysis of the (Fujita and Nishioka)	12	49
Ortho Effect, Quantitative Treatment of (Charton)	8	235
Ortho Esters, Hydrolysis of, Mechanism and Catalysis for (Cordes)	4	1
Ortho Substituent Effects (Charton)	8	235
Physical Properties and Reactivity of Radicals (Zahradnik and Carsky)	10	327
Pi Delocalization Effects, An Analysis of (Wells, Ehrenson, and Taft)	6	147
Planar Polymers, The Influence of Geometry on the Electronic Structure and Spectra of (Simmons)	7	1
Polar Delocalization Effects, An Analysis of (Wells, Ehrenson, and Taft)	6	147
Polarography, Physical Organic (Zuman)	5	81
Polar Substituent Effects (Reynolds)	14	165
Polyalkylbenzene Systems, Electrophilic Aromatic Substitution and Related Reactions in (Baciocchi and Illuminati)	5	1
Polymethine Approach, The, Free-Energy Relationships in Strongly Conjugated Substituents (Dähne and Hoffmann)	18	1

	VOL.	PAGE
Principle Component Analysis in Organic Chemistry, Application of (Zalewski)	18	77
Protonated Cyclopropanes (Lee)	7	129
Protonic Acidities and Basicities in the Gas Phase and in Solution: Substituent and Solvent Effects (Taft)	14	247
Proton-Transfer Reactions in Highly Basic Media (Jones)	9	241
Radiation Chemistry to Mechanistic Studies in Organic Chemistry, The Application of (Fendler and Fendler)	7	229
Radical Ions, The Chemistry of (Szwarc)	6	323
Saul Winstein: Contributions to Physical Organic Chemistry and Bibliography	9	1
Secondary Deuterium Isotope Effects on Reactions Proceeding Through Carbocations (Sunko and Hehre)	14	205
Semiempirical Molecular Orbital Calculations for Saturated Organic Compounds (Herndon)	9	99
Solutions under Pressure, Kinetics of Reactions in (le Noble)	5	207
Solvent Effects on Transition States and Reaction Rates (Abraham)	11	1
Solvent Isotope Effects, Mechanistic Deductions from (Schowen)	9	275
Solvolysis, in Water (Robertson)	4	213
Solvolysis Revisited (Blandamer, Scott, and Robertson)	15	149
Solvolytic Substitution in Simple Alkyl Systems (Harris)	11	89
Steric Effects, Quantitative Models of (Unger and Hansch)	12	91
Structural Principles of Unsaturated Organic Compounds: Evidence by Quantum Chemical Calculations (Dahne and Moldenhauer)	15	1
Structure, as Studied by Electron Spin Resonance Spectroscopy (Geske)	4	125
Structure–Energy Relationships, Reaction Coordinates and (Grunwald)	17	55
Structure-Reactivity and Hammett Relationships, Theoretical Interpretations of (Ehrenson)	2	195
Structure-Reactivity Relationships, Examination of (Ritchie and Sager)	2	323
Structure-Reactivity Relationships, in Homogeneous Gas-Phase Reactions (Smith and Kelley)	8	75
Structure-Reactivity Relationships, for Ortho Substituents (Charton)	8	235
Substituent Constants for Correlation Analysis, Electrical Effect (Charton)	13	119
Substituent Effects, in the Naphthalene Series (Wells, Ehrenson and Taft)	6	147
Substituent Effects on Chemical Shifts in the Sidechains of Aromatic Systems (Craik and Brownlee)	14	1
Substituent Effects in the Partition Coefficient of Disubstituted Benzenes: Bidirectional Hammett-type Relationships (Fujita)	14	75
Substituent Electronic Effects, The Nature and Analysis of (Topsom)	12	1

	VOL.	PAGE
Substitution Reactions, Electrophilic Aromatic (Berliner)	2	253
Substitution Reactions, Electrophilic Aromatic, in Polyalkylbenzene Systems (Baciocchi and Illuminati)	5	1
Substitution Reactions, Nucleophilic Aromatic (Ross)	1	31
Sulfur, Nucleophilic Substitution at (Ciuffarin and Fava)	6	81
Superbasic Ion Pairs: Alkali Metal Salts of Alkylarenes (Gau, Assadourian and Veracini)	16	237
Thermal Rearrangements, Mechanisms of (Smith and Kelley)	8	75
Thermal Unimolecular Reactions (Wilcott, Cargill and Sears)	9	25
Thermolysis in Gas Phase, Mechanisms of (Smith and Kelley)	8	75
Thermodynamics of Molecular Species (Grunwald)	17	31
Treatment of Steric Effects (Gallo)	14	115
Trifluoromethyl Group in Chemistry and Spectroscopy Carbon-Fluorine Hyperconjugation, The (Stock and Wasielewski)	13	253
Ultra-Fast Proton-Transfer Reactions (Grunwald)	3	317
Vinyl and Allenyl Cations (Stang)	10	205
Water, Solvolysis in (Robertson)	4	213
Y_x Scales of Solvent Ionizing Power (Bentley and Llewellyn)	17	121

RANDALL LIBRARY-UNCW

3 0490 0376553 2